T0180518

Lecture Notes in Computer Science

Lecture Notes in Artificial Intelligence   13561

Founding Editor

Jörg Siekmann

Series Editors

Randy Goebel, *University of Alberta, Edmonton, Canada*
Wolfgang Wahlster, *DFKI, Berlin, Germany*
Zhi-Hua Zhou, *Nanjing University, Nanjing, China*

The series Lecture Notes in Artificial Intelligence (LNAI) was established in 1988 as a topical subseries of LNCS devoted to artificial intelligence.

The series publishes state-of-the-art research results at a high level. As with the LNCS mother series, the mission of the series is to serve the international R & D community by providing an invaluable service, mainly focused on the publication of conference and workshop proceedings and postproceedings.

Amy Eguchi · Nuno Lau ·
Maike Paetzel-Prüsmann ·
Thanapat Wanichanon (Eds.)

RoboCup 2022: Robot World Cup XXV

 Springer

Editors
Amy Eguchi 🆔
University of San Diego
La Jolla, CA, USA

Nuno Lau 🆔
University of Aveiro
Aveiro, Portugal

Maike Paetzel-Prüsmann 🆔
Disney Research
Zurich, Switzerland

Thanapat Wanichanon 🆔
Mahidol University
Bangkok, Thailand

ISSN 0302-9743   ISSN 1611-3349 (electronic)
Lecture Notes in Artificial Intelligence
ISBN 978-3-031-28468-7   ISBN 978-3-031-28469-4 (eBook)
https://doi.org/10.1007/978-3-031-28469-4

LNCS Sublibrary: SL7 – Artificial Intelligence

© The Editor(s) (if applicable) and The Author(s), under exclusive license
to Springer Nature Switzerland AG 2023
This work is subject to copyright. All rights are reserved by the Publisher, whether the whole or part of the material is concerned, specifically the rights of translation, reprinting, reuse of illustrations, recitation, broadcasting, reproduction on microfilms or in any other physical way, and transmission or information storage and retrieval, electronic adaptation, computer software, or by similar or dissimilar methodology now known or hereafter developed.
The use of general descriptive names, registered names, trademarks, service marks, etc. in this publication does not imply, even in the absence of a specific statement, that such names are exempt from the relevant protective laws and regulations and therefore free for general use.
The publisher, the authors, and the editors are safe to assume that the advice and information in this book are believed to be true and accurate at the date of publication. Neither the publisher nor the authors or the editors give a warranty, expressed or implied, with respect to the material contained herein or for any errors or omissions that may have been made. The publisher remains neutral with regard to jurisdictional claims in published maps and institutional affiliations.

This Springer imprint is published by the registered company Springer Nature Switzerland AG
The registered company address is: Gewerbestrasse 11, 6330 Cham, Switzerland

Preface

RoboCup 2022 was successfully held in-person (with limited online participation options) after two unusual years – under unprecedented circumstances. RoboCup 2020 was canceled due to the COVID-19 pandemic. However, in 2021, each league's commitments and efforts with creative solutions led to the great success of the first fully online RoboCup competition, which brought the community closer together. In 2022, although teams from some regions had difficulty traveling to Bangkok, Thailand, many RoboCuppers gathered to participate in the in-person event, which made the community stronger by building onto the online experience from 2021. Many leagues continued to utilize the tools and strategies developed for the online competition in 2021 to strengthen the league experience.

As a challenging multidisciplinary endeavor, RoboCup continues to contribute to the advancement in robotics, and the Symposium highlights teams' accomplishments and robotics research that translate to better, faster, safer, more capable robots - and competition wins. This proceedings presents the science behind the advances in robotics, including the key innovations that led the winning teams to their successes, and the outcomes of research inspired by challenges across the different leagues at RoboCup.

The RoboCup 2022 Symposium received a total of 28 submissions for the Main and Development Tracks. The submissions were reviewed by the Program Committee of 63 members, receiving at least 3 reviews per paper. The committee carefully weighed the merits and limitations of each paper, and accepted 16 papers to be presented at the Symposium, for an overall acceptance rate of 57%. In addition, the Symposium proceedings includes 12 papers from the winners of the RoboCup 2022 competitions under the Champions Track. Every Champion Track paper had at least 2 positive reviews from experts in the corresponding league involving more members who were not on the Symposium Program Committee.

For the first time, the RoboCup Symposium hosted two satellite events. The "Open Humanoid Competition" was a hybrid workshop organized by the RoboCup Humanoid League to discuss a pathway to making the league more accessible for novice teams. The Rescue Simulation League hosted an online workshop, entitled "2022 Workshop on Artificial Intelligence and Simulation for Natural Disaster Management", aiming to bring together and build a community of researchers, disaster responders, and policy-makers interested in developing and applying AI and simulation techniques for natural disaster management.

Among the 16 accepted research papers, four papers were nominated as best paper award finalists. The awards committee evaluated the finalists based on the paper as well as their associated reviews and presentations, and selected one best paper:

– Maximilian Giessler, Marc Breig, Virginia Wolf, Fabian Schnekenburger, Ulrich Hochberg and Steffen Willwacher — "Gait Phase Detection on Level and Inclined Surfaces for Human Beings with an Orthosis and Humanoid Robots".

The RoboCup 2022 Symposium was delighted to host three Keynote Speakers:

Satoshi Tadokoro (Graduate School of Information Sciences, Tohoku University): "Search in Rubble Piles - ImPACT Tough Robotics Challenge"
Angelica Lim (Rosie Lab, School of Computing Science at Simon Fraser University): "Social Signals in the Wild: Multimodal Machine Learning for Human-Robot Interaction"
Manukid Parnichkun (Asian Institute of Technology): "Driverless Car Technologies"

We thank the members of the Program Committee and the additional reviewers for their time and expertise to help uphold the high standards of the Symposium Technical Program, as well as the members of the awards committee for their work in selecting the best paper award. This event would not have been possible without the tireless efforts of the Organizing Committee including the Local Organizing Committee (LOC) members in Thailand. We appreciate the enthusiastic support and participation of RoboCuppers across the world, both in-person and remote, and the technical and organizing committees of every league. Finally, our sincere gratitude goes to Jackrit Suthakorn, the RoboCup 2022 General Chair; Thanapat Wanichanon and Nantida Nillahoot, Program Chairs; Pattaraporn Posoknistakul, Secretariat, and all the Local Organizing Committee Chairs. The Symposium Co-chairs greatly appreciated all the support that we received from the LOC members and enjoyed this wonderful opportunity to work collaboratively together to help make the event another success.

February 2023

Amy Eguchi
Nuno Lau
Maike Paetzel-Prüsmann
Thanapat Wanichanon

Organization

Symposium Chairs

Amy Eguchi   University of California San Diego, USA
Nuno Lau   University of Aveiro, Portugal
Maike Paetzel-Prüsmann   Disney Research, Switzerland
Thanapat Wanichanon   Mahidol University, Thailand

Program Committee

Hidehisa Akiyama   Okayama University of Science, Japan
Emanuele Antonioni   Sapienza University of Rome, Italy
Minoru Asada   Osaka University, Japan
José Luis Azevedo   University of Aveiro, Portugal
Sven Behnke   University of Bonn, Germany
Reinaldo A. C. Bianchi   Centro Universitario FEI, Brazil
Joydeep Biswas   University of Texas at Austin, USA
Roberto Bonilla   JP Morgan, USA
Ansgar Bredenfeld   Dr. Bredenfeld UG, Germany
Xiaoping Chen   University of Science and Technology of China,
                           China
Xieyuanli Chen   University of Bonn, Germany
Esther Colombini   Unicamp, Brazil
Christian Deppe   Festo Didactic SE, Germany
Klaus Dorer   Hochschule Offenburg, Germany
Vlad Estivill-Castro   Universitat Pompeu Fabra, Spain
Farshid Faraji   Bonab Azad University, Iran
Alexander Ferrein   FH Aachen University of Applied Sciences,
                           Germany
Thomas Gabel   Frankfurt University of Applied Sciences,
                           Germany
Katie Genter   Red Ventures, USA
Reinhard Gerndt   Ostfalia University of Applied Sciences, Germany
Valentin Gies   University of Toulon, France
Justin Hart   University of Texas at Austin, USA
Dirk Holz   Google, USA
Luca Iocchi   Sapienza University of Rome, Italy

Masahide Ito	Aichi Prefectural University, Japan
Tetsuya Kimura	Nagaoka University of Technology, Japan
Irene Kipnis	Golda High School, Israel
Pedro U. Lima	University of Lisbon, Portugal
Luis José López Lora	Federación Mexicana de Robótica, Mexico
Patrick MacAlpine	University of Texas at Austin, USA
Sebastian Marian	Elrond Network, Romania
Julia Maurer	Mercersburg Academy, USA
Alex Mitrevski	Hochschule Bonn-Rhein-Sieg, Germany
Noriaki Mitsunaga	Osaka Kyoiku University, Japan
Alexander Moriarty	Intrinsic, USA
Tomoharu Nakashima	Osaka Metropolitan University, Japan
Daniele Nardi	Sapienza University of Rome, Italy
Luis Gustavo Nardin	École des Mines de Saint-Étienne, France
Itsuki Noda	Hokkaido University, Japan
Asadollah Norouzi	Singapore Polytechnic, Singapore
Oliver Obst	Western Sydney University, Australia
Hiroyuki Okada	Tamagawa University, Japan
Tatiana F.P.A.T. Pazelli	Federal University of Sao Carlos, Brazil
Eurico Pedrosa	University of Aveiro, Portugal
Paul G. Plöger	Bonn-Rhein-Sieg University of Applied Science, Germany
Daniel Polani	University of Hertfordshire, UK
Michael Quinlan	Everyday Robots, USA
Luis Paulo Reis	University of Porto, Portugal
A. Fernando Ribeiro	University of Minho, Portugal
Alessandra Rossi	University of Hertfordshire, UK
Raymond Sheh	Georgetown University, USA
Marco A. C. Simões	Universidade do Estado da Bahia, Brazil
Gerald Steinbauer	Graz University of Technology, Austria
Christoph Steup	Otto-von-Guericke University, Germany
Frieder Stolzenburg	Harz University of Applied Sciences, Germany
Komei Sugiura	Keio University, Japan
Toko Sugiura	Toyota National College of Technology, Japan
Jackrit Suthakorn	Mahidol University, Thailand
Jeffrey Tan	MyEdu AI Robotics Research Centre, Malaysia
Flavio Tonidandel	Centro Universitario da FEI, Brazil
Wataru Uemura	Ryukoku University, Japan
Rudi Villing	Maynooth University, Ireland
Arnoud Visser	University of Amsterdam, Netherlands
Ubbo Visser	University of Miami, USA
Oskar von Stryk	TU Darmstadt, Germany

Timothy Wiley	RMIT University, Australia
Aaron Wong	University of Newcastle, UK
Nader Zare	Dalhousie University, Canada

Additional Reviewers

Amini, Arash
Brizi, Leonardo
Gies, Valentin
Krause, Stefanie
Lekanne Gezegd Deprez, Hidde
Luo, Shengxin
Mosbach, Malte
Niemueller, Tim
Pasternak, Katarzyna
Suriani, Vincenzo
van Dijk, Daniel
Wong, Aaron

Contents

Main Track

Main Track

Object Recognition with Class Conditional Gaussian Mixture Model - A Statistical Learning Approach

Wentao Lu[1]([✉]), Qingbin Sheh[2], Liangde Li[3], and Claude Sammut[1]

[1] The University of New South Wales, Sydney, NSW 2052, Australia
{wentao.lu,c.sammut}@unsw.edu.au
[2] Kingsoft Office Software, Beijing, China
[3] University of California San Diego, La Jolla, CA, USA
lil009@ucsd.edu

Abstract. Object recognition is one of the key tasks in robot vision. In RoboCup SPL, the Nao Robot must identify objects of interest such as the ball, field features *et al*. These objects are critical for the robot players to successfully play soccer games. We propose a new statistical learning method, *Class Conditional Gaussian Mixture Model* (ccGMM), that can be used either as an object detector or a false positive discriminator. It is able to achieve a high recall rate and a low false positive rate. The proposed model has low computational cost on a mobile robot and the learning process takes a relatively short time, so that it is suitable for real robot competition play.

Keywords: Mixture model · Object recognition · Statistical learning

1 Introduction

In RoboCup 2016, SPL introduced a natural lighting rule that makes the competition environment more similar to a real soccer game. This significantly increases the difficulty for the robot to effectively and efficiently detect object of interest on the field. One way to address the natural lighting challenge is to binarize images taken by the robot's cameras. Adaptive thresholding binarizes an image using a sliding window to generate a monochrome image. While adaptive thresholding reduces variations due to different lighting across the field, it groups pixels into only two classes, black or white. Thus, to find such as balls or field lines in SPL are black and white, we need an effective algorithm to perform object recognition with this limited information.

In this paper, we propose a statistical learning algorithm, named *Class Conditional Gaussian Mixture Model* (ccGMM) to solve common vision recognition problems in RoboCup soccer. A Gaussian Mixture Model [10] is a probabilistic model for representing groups of sample data, called *components*, where a component is a statistical term to represent a sub-sample of the overall data

© The Author(s), under exclusive license to Springer Nature Switzerland AG 2023
A. Eguchi et al. (Eds.): RoboCup 2022, LNAI 13561, pp. 3–13, 2023
https://doi.org/10.1007/978-3-031-28469-4_1

population. Each component is characterized by its weighted mean and variance. The paper is organised as follows: the background of mixture models and the related works are discussed in Sect. 2; Sect. 3 will focus on the statistical approach to estimate the best number of components; In Sect. 4, the applications and experiments will be demonstrated to explain the performance of the proposed work.

2 Background

2.1 Mixture Models

A mixture model is a probabilistic model for representing a sub-population of the overall population in unsupervised learning. Mathematically, we can represent a mixture model by Eq. 1 [5], where we first sample the latent variable z and then sample the observation x from a distribution given z.

$$p(z, x) = p(z)p(x|z) \tag{1}$$

If the prior probability $p(x|z)$ is a Gaussian Distribution, this model is called Gaussian Mixture Model.

Equation 2 shows a weighted univariate Gaussian mixture model.

$$p(\theta) = \sum_{1}^{K} w_i \mathcal{N}(\mu_i, \sigma_i^2), where \sum_{1}^{K} w_i = 1 \tag{2}$$

Generally, binarized images in robot vision are 2-dimensional arrays. Hence, to apply a GMM, a multivariate GMM is needed, where each component (i.e. sub-population) is represented by the weighted mean vector μ_i and covariance matrix Σ_i. Equation 3 shows the multivariate version of the GMM.

$$p(\theta) = \sum_{1}^{K} w_i \mathcal{N}(\mu_i, \Sigma_i), where \sum_{1}^{K} w_i = 1 \tag{3}$$

In machine learning, a GMM is used as a form of unsupervised learning where the task is to cluster the data, without labels, into groups by given metrics. The GMM does not perform prediction, as is in supervised learning. However, in this work, we follow training the GMM with a *Maximize A Posterior* algorithm. This combined model is called a *Class Conditional Gaussian Mixture Model* as it does clustering and prediction in one model. The details of the method are discussed in Sect. 3.

2.2 Related Work

Gaussian Mixture Models are widely used for image categorisation. Greenspan and Pinhas [4] use a GMM to extract coherent image regions for image matching. Compared with our work, both use a GMM plus a posterior process to build the final algorithm, however, the main differences are: firstly, they use the GMM as an image descriptor extractor, while we use the GMM to represent images; second, they use the trained GMM to perform image matching while we do prediction. Ribeiro and Gonzaga's [4] give an analysis of how different real-time GMMs perform in background subtraction. Similar to our approach, they use a pixel-wise model to generate the Gaussian components and since their main focus is on video processing, real-time processing is also an essential requirement. The main difference is that background subtraction is a typical unsupervised learning application that groups coherent pixels into certain cluster, in their case, the hand regions.

In Robocup, there is also an application of GMMs in robot vision. Team TJArk propose a GMM-based approach to perform field colour recognition [11]. In their work, they use a GMM for green pixel recognition to find the field area. Similar to background subtraction, the GMM maintains components of different colours and then groups similar pixels into clusters.

3   Algorithm Design

3.1   Order Estimation

The first critical information we want to know is what is the best number of components we create. There are different methods for achieving this. Brute force search can be used. The main disadvantage of this approach is that it uses no prior knowledge and takes a long time to find the best number of components within a preset range. Furthermore, this approach can not guarantee that we can find the globally optimum number of components. Alternatively, we can perform a statistical estimation to find the estimated lower and upper bound of the best number of components that can be used in model selection. We test three different methods and discuss them below.

Algebraic Moment-Based Method. We first apply the method proposed by Dacunha-Castelle and Gassiat [3]. Considering the components in our application are multivariate, we can represent our model as in Eq. 4, where Σ_i^{-1}, $i = 1, \ldots, r$ are r different covariance matrices.

$$Q = \sum_1^r \pi_i G(\Sigma_i^{-1} \cdot) \tag{4}$$

After the data have been prepossessed, we want to know what is the optimal value for r, the number of components. We take the result of the moment-based estimation as in Proposition 2 of Dacunha-Castelle and Gassiat. We select $q = (r(r-1)/2 + 1))$ different unitary vectors $v_i, i = 1, \ldots, q$ and the estimators of their algebraic moments can be calculated from Eqs. 5 to 7.

$$X_t = S_t \cdot Y_t \tag{5}$$

Y_t is a random vector with distribution G and S_t is a random matrix with distribution μ. Then we can get

$$E(X_1 \cdot X_1^T) = E(S_1^2) \tag{6}$$

$$E((X_1 \cdot X_1^T)^k) = E(S_1^{2k})E((Y_1 \cdot Y_1^T)^k), \ for \ every \ k \in \mathbb{Z}^+. \tag{7}$$

The estimator of order $\hat{r}_n$ can be derived from the following function.

$$\hat{r}_n = \arg\min_p \sum_{i=1}^q |L(\hat{c}_n^p(v_i), p)| + A(p)l(n) \tag{8}$$

where $\hat{c}_n^p$ are the algebraic moments derived from Eqs. 5 to 7, $L(., p) = det H(., p)$ is a function defined on Hankel matrix [6,8] and $A(p)l(n)$ is a penalty term where $A(p)$ is a positive strictly increasing function and $l(n)$ is a positive function of n such that $\lim_{n \to +\infty} l(n) = 0$.

Maximum Penalized Likelihood Method. Assume we have a mixture model set G_Q, proposed by Keribin [7], for any order $r \in 1, \ldots, Q$ and $g \in G_r$, we can define the log sum function

$$l_n(g) = \sum_{i=1}^n \log g(X_i) \tag{9}$$

and the maximum likelihood statistic function

$$T_{n,r} = \sup_{g \in G_r} l_n(r) \tag{10}$$

Now we can find the boundary for the order estimator by calculating

$$\hat{q}_n = \arg \sup_{p \in \{1, \ldots, Q\}} (T_{n,p} - a_{n,p}) \tag{11}$$

where $a_{n,r}$ is some penalty function.

Likelihood Ratio Test Method. McLachlan and Rathnayake [9], proposed the Likelihood Ratio Test(LRTS) to find the smallest value of the number of components or the lower bound.

Again, we assume there are r components in the model, shown in Eq. 4. As described in their work, we first test the null hypothesis

$$H_0 : r = r_0 \tag{12}$$

against

$$H_1 : r = r_1 \tag{13}$$

for some $r_1 > r_0$. We define the likelihood ratio as

$$\lambda = \frac{\mathcal{L}(\Sigma_{r_0})}{\mathcal{L}(\Sigma_{r_1})} \qquad (14)$$

Now we can play the log trick and get

$$- 2 \log \lambda = 2\{\log \mathcal{L}(\Sigma_{r_1}) - \log \mathcal{L}(\Sigma_{r_0})\} \qquad (15)$$

We reject H_0 if the left-had side of Eq. 15 is sufficiently large. We keep doing the test by letting $r_1 = r_0 + 1$ until the log likelihood stops increasing.

3.2   Algorithm Overview

Due to the fact that most false positives for an object in SPL share some common features and objects of interest can show up in different orientations and scales, a trained ccGMM preserves two same length lists of Gaussian Mixtures of true positives and false positives. For example, T-junctions are often misclassified as corners; the ball is easily confused with the penalty spot; corners also can appear significantly different due to the robot's pose *et al.*

To deal with these challenges, two same length lists of Gaussian components are created to represent common objects on the soccer field. The best number of components is estimated using different statistical methods and then confirmed using the Grid Search algorithm.

Algorithm 1. Train

$X\_train, Y\_train, X\_test, Y\_test \leftarrow random\_split(data)$
$best\_accuracy \leftarrow 0$
while *Grid Search not finished* **do**
    $X\_train \leftarrow PCA(X\_train)$
    $TP\_components \leftarrow GMM(X\_train[label = True], \; Y\_train[label = True])$
    $FP\_components \leftarrow GMM(X\_train[label = False], \; Y\_train[label = False])$
    $accuracy \leftarrow 1 \cdot \{Inference(X\_test) = Y\_test\}/ \; length(X\_test)$
    if $accuracy > best\_accuracy$ **then**
        $best\_accuracy \leftarrow accuracy$
        $best\_model \leftarrow this\_model$
    end if
end while
return $best\_model$

Algorithm 1 shows the training method for the proposed model. Training is similar to the traditional Gaussian Mixture model where we use Principle Component Analysis(PCA) to reduce data dimensionality. We then perform a Grid Search by setting the lower and upper bounds derived from order estimation. During the Grid Search, in each iteration, we use the current hyper-parameter setting to get the trained GMM for both true positives and false positives for

---

Algorithm 2. Inference

Input $log\_weights,\ covariance,\ log\_prior,\ image\_data,\ PCA\_model, Precision\_matrix$:
$P, num\_components$
$X \leftarrow PCA\_model.transform(image\_data)$
$max\_likelihood \leftarrow -\infty$
$argmax\_index \leftarrow 0$
$i \leftarrow 0$
while $i < 2$ **do**
   $max\_ll \leftarrow -\infty$
   $j \leftarrow 0$
   $likelihood \leftarrow 0$
   $log\_likelihood \leftarrow 0$
   $component\_loglikelihood = Array[2]$
   while $j < num\_components$ **do**
      $X \leftarrow X - \tilde{X}$
      $logprob \leftarrow cholesky\_decomp(covariance[i][j]\ -\ prior * (XPX^{\top})[0])$
      $component\_loglikelihood[j] \leftarrow log\_weights[i][j]\ +\ logprob$
   end while
   if $component\_loglikelihood[j] > max\_ll$ **then**
      $max\_ll \leftarrow component\_loglikelihood[j]$
   end if
   while $j < num\_components$ **do**
      $likelihood\ += exp(component\_loglikelihood[j]\ -\ max\_ll)$
   end while
   $log\_likelihood\ +=\ max\_ll + \log(likelihood)$
   if $log\_likelihood > max\_likelihood$ **then**
      $max\_likelihood \leftarrow log\_likelihood$
      $argmax\_index \leftarrow i$
   end if
end while

---

a certain object. We use the Expectation Maximization algorithm to perform the training where the algorithm updates parameters in the model and tries to maximize the log-likelihood. At the end of each iteration, we evaluate the current model by calculating the metrics using the Inference algorithm. When the Grid Search meets the upper bound, the model with the best metrics will be selected.

Algorithm 2 is the inference process used both in training and prediction. Unlike the normal GMM, which is in unsupervised learning, ccGMM is a supervised learning model that performs classification. The main difference is that when we make the inference, we calculate the posterior likelihood for each component in the model, given the test data. We then perform *Maximize A Posterior* estimation to get the prediction. Hence, the method is called class conditional GMM. To avoid underflow, we play the log trick here to use the posterior log-likelihood instead of likelihood. Cholesky decomposition [1] is used to accelerate matrix decomposition.

4   Applications and Experiments

4.1   Field Feature Recognition

Team rUNSWift uses ccGMM as a false positives discriminator for detecting field features on the SPL soccer field.

Field features are the most important measurements for robot localisation and are found by probabilistic estimation, including variations of the Kalman filter or partial filter. To achieve minimum uncertainty, the localisation system requires accurate measurements that must not have many false positives. The rUNSWift code has a pipeline to detect possible field features in a camera frame.

The first part of the pipeline includes several handcrafted vision algorithms that find regions that contain candidate field features. Before introducing the ccGMM algorithm, we used the candidate regions directly for localisation, resulting in many false positives, greatly affecting the performance of the localisation system. The motivation for applying the ccGMM algorithm in the localisation system was to reduce the number of false positives.

The ccGMM is now the second part of the pipeline, which acts as a discriminator to remove false positives. Since there are many different kinds of field features, we maintain several ccGMM models for the important features.

To test how many false positives are removed by ccGMM, we place the robot at three different locations on the field and performs a quick scan for 60 s. We then compare how many candidate field features are accepted by the handcrafted algorithm but are rejected by the ccGMM discriminator.

A second qualitative experiment is designed to analyse the quality of the ccGMM discriminator. The robot performs normal game play behaviours on the field, randomly sampling field features rejected or accepted by the ccGMM. Manual qualitative analysis is done after the session.

As Table 1 and Fig. 1 show, the average rejection rates for T-junctions and corners are 92.7% and 87.0% respectively. This high rejection rate is mostly due to the handcrafted candidate region proposal algorithm counting pixels in extremely low resolution. In the next qualitative experiment, we analyse the sample rejected cases and discuss why so many are rejected.

Figures 2 and 3 show some accepted and rejected corners. The rejected candidates include shapes, such as robot feet, lines and the hardest case, centre curves. However, limited by the robot's perceptive and low resolution, some corners are still too similar to curves, resulting in misclassification. With most of the false positives removed, the robot's localisation remains stable most of the time, even in extreme lighting conditions, as reported by team rUNSWift 2019 [2].

Table 1. Static robot field feature experiment

Trial	Pose(X, Y, Theta)	T_Rejected	T_Total	Corner_Rejected	Corner_Total
1	$(-4500, 0, 0)$	284	286	3700	4588
2	$(-4500, 0, 0)$	666	680	3761	4335
3	$(-4500, 0, 0)$	595	605	4058	4413
4	$(2000, 0, 0)$	486	524	2020	2407
5	$(2000, 0, 0)$	380	414	1819	2350
6	$(2000, 0, 0)$	410	444	2261	2808
7	$(3000, -1500, 0)$	633	684	4435	4788
8	$(3000, -1500, 0)$	284	307	3831	4031
9	$(3000, -1500, 0)$	269	350	3893	4127

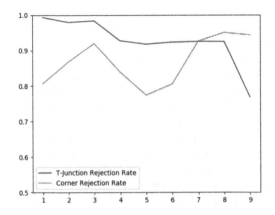

Fig. 1. Rejection rate

Fig. 2. Accepted corners

Fig. 3. Rejected corners

4.2   Bottom Camera Ball Recognition

Team rUNSWift also uses the ccGMM method on the Nao V6's bottom camera and both Nao V5's cameras for ball recognition. On the V6, we implemented a neural network based recognition system for the top camera. This neural network consists of three convolutional layers and two fully connected layers with the same input size as the ccGMM. Comparing the approaches, several advantages and disadvantages were apparent. First, the processing time is lower for ccGMM than for our convolutional neural network (CNN). While the CNN has a longer detection range, the ccGMM works better in low-resolution, with more robust detection. Considering these facts, we chose to use ccGMM on the bottom camera with half resolution (640 × 480) for ball recognition. For comparison, we conduct an experiment where we set up a single Nao robot to track the ball in different field settings.

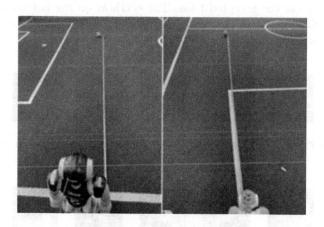

Fig. 4. Two standing position

Figure 4 shows two different standing position settings for the experiment. In this experiment, the observer robot will stand on the boundary line to perform normal stand behaviour while trying to track the ball. The ball observation range is recorded while the robot can consistently see and track the ball. In this case, we define the consistent tracking as not losing the ball for consecutive 10 frames. Also, 0 to 3 opponent robots are randomly placed near the ball during the experiment. The experiment is terminated when either the proposed ccGMM or the CNN is called 1000 times after the ball is placed at the edge of observer robot's consistent tracking range. Both observation range and running time are reported as the overall average shown in Table 2.

Table 2. Experiment result

Algorithm	Time/Call (us)	Consistent observation range (m)
ccGMM	**158**	3.56
CNN	739	**3.92**

Fig. 5. Accepted balls

Figure 5 shows the accepted balls. The ccGMM on the bottom camera has almost zero false positives and works consistently with the ball on field line or at the robot's feet.

Fig. 6. Mean Components for True Positive Balls(Weights: 0.18, 0.19, 0.19, 0.21, 0.23)

Fig. 7. Mean Components for False Positive Balls(Weights: 0.15, 0.18, 0.18, 0.22, 0.27)

Figures 6 and 7 show a sample mean components for a trained ccGMM for ball recognition where we set *number of components* = 5. Weights for each component are included.

5   Conclusion

We proposed a new statistical learning method for real-time object recognition. The *Class Conditional Gaussian Mixture Model* (ccGMM) is a novel combination of aspects of unsupervised and supervised learning. The method requires only limited computational power and works robustly for different light conditions, and on low-resolution images in the RoboCup SPL competition. In further research, we will extend and generalise the method to be applicable in a wider variety of robotics vision problems.

References

1. Cholesky decomposition. https://doi.org/10.1093/oi/authority.20110803095609584. https://www.oxfordreference.com/view/10.1093/oi/authority.20110803095609584
2. Ashar, J., et al.: Runswift team report (2019). https://github.com/UNSWComp uting/rUNSWift-2019-release/raw/master/rUNSWift_Team_Report.pdf
3. Dacunha-Castelle, D., Gassiat, E.: The estimation of the order of a mixture model. Bernoulli 279–299 (1997)
4. Greenspan, H., Pinhas, A.T.: Medical image categorization and retrieval for PACS using the GMM-KL framework. IEEE Trans. Inf. Technol. Biomed. $11(2)$, 190–202 (2007). https://doi.org/10.1109/TITB.2006.874191
5. Grosse, R., Srivastava, N.: Lecture 16: Mixture models. https://www.cs.toronto.edu/~rgrosse/csc321/mixture_models.pdf
6. Karlin, S., Studden, W.: Tchebycheff Systems: With Applications in Analysis and Statistics. Pure and Applied Mathematics. Interscience Publishers (1966). https://books.google.com.au/books?id=Rrc-AAAAIAAJ
7. Keribin, C.: Consistent estimation of the order of mixture models. Sankhyā: Indian J. Stat. Series A 49–66 (2000)
8. Krein, M.G., Nudelman, A.A.: The Markov moment problem and extremal problems: ideas and problems of P. L. Cebysev and A. A. Markov and their further development/by M. G. Krein and A. A. Nudelman; [translated from the Russian by Israel Program for Scientific Translations, translated by D. Louvish]. American Mathematical Society Providence, R.I (1977)
9. McLachlan, G.J., Rathnayake, S.: On the number of components in a Gaussian mixture model. Wiley Int. Rev. Data Min. Knowl. Disc. $4(5)$, 341–355 (2014)
10. Reynolds, D.: Gaussian mixture models. In: Li, S.Z., Jain, A. (eds.) Encyclopedia of Biometrics, pp. 659–663. Springer, Boston (2009). https://doi.org/10.1007/978-0-387-73003-5_196
11. TJArk: Team description paper & research report (2017). https://github.com/TJArk-Robotics/coderelease_2017/blob/master/TJArkTeamResearchReport2017.pdf

Instance-Based Opponent Action Prediction in Soccer Simulation Using Boundary Graphs

Thomas Gabel[(✉)] and Fabian Sommer

Faculty of Computer Science and Engineering, Frankfurt University of Applied
Sciences, 60318 Frankfurt am Main, Germany
{tgabel,fabian.sommer}@fb2.fra-uas.de

Abstract. The ability to correctly anticipate an opponent's next action
in real-time adversarial environments depends on both, the amount of
collected observations of that agent's behavior as well as on the capability
to incorporate new knowledge into the opponent model easily. We present
a novel approach to instance-based action prediction that utilizes graph-
based structures for the efficiency of retrieval, that scales logarithmically
with the amount of training data, and that can be used in an online and
anytime manner. We apply this algorithm to the use case of predicting
a dribbling agent's next action in Soccer Simulation 2D.

1 Introduction

Opponent modeling and action prediction have a long history in robotic soccer.
The ability to anticipate what an opponent player is going to do in the next time
step and reacting with appropriate counter measures can bring about significant
advantages to one's own agents. In this paper, we extend our previous work
[8] on predicting the low-level behavior of agents in Soccer Simulation 2D into
a direction that makes it scalable and practically applicable under the hard
real-time constraints that are imposed in this domain. Hitherto, we approached
the task of predicting an opponent's next action in an instance-based manner
by explicitly storing all training instances in memory and then (e.g. during a
match) searching linearly for the nearest neighbor to the current situation and
using that neighbor's class label as the predicted next opponent action.

Unfortunately, it is well-known that instance-based classification approaches,
when being applied in the described naive manner, scale poorly (usually linearly)
with the amount of training data. As a consequence, when intending to apply
these ideas in our soccer simulation competition team, we arrive at a set of
challenging requirements:

a) *Instance-based:* Instance-based learning is a lazy learning approach; new
   instances shall, if necessary, be memorized easily.
b) *Real-time capable:* There are hard real-time constraints in robotic soccer. Thus, when searching for the nearest neighbor from the set of stored
   instances, hard time limits must be respected.

© The Author(s), under exclusive license to Springer Nature Switzerland AG 2023
A. Eguchi et al. (Eds.): RoboCup 2022, LNAI 13561, pp. 14–26, 2023
https://doi.org/10.1007/978-3-031-28469-4_2

c) *Incremental:* The approach should be applicable online, which means that it must be possible to incorporate new experience on the fly during the application without the need to perform some computationally heavy relearning.

d) *Anytime:* Usually, the available computational budget varies from time to time. Thus, it would be highly desirable to have an anytime prediction algorithm whose accuracy improves with more computational power.

e) *Simplicity:* The desired prediction algorithm shall be simple to implement and have no dependency on certain libraries or mighty learning frameworks such that it can be utilized easily on any (competition) machine.

Needless to say that any prediction algorithm with linear time requirements in the number n of training examples is ruled out as it would perform too poorly and not scale for larger dataset sizes. Hence, for an instance-based prediction algorithm ideally logarithmic complexity is desired or at least a dependency on n according to some power law with a power value significantly below one.

Our contribution in this paper is twofold. On the one hand, we propose a novel instance-based classification approach that fulfills the mentioned five requirements. At the heart of this approach is the construction of an index structure to efficiently guide the search for most similar instances that we call a *Boundary Graph*. We build up the graph structure from training data, which means that its topology is not fixed a priori. It is also worth noting that the construction process can be applied in an online setting, i.e. no batch access to the full dataset of instances is needed and, hence, the graph index structure can be extended as more and more training examples come in. Both, the build-up as well as the employment of that graph-based index structure are inherently stochastic – a fact that we found to substantially improve the robustness of the approach as well as to reduce its dependency on other factors like the order of presentation of instances during learning.

On the other hand, we empirically evaluate the performance of the delineated approach for the use case of predicting a dribbling opponent agent's next action in soccer simulation. Knowing the opponent's next action with high certainty before it is executed by the opponent may enable our agents to simultaneously compute the best possible answer to that future action and, hence, improve our team's playing strength.

We start by providing background knowledge and reviewing related work in Sect. 2. While Sect. 3 presents the mentioned boundary graph-based approach in full detail, in Sect. 4 we return to robotic soccer simulation, explain how to utilize the proposed approach for the dribble action use case and present corresponding empirical findings.

2   Background and Related Work

In the following, we outline the basics that are needed to understand our approach as well as the application use case it is intended for and discuss relevant related work.

2.1   Robotic Soccer Simulation

In RoboCup's 2D Simulation League, two teams of simulated soccer-playing agents compete against one another using the Soccer Server [12] as real-time soccer simulation system. The Soccer Server allows autonomous software agents to play soccer in a client/server-based style: It simulates the playing field, communication, the environment and its dynamics, while the player clients connect to the server and send their intended actions (e.g. a parameterized kick or dash command) once per simulation cycle to the server. The server takes all agents' actions into account, computes the subsequent world state and provides all agents with information about their environment.

So, decision making must be performed in real-time or, more precisely, in discrete time steps: Every 100 ms the agents can execute a low-level action and the world-state will change based on the individual actions of all players. Speaking about low-level actions, we stress that these actions themselves are "parameterized basic actions" and the agent can execute only one of them per time step:

- $dash(x, \alpha)$ – lets the agent accelerate by relative power $x \in [0, 100]$ into direction $\alpha \in (-180°, 180°]$ relative to its body orientation
- $turn(\alpha)$ – turn the body by $\alpha \in (-180°, 180°]$ where, however, the Soccer Server reduces α depending on the player's current velocity (inertia moment)
- $kick(x, \alpha)$ – kick of the ball (only, if the ball is within the player's kick range) by relative power $x \in [0, 100]$ into direction $\alpha \in (-180°, 180°]$
- There exist a few further actions (like tackling, playing foul, or, for the goal keeper, catching the ball) whose exact description is beyond scope.

It is clear that these basic actions must be combined cleverly in consecutive time steps in order to create "higher-level actions" like intercepting balls, playing passes, marking players, or doing dribblings.

2.2   Related Work on Opponent Modeling

Opponent modeling enables the prediction of future actions of the opponent. In doing so, it also allows for adapting one's own behavior accordingly. Instance-based approaches have frequently been used as a technique for opponent modeling in multi-agent games [5], including the domain of robotic soccer [2,6].

In [15], the authors make their simulated soccer agents recognize currently executed higher-lever behaviors of the ball leading opponent. These include passing, dribbling, goal-kicking and clearing. These higher-level behaviors correspond to action sequences that are executed over a dozen or more time steps. The authors of [14] deal with the instance-based recognition of skills (shoot-on-goal skill) executed by an opponent soccer player, focusing on the adjustment of the distance metrics employed. In [8] we argued that opponent modeling is useful for counteracting adversary agents, but that we disagree with the authors of [14] claiming that "in a complex domain such as RoboCup it is infeasible to predict an agent's behavior in terms of primitive actions". Instead we have shown prototypically in [8] that a low-level action prediction can be achieved during an on-going play using instance-based methods. We grasp this prior work of ourselves now, addressing the crucial point that we omitted to handle in that paper: Instance-based learning algorithms learn by remembering instances, which is why, for certain applications, specifically data intensive ones, retrieval times over the set of stored instances can quickly become the system's bottleneck. This issue is a showstopper in a real-time application domain like robotic soccer simulation.

2.3   Related Work on Index Structures for Efficient Retrieval

Index structures in instance-based methods are supposed to more efficiently guide the search for similar instances. Before the actual retrieval utilizing an index structure can take place that structure must be created. Tree-based structures have often been employed to speed up access to large datasets (e.g. geometric near-neighbor access trees [4] or nearest vector trees [10]). Tree-based algorithms that also feature online insertion capabilities include cover trees [3], boundary trees [11] (see below), or kd-trees [16] where the latter have the advantage of not requiring full distance calculations at tree nodes.

Boundary Trees [11] are a powerful tree-based index structure for distance-based search. They consist of nodes representing training instances connected by edges such that any pair of parent and child node belongs to different classes[1]. This fact is eponymous as with each edge traversal a decision boundary is crossed.

Given a boundary tree T and a new query q, the tree is traversed from its root by calculating the distance between q and all children of the current node, moving to and traversing successively that child which has the lowest distance to q. Boundary trees use a parameter $k \in [1, \infty]$ that determines the maximal number of children any node is permitted to have. The retrieval is finished, if a leaf has been reached or if the current (inner) node v has less than k children and the distance between q and v is smaller than the distance between q and all children of v. This way, a "locally closest" instance x^* to the query is found, meaning that neither the parent(s) of x^* nor the children of x^* are more similar.

The tree creation procedure for boundary trees is inspired by the classical IB2 algorithm [1]. The next training instance x_i is used as query using the so far existing boundary tree T_{i-1}. If the result of the tree-based retrieval returns an instance x^* whose class label does not match the class label of x_i (i.e. x_i could not be "solved" using T_{i-1}), then x_i is added as a new child node of x^*.

In [11], Mathy et al. propose to extend the described approach to an ensemble of boundary trees, which they name a boundary forest (BF). Essentially, they train an ensemble of (in that paper usually 50) boundary trees on shuffled versions of the training data set and employ different kinds of voting mechanisms (e.g. majority voting or Shepard weighted average [13]) using the retrieval results of the boundary trees. The Boundary Graph approach we are presenting in the next section takes some inspiration from boundary trees which is why we also use them as a reference method in our empirical evaluations.

---

[1] While the definition given here focuses on classification tasks, a straightforward generalization to other tasks like regression or mere retrieval can easily be made.

3   Boundary Graphs

It is our goal to develop an instance-based technique that covers both, a method to decide which instances to store in memory and which not as well as algorithms to build up and employ an index structure that facilitates an efficient retrieval. Boundary graphs (BG) in combination with techniques to create and utilize them, represent the backbone of our approach.

3.1   Notation

In what follows, we assume that each instance $x \in \mathbb{R}^D$ is a D-dimensional tuple of real values and has a label $l(x) \in \mathcal{L} \subset \mathbb{R}^m$ attached to it (where in case of a classification task $\mathcal{L}$ is simply the enumeration of class labels). Distance between instances is measured using a distance metric $d : \mathbb{R}^D \times \mathbb{R}^D \to \mathbb{R}^+$ that for any two instances returns a non-negative real number $d(x, y)$. Note that we do not impose any further requirements on d throughout the rest of the paper, except that, for ease of presentation, we assume it to be symmetric. Furthermore, we need a metric function $d_l : \mathcal{L} \times \mathcal{L} \to \mathbb{R}^+$ to assess the difference of label vectors.

For a given set of training instances $\mathcal{X} = \{x_1, \ldots, x_n\}$, a *Boundary Graph* $\mathcal{B} = (V, E)$ is an undirected graph without loops with a set of nodes $V \subseteq \mathcal{X}$ and a set of edges

$$E \subseteq \{(x_i, x_j) | x_i, x_j \in V \text{ and } i \neq j\}, \tag{1}$$

where, by construction, each edge from E connects only instances with differing labels. This means, for each $(x_i, x_j) \in E$ it holds

$$d_l(l(x_i), l(x_j)) > \varepsilon \tag{2}$$

where $\varepsilon > 0$ is a threshold that defines when two label vectors are considered to be different. The definition given so far and the relations in Formula 1 and 2 are not finalized, most specifically since Eq. 1 gives just a subset specification. We are going to concretize this specification in the next paragraphs, emphasizing upfront that the boundary graphs we are creating will be a sparse representation of the case data and, thus, contain only a tiny fraction of the edges that would be allowed to be contained in E according to Eqs. 1 and 2.

3.2   Querying a Boundary Graph

Given a query $q \in R^D$ and a boundary graph $\mathcal{B} = (V, E)$, the retrieval algorithm moves *repeatedly* through the graph structure, calculating the distance between q and the current node $x \in V$ as well as between q and the neighbors of x, i.e. for all $v \in V$ for which an edge $(x, v) \in E$ exists. It successively and greedily "moves" onwards to the node with the lowest distance to q until some minimum x^* has been reached, which means that $d(q, x^*) \leq d(q, x) \forall (x^*, x) \in E$.

Importantly, this procedure is repeated for r times, where the starting node is selected randomly from V each time. Hence, r determines the number of random

---

BG_PREDICT(q, $\mathcal{B}$, r)
Input: query $q \in \mathbb{R}^D$,
    boundary graph $\mathcal{B} = (V, E)$,
    number r of random retrieval
    restarts,
    amalgamation function $\mathcal{A}$
Output: BG-based prediction $\mathcal{R}(q)$
1:   // *retrieval*
2:   $\mathcal{N}_q^r \leftarrow$ BG_RETRIEVE($q, \mathcal{B}, r$)
3:   // *prediction (cf. Eqn. 4-6)*
4:   $\mathcal{R}(q) \leftarrow \mathcal{A}(\mathcal{N}_q^r)$
5:   **return** $\mathcal{R}(q)$

BG_RETRIEVE(q, $\mathcal{B}$, r)
Input: query $q \in \mathbb{R}^D$,
    boundary graph $\mathcal{B} = (V, E)$ with $V \neq \emptyset$,
    number r of random retrieval restarts
Output: r-dimensional vector $\mathcal{N}_q^r$ of
    potential nearest neighbors
1:   $\mathcal{N}_q^r \leftarrow r$-dimensional vector
2:   **for** $i = 1$ **to** r **do**
3:     $x^\star \leftarrow$ random node from V
4:     $stop \leftarrow false$
5:     **while** $stop = false$ **do**
6:       $x \leftarrow \arg\min_{v \in V \text{ s.t. } (x^\star, v) \in E} d(q, v)$
7:       **if** $d(q, x) < d(q, x^\star)$
8:       **then** $x^\star \leftarrow x$ **else** $stop \leftarrow true$
9:     $\mathcal{N}_q^r[i] \leftarrow x^\star$
10: **return** $\mathcal{N}_q^r$

Algorithm 1: Boundary Graph-Based Prediction and Retrieval

---

retrieval starting points from which the distance-guided search is initiated. Consequently, as retrieval result a vector $\mathcal{N}_q^r = (n_1, \ldots, n_r)$ of r estimated nearest neighbors is obtained.

Algorithmically, we embed the delineated step (function BG_RETRIEVE in Algorithm 1) into the superjacent function BG_PREDICT for boundary graph-based prediction which, effectively, performs both, the retrieval task and the prediction on top of it. The entries of the vector of r nearest neighbor estimates are combined to form an overall prediction $\mathcal{R}(q)$ using some amalgamation function $\mathcal{A}$, such that

$$\mathcal{R}(q) = \mathcal{A}(\mathcal{N}_q^r) = \mathcal{A}((n_1, \ldots, n_r)). \tag{3}$$

For classification tasks, we might use a simple majority vote

$$\mathcal{A}((n_1, \ldots, n_r)) \in \arg\max_{t \in \mathcal{L}} |\{n_j | l(n_j) = t, j = 1, \ldots, r\}| \tag{4}$$

or an inverted distance-weighted voting scheme, like

$$\mathcal{A}((n_1, \ldots, n_r)) \in \arg\max_{t \in \mathcal{L}} \sum_{j=1}^{r} \begin{cases} 1/d(q, n_j) & \text{if } l(n_j) = t \\ 0 & \text{else} \end{cases}. \tag{5}$$

In a similar manner, for regression tasks the estimated value becomes [13]

$$\mathcal{A}((n_1, \ldots, n_r)) = \frac{\sum_{j=1}^{r} l(n_j)/d(n_j, q)}{\sum_{j=1}^{r} 1/d(n_j, q)}. \tag{6}$$

A pseudo-code summary of the entire retrieval and prediction approach using a BG is given in Algorithm 1. For the empirical case study presented below we stick to a simple majority vote according to Eq. 4 and employ a normalized L_1 norm as distance measure d. Before, however, we can utilize a BG, we must build it up which is why we focus on the construction of boundary graphs next.

BG_CONSTRUCT($\mathcal{X}$, r)	BG_TRAIN($\mathcal{B}$, x, r)
Input: set of train instances $\mathcal{X} = \{x_1, \ldots, x_n\}$, number r of random retrieval restarts	**Input:** single (new) instance x, boundary graph $\mathcal{B} = (V, E)$, number r of random retrieval restarts
Output: boundary tree $\mathcal{B}$	**Requires (global variables):** amalgamation function $\mathcal{A}$, metric d and d_l, label discrimination threshold ε
1: $\mathcal{B} \leftarrow (\emptyset, \emptyset)$	**Output:** (possibly extended) boundary graph $\mathcal{B}$
2: // loop over all instances	1: if $V = \emptyset$ then $V \leftarrow V \cup x$
3: **for** $i = 1$ **to** n **do**	2: **else**
4: $\mathcal{B} \leftarrow$ BG_TRAIN($\mathcal{B}, x_i, r$)	3: $\mathcal{N}_x^r \leftarrow$ BG_RETRIEVE($x, \mathcal{B}, r$)
5: **return** $\mathcal{B}$	4: **for** $i = 1$ **to** r **do**
	5: $\delta \leftarrow d_l(l(x), l(\mathcal{N}_x^r[i]))$
	6: **if** $\delta > \varepsilon$ **then**
	7: $V \leftarrow V \cup x$
	8: $E \leftarrow E \cup (x, \mathcal{N}_x^r[i])$
	9: **return** (V, E)

Algorithm 2: Construction of and Retain Procedure for Boundary Graphs

3.3   Graph Construction

We assume that the instances x_1 to x_n from the set of training instances $\mathcal{X}$ are presented to the boundary graph construction algorithm successively. Given a single training instance x_i, the algorithm first queries the boundary graph $\mathcal{B}_{i-1} = (V_{i-1}, E_{i-1})$ which has been trained for the preceding $i - 1$ training instances, yielding a vector $\mathcal{N}_{x_i}^r = (n_1, \ldots, n_r)$ of r possible nearest neighbors. The algorithm then iterates over these n_j ($j = 1, \ldots, r$) and, if $d_l(l(x_i), l(n_j)) > \varepsilon$ (i.e. n_j does "not solve" x_i, which in the case of classification tasks boils down to $l(x_i) \neq l(n_j)$), then x_i is added as a new node to V_{i-1} and a (bidirectional) edge (x_i, n_j) is added to E_{i-1}. The resulting, extended boundary graph is accordingly denoted as $\mathcal{B}_i$. To sum up, training instances are added as nodes to the graph (including connecting edge), if the algorithm stochastically discovers a random retrieval starting point for which the currently existing boundary graph's prediction would be wrong and where, hence, a correction is needed.

Again, a pseudo-code summary of the algorithm to constructively building up a boundary graph for a sequence of training instances $\mathcal{X}$ is provided in Algorithm 2, denoted as BG_CONSTRUCT. Note that the algorithm can be easily deployed in an online setting where new instances arrive during runtime by simply calling the BG_TRAIN function given in the right part of Algorithm 2. Additionally, Fig. 1 visualizes exemplary boundary graphs for two synthetic two-dimensional two-class problem.

Constructing vs. Applying the Graph Structure. As we will show below, the algorithms described have a logarithmic retrieval complexity in the amount of training instances n for the opponent action prediction dataset we use subsequently. Accordingly, training time scales mildly as well because each train step

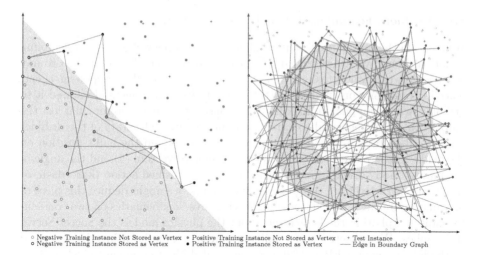

○ Negative Training Instance Not Stored as Vertex   • Positive Training Instance Not Stored as Vertex   + Test Instance
○ Negative Training Instance Stored as Vertex   • Positive Training Instance Stored as Vertex   —— Edge in Boundary Graph

Fig. 1. Two Exemplary Visualizations of Boundary Graphs for Synthetic Domains: Out of the 80 training instances (left), 19 are included in the graph's set of vertices (11 from the negative class, 8 from the positive one), when trained using $r = 9$. Among that set of nodes, there are 88 possible edges that would cross the decision boundary. From those, 40 are included in the graph's set of edges. The boundary graph for the "doughnut" domain (right) has been constructed using $r = 3$. The graph stores 203 out of the 400 training instances and connects them by 534 edges.

essentially includes a retrieve step, rendering the complexity of training to be a loop of n repetitions wrapped around the retrieval procedure ($O(n \log n)$).

The boundary graph approach has the favorable characteristic to be an *any-time retrieval algorithm*. By handling the parameter r of random retrieval starting points within the graph differently during training (r_t) and the application (r_a) of the learned graph, i.e. separating $r_t := r$ from r_a ($r_a \neq r$), one can gain a significant performance boost by letting $r_a > r_t$, given that a sufficient amount of time is available for the system to respond to a query. This is a desirable property in real-time and online application settings since the accuracy of the retrieval grows with r_a as we will delineate in the next section.

4 Empirical Evaluation

Our empirical investigations on the boundary graph approach were primarily driven by our target application problem of predicting an opponent soccer player's next low-level action. We focus on a dribbling opponent, leaving the investigation of other opponent behaviors for future work. We first more introduce the task at hand more precisely and then present achieved classification results including an analysis of our algorithms' scaling behavior. In a separate paper [9], we present detailed results on the performance of our proposed algorithms for a variety of classical benchmark datasets beyond the realm of robotic soccer.

4.1   Problem Formulation and Data Collection

We focus on the task of predicting a ball leading opponent player's next dribbling action, although we stress that our approach could generally be applied to any other action prediction task as well. Therefore, we selected an opponent agent, placed it randomly on the field with the ball in its kick range and allowed it to dribble towards our goal for maximally 20 consecutive time step (or till it lost the ball) when it was relocated to a new random position. The state of the dribbler was described by a 9-tuple consisting of the player's and the ball's position on the field (4), the player's and ball's current velocity vectors (4) and the angle of the player's body orientation (1). The actually performed action (kick, dash, or turn) was extracted from the game log file, though we might deduce this piece of information during a match, too, by applying inverse kinematics on two consecutive states exploiting the knowledge about the physics models the Soccer Server [12] applies. We collected half a million training examples using the described methodology using a FRA-UNIted agent as dribbling opponent which utilizes its established dribbling behavior that was trained with reinforcement learning [7]. The class distribution in this dataset features 7.2% turning, 20.4% kicking, and 72.4% dashing actions such that a naive classifier that always predicts the majority class would yield an error of 27.6%. Note that in the context of the evaluation presented here, we solely focused on the classification of the *type* of the performed action, not on its real-valued parameter(s) (cf. Sect. 2.1).

We compare our boundary graph approach to the classical nearest neighbor algorithm (which linearly iterates over all stored training instances) as well as to the boundary forest approach (BF) from the literature (cf. Sect. 2.3). We measure performance in terms of the achieved classification error on an independent test set as well as in terms of required real-time (on a contemporary 3 GHz CPU, single-core, i.e. without any parallelization[2]).

4.2   Results

Table 1 summarizes the remaining classification errors when predicting the opponent's low-level dribble actions for different training set sizes n. All numbers reported are averages over 100 repetitions of the experiment using different random number seedings. As expected, the nearest neighbor classifier turns out to be a simple, but computationally prohibitive baseline. When opposing boundary forests and boundary graphs, it is advisable to compare settings that are conceptually similar, viz when the number t of trees and the number r of random retrieval restarts match. The result table reports results for $r = t \in \{50, 100\}$ and shows that boundary graphs slightly, but consistently outperform the forest approach except for small training set sizes where, however, all approaches "fail" since their accuracy is not so far off the error of the naive classifier (27.6%) that just predicts the majority class.

---

[2] We emphasize that all discussed approaches are easily parallelizable and that computation times could, thus, be reduced dramatically given the appropriate hardware.

Table 1. Classification errors and belonging standard errors of the discussed algorithms in percent subject to different amounts of training data for 100 experiment repetitions. Better-performing algorithms (between BF and BG only) are highlighted in bold.

n	k-NN ($k = 1$)	BF ($t = 50$)	BG ($r = 50$)	BF ($t = 100$)	BG ($r = 100$)
400	27.80 ± 0.48	$\mathbf{28.26 \pm 0.44}$	29.41 ± 0.47	$\mathbf{27.64 \pm 0.45}$	29.44 ± 0.45
1600	23.92 ± 0.19	$\mathbf{25.12 \pm 0.21}$	25.22 ± 0.20	$\mathbf{24.71 \pm 0.21}$	25.49 ± 0.20
6400	22.10 ± 0.10	22.14 ± 0.11	$\mathbf{21.88 \pm 0.11}$	$\mathbf{21.51 \pm 0.11}$	21.92 ± 0.11
25600	18.77 ± 0.06	18.55 ± 0.06	$\mathbf{18.07 \pm 0.05}$	$\mathbf{17.94 \pm 0.05}$	18.19 ± 0.05
51200	16.64 ± 0.02	16.31 ± 0.04	$\mathbf{15.72 \pm 0.03}$	15.75 ± 0.03	$\mathbf{15.72 \pm 0.02}$
102400	13.46 ± 0.03	13.27 ± 0.02	$\mathbf{12.62 \pm 0.02}$	12.78 ± 0.03	$\mathbf{12.61 \pm 0.02}$
204800	8.70 ± 0.01	9.25 ± 0.02	$\mathbf{8.77 \pm 0.01}$	8.79 ± 0.01	$\mathbf{8.64 \pm 0.01}$
409600	5.42 ± 0.01	5.90 ± 0.01	$\mathbf{5.43 \pm 0.01}$	5.55 ± 0.01	$\mathbf{5.22 \pm 0.01}$

The left part of Fig. 2 visualizes the scaling behavior of boundary graphs (black) and boundary forests (gray), reporting the average number of milliseconds required to answer a single test query, i.e. to predict the opponent's next dribble action, subject to different amounts of training data that has been processed to generate the BF/BG. Apparently, boundary graphs need about a third more computational effort compared to their same-sized tree-based counterparts, but achieve lower classification errors as discussed in the preceding paragraph. It is worth noting that we have set the value of the BF parameter k (cf. Sect. 2.3) to infinity during all our experiments. Setting k to a finite value would further reduce the computational requirements of that algorithm, but at the same time impair its performance even more as delineated by [11].

Another interesting observation is that a boundary graph ($r = 50$), which has been constructed using $n = 409.6k$ instances, stores about 60% of them as vertices in the graph (space complexity grows linearly with n). Yet, during a BG-based retrieval for a single test query q the distance calculation (which, essentially, represents the computational bottleneck) between q and stored instances must, effectively, be done for only $\approx 1.8\%$ of the n given training instances.

After all, the chart shows that any of the graph- or forest-based approaches have a logarithmic time complexity and could very well be deployed practically by a soccer-playing agent since the retrieval time of less than 40 ms (on the mentioned hardware) would fit well into a soccer simulation time step (even without any parallelization). Since the addition of a single new instance requires basically one retrieval plus a loop over r (which has constant effort in n), it requires roughly the same amount of computation as processing a test query and, thus, even an online extension of a boundary graph during a running match is feasible, for example when observing the current opponent dribbling. By contrast, the nearest neighbor classifier (also shown in the chart) has linear complexity and requires more than 50 ms already for 8000 stored instances (and even 3000 ms per test query for $n = 409.6k$) which renders this algorithm practically useless.

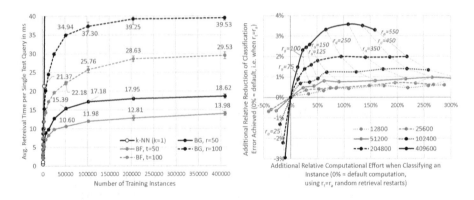

Fig. 2. Scaling Behavior of Boundary Graphs: See the text for details.

An outstanding characteristic of the boundary graph algorithm is its any-time behavior. By increasing the number of random retrieval restarts during the application phase (for example in a match, when in a specific time step less other computations are done for whatever reason) the accuracy of the prediction increases. This relationship is visualized in the right part of Fig. 2 for a boundary graph that has been *constructed* using $r_t = 50$ random retrieval restarts. So, the point of origin represents the "default" setting where $r_t = r_a = 50$. For positive x values the computational budget during application (not during training, i.e. $r_t = 50$ is not altered) has been increased expressing the *relative* extra effort on top of the default in terms of additional real-time. Likewise, negative x values denote that less computational power is invested into the retrieve process during testing. The ordinate shows the impact of the described variation of r_a in terms of *relative* gain/loss in classification performance compared to what is achieved with the default setting. So, for example, for $n = 409.6k$ training instances (here, processing a test query for $r_a = 50$ needs 18.6 ms on average) we observe that by doubling the retrieval time (+100%, i.e. 37.2 ms, corresponding to $r_a \approx 350$) the originally achieved classification error can be reduced by ca. 3.5%.

5   Conclusion

We have proposed boundary graphs as a useful and scalable tool for instance-based prediction. Although we have focused solely on its use for classification throughout this paper, the approach is general enough to cover other tasks like regression or mere instance retrieval as well. We provided algorithms for creating and utilizing boundary graphs and applied them successfully for the prediction of the next low-level action of a dribbling simulated soccer player. In so doing, we found that this approach scales very well and is applicable under hard real-time constraints even with large sets of training data which are required for high-quality predictions. Our next steps include the employment of this approach for determining the real-valued parameters of the predicted action which, of course, represents a regression task. Moreover, we also intend to evaluate in depth the performance of boundary graphs for other established benchmark datasets beyond the realm of robotic soccer.

References

1. Aha, D., Kibler, D., Albert, M.: Instance-based learning algorithms. Mach. Learn. **6**, 37–66 (1991)
2. Ahmadi, M., Keighobadi-Lamjiri, A., Nevisi, M., Habibi, J., Badie, K.: Using a two-layered case-based reasoning for prediction in soccer coach. In: Proceedings of the International Conference of Machine Learning; Models, Technologies and Applications (MLMTA 2003), pp. 181–185. CSREA Press (2003)
3. Beygelzimer, A., Kakade, S., Langford, J.: Cover tree for nearest neighbor. In: Proceedings of the Twenty-Third International Conference on Machine Learning (ICML), Pittsburgh, USA, pp. 97–104. ACM Press (2006)
4. Brin, S.: Near neighbors search in large metric spaces. In: Proceedings of the Twenty-First International Conference on Very Large Data Bases (VLDB), Zurich, Switzerland, pp. 574–584. Morgan Kaufmann (1995)
5. Denzinger, J., Hamdan, J.: Improving modeling of other agents using stereotypes and compactification of observations. In: Proceedings of 3rd International Conference on Autonomous Agents and Multiagent Systems, New York, pp. 1414–1415 (2004)
6. Fukushima, T., Nakashima, T., Akiyama, H.: Online opponent formation identification based on position information. In: Akiyama, H., Obst, O., Sammut, C., Tonidandel, F. (eds.) RoboCup 2017. LNCS (LNAI), vol. 11175, pp. 241–251. Springer, Cham (2018). https://doi.org/10.1007/978-3-030-00308-1_20
7. Gabel, T., Breuer, S., Roser, C., Berneburg, R., Godehardt, E.: FRA-UNIted - Team Description 2017 (2018). Supplementary material to RoboCup 2017: Robot Soccer World Cup XXI
8. Gabel, T., Godehardt, E.: I know what you're doing: a case study on case-based opponent modeling and low-level action prediction. In: Proceedings of the Workshop on Case-Based Agents at the International Conference on Case-Based Reasoning (ICCBR-CBA 2015), Frankfurt, Germany, pp. 13–22 (2015)
9. Gabel, T., Sommer, F.: Case-based learning and reasoning using layered boundary multigraphs. In: Keane, M.T., Wiratunga, N. (eds.) ICCBR 2022. LNCS, vol. 13405, pp. 193–208. Springer, Cham (2022). https://doi.org/10.1007/978-3-031-14923-8_13
10. Lejsek, H., Jonsson, B., Amsaleg, L.: NV-tree: nearest neighbors in the billion scale. In: Proceedings of the First ACM International Conference on Multimedia Retrieval (ICMR), Trento, Italy, pp. 57–64. ACM Press (2011)
11. Mathy, C., Derbinsky, N., Bento, J., Rosenthal, J., Yedidia, J.: The boundary forest algorithm for online supervised and unsupervised learning. In: Proceedings of the 29th AAAI Conference on Artificial Intelligence, Austin, USA, pp. 2864–2870. AAAI Press (2015)
12. Noda, I., Matsubara, H., Hiraki, K., Frank, I.: Soccer server: a tool for research on multi-agent systems. Appl. Artif. Intell. **12**(2–3), 233–250 (1998)
13. Shepard, D.: A 2-dimensional interpolation function for irregularly-spaced data. In: Proceedings of the 23rd ACM National Conference, pp. 517–524. ACM (1968)

14. Steffens, T.: Similarity-based opponent modelling using imperfect domain theories. In: Proceedings of the IEEE Symposium on Computational Intelligence and Games (CIG05), Colchester, UK, pp. 285–291 (2005)
15. Wendler, J., Bach, J.: Recognizing and predicting agent behavior with case based reasoning. In: Polani, D., Browning, B., Bonarini, A., Yoshida, K. (eds.) RoboCup 2003. LNCS (LNAI), vol. 3020, pp. 729–738. Springer, Heidelberg (2004). https://doi.org/10.1007/978-3-540-25940-4_72
16. Wess, S., Althoff, K.-D., Derwand, G.: Using k-d trees to improve the retrieval step in case-based reasoning. In: Wess, S., Althoff, K.-D., Richter, M.M. (eds.) EWCBR 1993. LNCS, vol. 837, pp. 167–181. Springer, Heidelberg (1994). https://doi.org/10.1007/3-540-58330-0_85

Trajectory Prediction for SSL Robots Using Seq2seq Neural Networks

Lucas Steuernagel$^{(\boxtimes)}$ [ID] and Marcos R. O. A. Maximo [ID]

Autonomous Computational Systems Lab (LAB-SCA), Computer Science Division,
Aeronautics Institute of Technology, Praça Marechal Eduardo Gomes 50,
Vila das Acacias, 12228-900 Sao Jose dos Campos, SP, Brazil
lucas.tnagel@gmail.com, mmaximo@ita.br
http://www.itandroids.com.br/en/

Abstract. The RoboCup Small Size League employs cylindrical robots of 15 cm height and 18 cm diameter. Presently, most teams utilize a Kalman predictor to forecast the trajectory of other robots for better motion planning and decision making. The predictor is limited for such task, for it typically cannot generate complex movements that take into account the future actions of a robot. In this context, we introduce an encoder-decoder sequence-to-sequence neural network that outperforms the Kalman predictor in trajectory forecasting. The network consists of a Bi-LSTM encoder, an attention module and a LSTM decoder. It can predict 15 future time steps, given 30 past measurements, or 30 time steps, given 60 past observations. The proposed model is roughly 50% more performant than a Kalman predictor in terms of average displacement error and runs in less than 2 ms. We believe that our new architecture will improve our team's decision making and provide a better competitive advantage for all teams. We are looking forward to integrating it with our software pipeline and continuing our research by incorporating new training methods and new inputs to the model.

Keywords: Trajectory prediction · Neural networks · Encoder-decoder · Small Size League · Sequence-to-sequence

1 Introduction

The Small Size League (SSL) is one of the robot soccer categories in the RoboCup competition, an international scientific community focused on intelligent robots. SSL robots are cylindrical with 18 cm diameter and 15 cm height. Games are very dynamic and competitive, so teams have constant pressure to improve their strategies and predict their opponent's behavior and trajectory.

Predicting the opponent's trajectory is an important step in trajectory planning, as it avoids collision between robots, and in decision making, as it helps us determinate the opponent team's strategy. SSL teams employ a Kalman filter to filter measurements from the SSL vision server and estimate robots' position.

© The Author(s), under exclusive license to Springer Nature Switzerland AG 2023
A. Eguchi et al. (Eds.): RoboCup 2022, LNAI 13561, pp. 27–38, 2023
https://doi.org/10.1007/978-3-031-28469-4_3

It is possible to extend Kalman's algorithm into a predictor by propagating its state matrix over the last estimated state.

Such an approach has some limitations. First, the Kalman predictor cannot generate complex movements, like the ones seen in an SSL game, because it propagates the estimate using simple kinematic-based models that do not consider future actions taken by the robots. Second, it models only kinematics aspects of the robots, however, the robots' trajectory depends on the ball's position and the position of other robots surrounding the one we are analyzing.

In the RoboCup context, most efforts have been in predicting the opponent's behavior. Adachi et al. [1,2] propose a classification for all possible actions of robots during a game. They use their classes to cluster robots' movements into groups and identify their behavior, using similarities between the robots' trajectories. Likewise, in Erdogan, C., Veloso, M.M [7], researches define a "behavior" as a trajectory during the period a team has the ball under control. Then, they cluster behaviors according to their similarity. During a game, the system is supposed to identify which trajectory pattern a robot is executing among the classified ones. The authors, cite, however, that, although their algorithm could effectively detect the rival's behavior, it did not do so fast enough for the team to adapt.

All the aforementioned works have tried to predict an opponent's behavior. Our work aims at predicting the opponents' trajectories (velocities and positions) by abstracting robots' behavior in a time series forecasting neural network. The contribution of our work is adapting time series forecasting to predict trajectories, aiming at an efficiency superior to a Kalman predictor.

There has been multiple works regarding trajectory prediction of different types of agents. Park, S. et al. [10] introduces an encoder-decoder architecture to forecast the trajectory of multiple vehicles. They utilize an occupancy grid map, whose cells should contain a maximum of one road car, to reduce the prediction into a classification problem. Although the system showed promising results for the proposed experiments, the RoboCup SSL field is too big to be divided into cells that contain a single robot, rendering an occupancy grid map too computationally complex.

In addition, Capobianco, S. et al. [6] presents an encoder-decoder architecture that contains an attention module to aggregate information from the past and the future for vessel trajectory prediction. They also analyze the influence of adding the vessel's destination to improve prediction.

Regarding multi-agent prediction, Ivanovic, B. Pavone, M [8] proposed using a graph to account for the influence of neighboring pedestrians in the trajectory of the target agent, whose path we want to predict, in an attempt to obtain a better forecast. The model has been updated in Salzmann, T. [12] to include the influence of many types of agents in the trajectory of each other, providing a more accurate prediction for a multi-agent scenario.

Our work mingles the aforementioned techniques and ideas into the RoboCup SSL world. Our contribution is to introduce an encoder-decoder sequence-to-sequence (seq2seq) architecture using attention to successfully predict the trajectory of robots in this context.

This paper is organized as follows. Section 2 describes our data set preprocessing, Sect. 3 describes our proposed model, Sect. 4 talks about how we trained the network, and Sect. 5 discusses our testing methodology. Sections 6 and 7 presents our results and conclusions.

2   Dataset Cleansing

All SSL games are recorded and their data is publicly available at the Robocup's website[1]. The data is not completely trustworthy because some detection packets are missing and others are repeated. Packets may be missing due to network delays or a faulty camera. Repeated detection may happen if an object appears in the field of view of multiple cameras.

The detection rate 60 Hz. It means that the interval between two frames is 0.0166 s. After analyzing the data set, we found the time range for two frames to be consecutive. If the difference between the time of capture of two frames is in the range $[0.01, 0.022]$ s, we consider them consecutive. When the difference is less than 0.01, we discard the newest frame.

Nonetheless, whenever the difference is greater than 0.022, we might create a problem, because our time series would not have a constant time difference between elements. To solve this, we repeated the last valid measurement in our stream, using intervals of 1/60 s, until the time difference between the last inserted item and the detected one stays within our defined range.

To distinguish between repeated data and real data, we created a new attribute for the measurement packet: a boolean mask. It contains true for each real element and false for repeated ones. We leverage this mask to smooth the time series we are interested in analyzing. We employed the algorithm described in Barrat, S. et al. [4] to optimize the parameters of a Kalman smoother, which substitutes the repeated points for a reliable interpolation that considers the objects' dynamics. A smoother is reliable because we are dealing only with previously collected data, not a stream. For our smother, we consider a linear system with the dynamics shown in 1 and with the sensor measurements shown in 2.

$$x_{t+1} = Ax_t + \omega_t, \tag{1}$$

$$y_t = Cx_t + v_t, \tag{2}$$

where x_t is the state, ω_t is the process noise, y_t the sensor measurement, A is the state dynamics matrix and C is the output matrix [4]. Our optimizer calculates the best ω_t and v_t, having set A and C as follows:

$$A = \begin{pmatrix} 1 & 1 & 0 & 0 \\ 0 & 1 & 0 & 0 \\ 0 & 0 & 1 & 1 \\ 0 & 0 & 0 & 1 \end{pmatrix} \quad C = \begin{pmatrix} 1 & 0 & 0 & 0 \\ 0 & 0 & 1 & 0 \end{pmatrix}. \tag{3}$$

---

[1] https://ssl.robocup.org/game-logs/.

2.1   Robot's Trajectory and Velocity

From each data packet, we can extract the position of each robot in the field. We captured each x and y coordinate between a play and a stop signal from the game controller (virtual referee). We will reference henceforth such a period as play time. This way we avoid tracking robots that team members displace during time out.

The positioning data for robots that come from the SSL vision system (the RoboCup program that processes cameras' images), which transforms coordinates in pixel to coordinates in the field's system, has quantization noise: the robot's position oscillates between two adjacent pixels.

To bypass it, we utilized the algorithm described in Barrat, S. et al. [4] to optimize the parameters for a Kalman smoother. We set the learning rate to 1×10^{-2} and the regularization parameter to 1×10^{-10}. After running the optimizer for 25 iterations in a single trajectory, we smoothed all the trajectories from our data set.

The advantage of using Kalman to smooth data is that although our observations contain only the robot's x and y position, we can configure our state dynamics to track the speed in the x axis and in the y axis. After doing so, the Kalman smoother returns the velocity of the robot, in addition to its position. To streamline the development of a prediction algorithm, we chose to track the robot's speed in $mm/frame$ instead of mm/s.

2.2   Robots' Heading

The robots orientation angle (heading) is defined within the range $[0, 2\pi)$ and, when the robot's heading is close to 0 radians, the measurements jump too frequently from 0 to the surroundings of 2π.

To overcome this problem, we calculated the sine and cosine of the heading and smoothed the data using a Kalman smoother, whose parameters we optimized using the optimization algorithm we described in Sect. 2. We configured it with a regularization parameter of 1×10^{-10} and a learning rate of 1×10^{-4}. After running the algorithm for 25 iterations in a single sequence of headings, we had the parameters to smooth all sequences of sine and cosine we obtained from our data set. From the smoothed sine and cosine, we calculate the arc tangent to obtain the heading in radians again.

3   Neural Network

We chose an encoder-decoder architecture for our network because it has shown outstanding performance for sequence to sequence neural networks [13]. Our neural network has been implemented using Tensorflow and Keras. An overview of the architecture we developed is shown in Fig. 1a. It has an encoder which summarizes information [5] from the past into a fixed size tensor h_e and a decoder that predicts the future trajectory from the context vector z. Our model utilizes data from only a single robot to forecast its future position.

The decoder state initializer transforms LSTM hidden and cell states into initialization states for the decoder's LSTM cells. We also employed an additive attention mechanism to score the relationship between a summarization of the past and a prediction of the future [6]. The attention mechanism has also shown the best performance for such task, in comparison to other techniques [6].

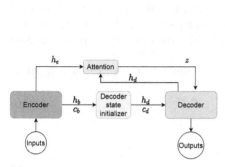

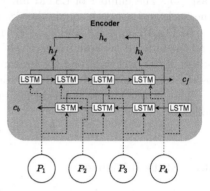

(a) Overview of the neural network we developed.

(b) Simplified overview of the model's encoder.

Fig. 1. Diagrams of the building blocks of our proposed architecture.

Considering the data we captured from the robots and the pre-processing we made, we can define a measurement P_t at time t as the following:

$$P_t = \begin{bmatrix} x_t\ y_y\ \dot{x}_t\ \dot{y}_t\ \psi_t \end{bmatrix}^\top \in \mathbb{R}^5, \tag{4}$$

where x_t and y_t represent the position of the robot at time t, $\dot{x}_t$ and $\dot{y}_t$ its velocity and ψ_t its heading.

If we want to use a look back window of size n to predict m time steps in the future from t, the input is going to be the following:

$$X = \begin{bmatrix} P_{t-n+1} \cdots P_{t-1}\ P_t \end{bmatrix} \in \mathbb{R}^{n\times 5}. \tag{5}$$

The neural network outputs a sequence of speeds the robot is going to have in the future m time steps. Let Q be the following:

$$Q_i = \begin{bmatrix} \dot{x}_i\ \dot{y}_i \end{bmatrix}^\top \in \mathbb{R}^2, \tag{6}$$

where the subscript i represents a future time step i, so the neural network prediction of m time steps ahead of t consists of:

$$Y = \begin{bmatrix} Q_{t+1}\ Q_{t+2} \cdots Q_{t+m,} \end{bmatrix} \in \mathbb{R}^{m\times 2}. \tag{7}$$

As the velocity we use is measured in millimeters per frame, we can use (8) and (9) to obtain the robot's future positions from the predicted velocity.

$$x_i = x_{i-1} + \dot{x}_{i-1}, \tag{8}$$
$$y_i = y_{i-1} + \dot{y}_{i-1}. \tag{9}$$

3.1   Encoder

Figure 1b depicts a simplified version of the model's encoder. For simplicity, the diagram shows an input of four time steps in the past. We use a BiLSTM architecture for the encoder, as it has shown great performance in summarization tasks [5]. The number of LSTM units (cells) for both the forward network and the backward network equals the size of the look back window.

In Fig. 1b, $c_b \in \mathbb{R}^n$ and $c_f \in \mathbb{R}^n$ represent the cell states for the backward and the forward network, respectively. We call $h_b \in \mathbb{R}^n$ and $h_f \in \mathbb{R}^n$ the hidden states of the backward and the forward networks, respectively. They are defined as the concatenation of the hidden state for each LSTM cell.

The encoded representation for the given input is h_e, which is defined as follows:

$$h_e = \begin{bmatrix} h_f \ h_b \end{bmatrix}^\top \in \mathbb{R}^{n \times 2}. \tag{10}$$

3.2   State Initializer

As demonstrated in Bahdanau et al. [3], we apply a transformation to initialize our decoder. Our approach, however, is slightly different than the one proposed in [3]. We apply the following equations to create the initial states $h_d \in \mathbb{R}^n$ (hidden state) and $c_d \in \mathbb{R}^n$ (cell state) for our decoder:

$$h_d = \tanh\left(W_1 h_b + b_1\right), \tag{11}$$

$$c_d = \tanh\left(W_2 c_b + b_2\right), \tag{12}$$

where W_1, W_2, b_1, and b_2 are trainable parameters.

3.3   Attention Aggregator

We use the additive attention mechanism as an aggregator for our encoder-decoder architecture. The inputs for our aggregator are h_e as the key and h_d for the query. It outputs a context vector $z \in \mathbb{R}^n$. We calculate the context vector z for each time step prediction, based on the decoder state h_d for the last prediction. We will refer henceforth to z as z_i as the context vector for the time step i.

The attention mechanism serves to score each context vector z_i with respect to the decoder's state h_d at time step i, while it is generating the output trajectory. It means we are "assigning a probability to each context of being attended by the decoder" [6].

3.4   Decoder

The decoder, depicted in Fig. 2a, has n LSTM cells, which are initialized with the output of the state initializer module (Subsect. 3.2). The initialization states are depicted in the figure as h_{init} and c_{init}.

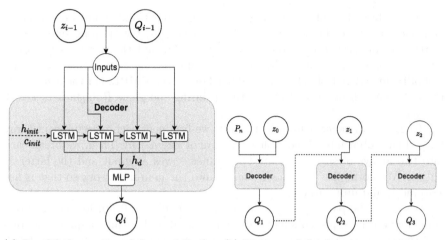

(a) Simplified overview of the model's decoder.

(b) Diagram of the decoder recursive prediction mechanism.

Fig. 2. Diagrams of the building blocks of our decoder.

The output of the forward LSTM network is h_d, which we built as the concatenation of all cells' hidden state. The h_d is then fed into a multilayer perceptron network (MLP) with linear activation to calculate Q_i, so the MLP has two neurons. We concatenate Q_{i-1} with z_{i-1} to serve as the input for the decoder.

The decoder works in a recursive manner, as shown in Fig. 2b. The figure depicts the prediction of three time steps. To predict the first one, we feed the decoder with $\dot{x}_n$ and $\dot{y}_n$ from P_n. z_0 represents the context vector obtained using the decoder's initialization states. For the time steps after the first one, we always utilize Q_{i-1} and z_{i-1} (calculated from h_d at time step i).

4   Training

We trained the neural network for 10 epochs with batches of 2.048 elements. We employed Adam as the optimizer, using its standard parameters [9]. We utilized an exponential learning rate decay that follows the implementation of Keras `ExponentialDecay` function with the following parameters: 10^{-3} as the initial learning rate, 1000 decay steps, 0.98 as the decay rate and the staircase parameter as false.

Our loss function $\mathbb{L}$ for training is the sum of the mean squared error and the mean absolute error. We multiply this sum by a factor of 100 to avoid working with small numbers for loss, which have shown to slow down training [11]. Our loss function is, then, described as follows:

$$\mathbb{L}(Y_{true}, Y_{pred}) = \frac{100}{K} \left[\sum_{k}^{K} \left(Y_{true}^{k} - Y_{pred}^{k} \right)^2 + \sum_{k}^{K} |Y_{true}^{k} - Y_{pred}^{k}| \right], \qquad (13)$$

where K is the number of samples, Y_{true} represent real values and Y_{pred} predicted ones. The superscript k indicates which sample we are analyzing.

We used two logs of RoboCup games to build our training and validation set and one for our testing set. In terms of percentages, the training set contains 78.03% of all data, the validation set has 8.67% and the testing set, 13.2%. The logs for training and validation were captured in 2019 during the games RobotTeam *versus* RoboDragons and ER-Force *versus* MRL. For the testing set, the data came from a 2019 log captured during the game RoboJackets *versus* nAMec.

To create an element for our data set, we fragment each measurement array for a single robot into all possible subarrays of size $n + m$. Then, we break each into two arrays of size n and m. The former serves as input and the latter as output. We normalize the data to be fed into the neural network so that is has zero mean and unit variance.

We chose the pairs $(n, m) = (30, 15)$ and $(n, m) = (60, 30)$ for our architecture. The former represents half a second of input and outputs a quarter second and the latter is an input of a second of measurements to predict the next half second. As the tuple (n, m) is variable, so is the exact number of elements in our dataset. For $n = 30$ and $m = 15$, we have a total of 325,504 elements, and for $n = 60$ and $m = 30$, we have 306,739. Despite this difference, the distribution of elements in each sub-set remains the same for every n and m.

5   Testing

To properly assess the effectiveness of our algorithm, we utilized three metrics. The mean absolute error (MAE), the average displacement error (ADE) and the final displacement error (FDE). The ADE is the result of 14 calculated for every predicted x and y coordinate. FDE is result of 14 considering only the position at the last time step of each predicted trajectory. In 14, x and y are the ground truth positions, x_{pred} and y_{pred} are predicted values and n is the number of samples we have for testing.

$$\frac{\sum \sqrt{(x - x_{pred})^2 + (y - y_{pred})^2}}{n}. \tag{14}$$

A lower ADE implies a lower drift from the ground truth and a lower FDE indicates a better prediction on the long term. These metrics, thus, provide us a concrete way to measure how bad we are predicting the trajectories.

We compared our model to a Kalman predictor and to multi-layer perceptron network. The Kalman predictor receives as input raw measurements from the robots' positions and updates its state matrices. The predictor has been initialized with the parameters we got from the optimizer described in Sect. 2. It first updates its state with the first n measurements for each trajectory. Afterwards, updates are done at each time step. After each update, we use its state dynamics matrix to forecast m time steps in the future, given the last valid measurement. We calculate the MAE, ADE and FDE from the results using the smoothed trajectories as the ground truth.

The multilayer perceptron (MLP) developed for the comparison, in turn, consists of four layers. Every layer has a ReLu activation function, except the last, which has no activation. The layers have 128, 1024, 128 and 2 m neurons, respectively. The MLP network receives the same input X as our proposed model. Its last layer predicts Q for each of the m future time steps.

To perform inference in the context of a Robocup game, the neural network must run in real time, so we also analyze the execution time of each proposed solution, calculating it for 100 inferences and getting the average. We discarded the first inference time before starting to count the time, as it involves loading libraries. We optimized the neural network using Tensorflow Lite default optimizer for inference in CPUs. We set a restriction of inference time to the maximum of 0.016 s, the interval between two vision frames.

We have also done a qualitative analysis of the predicted trajectory, i.e. how well it predict sharp turns, how noisy the prediction is and how trustworthy the results are.

6 Results

We present on Table 1 the results of the MAE, FDE and ADE for the proposed architecture, the MLP network and the Kalman predictor, using $(n, m) = (30, 15)$.

Table 1. Comparison between our architectures and other forecasting methods for $(n, m) = (30, 15)$.

Metric	Proposed model	MLP	Kalman predictor
MAE	53.95	67.73	155.70
ADE	43.64	53.89	123.60
FDE	4.80	6.36	16.46

Our proposed sequence-to-sequence network outperforms the Kalman predictor and the MLP network. We were able to decrease MAE, ADE and FDE, respectively in 65.35%, 64.69% and 70.83%. In comparison to a simple MLP architecture we improve our results, respectively, in 20.34%, 19.02% and 24.52%.

In Table 2 we present again the values of MAE, FDE and ADE for the methods we are comparing. This time we used $(n, m) = (60, 30)$.

Table 2. Comparison between our architectures and other forecasting methods for $(n, m) = (60, 30)$.

Metric	Proposed model	MLP	Kalman predictor
MAE	328.23	440.49	705.18
ADE	261.82	350.57	560.02
FDE	22.79	30.03	42.31

For $(n, m) = (60, 30)$, errors are naturally higher, as we are predicting more time steps into the future and all methods tend to diverge in longer predictions. However, our model still outperforms the Kalman predictor and the MLP network. With respect to the Kalman filter, we improve MAE, ADE and FDE in 53.45%, 53.32% and 46.13%, respectively. Considering the MLP network, the improvements are 25.48%, 25.31% and 24.10%.

Optimizing the $(n, m) = (30, 15)$ for inference using Tensorflow Lite default optimizer, we get an execution time of 0.6 ms for each sample in an Intel Core i7-7550U CPU. When we consider a batch prediction of eleven inputs (the number of robots in a team), we optimize the model again and achieve an execution time of 2.08 ms for each batch of eleven elements. The optimization increases the MAE, ADE and FDE values in 0.1%.

For the $(n, m) = (60, 30)$ model, the optimization for a single input gives an execution time of 1.85 ms in the aforementioned CPU model. When we optimize for the batch prediction of eleven elements, our inference time for batches is 11.51 ms. The same worsening in the performance scores has been observed here.

SSL robots have a diameter of 18 cm. The final displacement error between the real trajectory and the predicted one is less than 6 mm for a prediction of 15 time steps. In terms of dimensions, the error is a remarkable result. Considering the prediction of 30 time steps, the FDE is 22 mm.

We present in Fig. 3a many predictions on a curved trajectory. The black line represents the ground truth and the green lines represent forecasted trajectories from several consecutive inputs. It illustrates how the neural network updates its predictions based on newer information. We can notice that the algorithm works well when predicting the curvature radius of a turn. In addition, Fig. 3b highlights the fidelity of prediction while performing a turn.

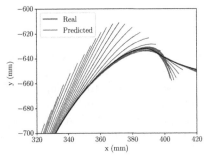
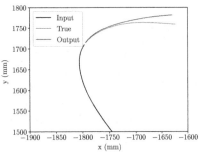

(a) Up close in a curved trajectory, containing consecutive predictions (in green), compared to the whole trajectory (black), using $(n, m) = (30, 15)$

(b) Single prediction for the $(n, m) = (60, 30)$ network (green), compared to the true trajectory (red).

Fig. 3. Plots of examples of predictions. (Color figure online)

It is worth mentioning that the network struggles with sharp turns. During a sharp turn, there is not enough information in the past about what the robot might do, no matter how big is the look back window. As soon as the robot start turning, the neural network quickly correct its prediction to the right direction.

The network with $(n, m) = (60, 30)$ provides us with more information about the future, nonetheless its short term prediction is not as accurate as that of the $(n, m) = (30, 15)$. If we calculate the MAE, ADE and FDE for the look back window of 60, considering only its first 15 predicted time steps for the error metrics we got a MAE of 64.32, an ADE of 52.08 and a FDE of 5.98.

7   Conclusions

We applied time series forecasting techniques to the context of RoboCup Small Size League to propose a sequence-to-sequence architecture that outperforms a Kalman predictor and a multi-layer perceptron network while meeting our inference time restrictions. The next step in our research will be integrating them to our software pipeline to identify whether we need a more precise prediction or more time steps in the future. A simple video of the network forecasting a trajectory is available at Youtube[2].

Future work can be done to improve our network. We can leverage the graph architecture introduced in [12] to model the whole opponent team and predict the trajectories of all robots. This is interesting because a single entity controls all the robots of a team in a SSL game. In addition, we can add new features to the prediction, like the kick and the dribble actions as they are important for the decision making of our team. We can also aggregate more information from the game to improve our prediction, for using the trajectory of the ball and of other robots as inputs might provide us with improved results.

As we are committed to contributing to the RoboCup community and to ensuring the transparency of our research, every code we used at this paper is available at our git repository[3]. It also includes a bash script to download the dataset from the RoboCup's official repository.

Aknowledgements. We would like to thank ITAndroids – Aeronautics Institute of Technology robotics team – for providing the research opportunity and supporting it. We are also grateful to Ana Carolina Lorena for the insightful comments when reviewing this work.

References

1. Adachi, Y., Ito, M., Naruse, T.: Classifying the strategies of an opponent team based on a sequence of actions in the RoboCup SSL. In: Behnke, S., Sheh, R., Sariel, S., Lee, D.D. (eds.) RoboCup 2016. LNCS (LNAI), vol. 9776, pp. 109–120. Springer, Cham (2017). https://doi.org/10.1007/978-3-319-68792-6_9
2. Adachi, Y., Ito, M., Naruse, T.: Online strategy clustering based on action sequences in RoboCupSoccer small size league. Robotics **8**(3) (2019). https://doi.org/10.3390/robotics8030058. https://www.mdpi.com/2218-6581/8/3/58

---

[2] https://youtu.be/KKBwuEjD72w.

[3] https://gitlab.com/itandroids/open-projects/trajectory-prediction-ssl-seq2seq.

3. Bahdanau, D., Cho, K., Bengio, Y.: Neural machine translation by jointly learning to align and translate. In: International Conference on Learning Representations, ICLR 2015, San Diego, USA (2015)
4. Barratt, S.T., Boyd, S.P.: Fitting a Kalman smoother to data. In: 2020 American Control Conference (ACC), Denver, USA, pp. 1526–1531 (2020). https://doi.org/10.23919/ACC45564.2020.9147485
5. Britz, D., Goldie, A., Luong, M.T., Le, Q.: Massive exploration of neural machine translation architectures. In: Proceedings of the 2017 Conference on Empirical Methods in Natural Language Processing, Copenhagen, Denmark, pp. 1442–1451. Association for Computational Linguistics (2017). https://doi.org/10.18653/v1/D17-1151. https://aclanthology.org/D17-1151
6. Capobianco, S., Millefiori, L.M., Forti, N., Braca, P., Willett, P.: Deep learning methods for vessel trajectory prediction based on recurrent neural networks. IEEE Trans. Aerosp. Electron. Syst. **57**, 4329–4346 (2021). https://doi.org/10.1109/TAES.2021.3096873
7. Erdogan, C., Veloso, M.M.: Action selection via learning behavior patterns in multi-robot domains. https://doi.org/10.1184/R1/6602957.v1
8. Ivanovic, B., Pavone, M.: The trajectron: probabilistic multi-agent trajectory modeling with dynamic spatiotemporal graphs. In: 2019 IEEE/CVF International Conference on Computer Vision (ICCV), Seoul, Korea, pp. 2375–2384. IEEE (2019). https://doi.org/10.1109/ICCV.2019.00246
9. Kingma, D.P., Ba, J.: Adam: a method for stochastic optimization. In: Bengio, Y., LeCun, Y. (eds.) 3rd International Conference on Learning Representations, ICLR 2015, San Diego, CA, USA, 7–9 May 2015, Conference Track Proceedings (2015). arxiv.org/abs/1412.6980
10. Park, S., Kim, B., Kang, C., Chung, C., Choi, J.: Sequence-to-sequence prediction of vehicle trajectory via LSTM encoder-decoder architecture. In: Proceedings of the 2018 IEEE Intelligent Vehicles Symposium, IV 2018, pp. 1672–1678. Institute of Electrical and Electronics Engineers Inc. (2018). https://doi.org/10.1109/IVS.2018.8500658. Publisher Copyright: 2018 IEEE.; Null; Conference date: 26–30 September 2018
11. Pascanu, R., Mikolov, T., Bengio, Y.: On the difficulty of training recurrent neural networks. In: Proceedings of the 30th International Conference on International Conference on Machine Learning, ICML 2013, vol. 28, pp. III-1310–III-1318. JMLR.org (2013)
12. Salzmann, T., Ivanovic, B., Chakravarty, P., Pavone, M.: Trajectron++: dynamically-feasible trajectory forecasting with heterogeneous data. In: Vedaldi, A., Bischof, H., Brox, T., Frahm, J.-M. (eds.) ECCV 2020. LNCS, vol. 12363, pp. 683–700. Springer, Cham (2020). https://doi.org/10.1007/978-3-030-58523-5_40
13. Sutskever, I., Vinyals, O., Le, Q.V.: Sequence to sequence learning with neural networks. In: Proceedings of the 27th International Conference on Neural Information Processing Systems, NIPS 2014, vol. 2, pp. 3104–3112. MIT Press, Cambridge (2014)

Gait Phase Detection on Level and Inclined Surfaces for Human Beings with an Orthosis and Humanoid Robots

Maximilian Gießler$^{(\boxtimes)}$ (iD), Marc Breig, Virginia Wolf, Fabian Schnekenburger, Ulrich Hochberg, and Steffen Willwacher

Offenburg University of Applied Sciences, 77652 Offenburg, Germany
maximilian.giessler@hs-offenburg.de

Abstract. In this paper, we propose an approach for gait phase detection for flat and inclined surfaces that can be used for an ankle-foot orthosis and the humanoid robot *Sweaty*. To cover different use cases, we use a rule-based algorithm. This offers the required flexibility and real-time capability. The inputs of the algorithm are inertial measurement unit and ankle joint angle signals. We show that the gait phases with the orthosis worn by a human participant and with *Sweaty* are reliably recognized by the algorithm under the condition of adapted transition conditions. E.g., the specificity for human gait on flat surfaces is 92 %. For the robot *Sweaty*, 95 % results in fully recognized gait cycles. Furthermore, the algorithm also allows the determination of the inclination angle of the ramp. The sensors of the orthosis provide 6.9° and that of the robot *Sweaty* 7.7° when walking onto the reference ramp with slope angle 7.9°.

Keywords: Orthosis · Gait phase detection · Inclined surface · Inertial measurement unit · Humanoid robot · *Sweaty*

1 Introduction

Gait phase detection is a frequently discussed area in biomechanics. Gait phase detections (e.g. detecting initial and final contact of the feet with the ground) using spatially fixed force plates, motion capture systems, or the combination of both systems currently represent the reference [1,3,9,12]. In this work, we will study the scope of application of a gait phase detection approach both for humans handicapped in their locomotion and for humanoid robots. Our focus is specifically on bipedal walking on inclined surfaces.

An actuated orthosis can allow people with impaired dorsiflexion of the ankle to walk again on inclined surfaces [2,3,5]. Figure 1(b) shows an example of such an orthosis. The sensors built into the orthosis use an algorithm to detect gait phases and the angle of inclination of the walking surface [13].

For humanoid robots, there are also several possible applications. Knowledge of the gait phase for each leg can be an input for stability-maintaining control

© The Author(s), under exclusive license to Springer Nature Switzerland AG 2023
A. Eguchi et al. (Eds.): RoboCup 2022, LNAI 13561, pp. 39–49, 2023
https://doi.org/10.1007/978-3-031-28469-4_4

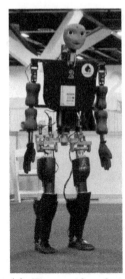
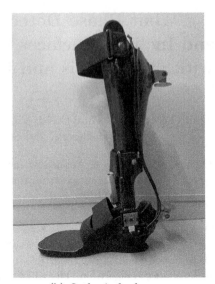

(a) Humanoid robot
Sweaty

(b) Orthosis for human

Fig. 1. Humanoid robot *Sweaty* (a), which was originally designed to compete in the *RoboCup Soccer* and the orthosis (b).

algorithms. E.g., we could determine a leg with which to perform a reflex like quick motion (e.g., a forward lunge) for stabilization. Similarly, we can assign either a position- or force-controlled control algorithm to specific gait phases. We can use the gait phase detection as a transition condition between position or force-control [6]. In analogy to the use by people with a handicap, we can also estimate the slope angle of the walking surface. This allows an adjustment of the ankle joint angle and the leg length for bipedal walking on sloped surfaces.

Both application scopes indicate a positive influence of gait phase detection for walking on inclined surfaces. To detect gait phases in different application areas, we propose an algorithm that uses sensor signals of an inertial measurement unit (IMU) and the angle of the ankle joint as inputs. We focus on an algorithm with a low time delay. We validate the algorithm both by measurements with a self-developed orthosis and by simulated and physical data of the humanoid robot *Sweaty* (cf. Fig. 1 (a)).

2   State of the Art

There are many published methods for gait phase detection in the literature. On the one hand, these differ in the number of detectable gait phases. Between two and eight phases can be detected [14]. Likewise, different sensors are used such as force sensors [13], electromyography (EMG) [1] and IMU sensors [6,11].

Various detection algorithms e.g. threshold methods [4,8,10] or machine learning [5,7,11] are also used.

IMUs are widely used because of the measurable quantities, weight, low energy consumption and mounting options on different parts of the body [14]. From Vu et al. [14] it appears that better results are obtained when placing the IMU on the shank compared to mounting it on the thigh or hip.

Furthermore, IMUs are also combined with machine learning (ML) methods such as long short-term memory [11], hidden Markov models [7,11], or neural networks [5]. Vu et al. show in [14] that these algorithms are also suitable for online applications, but these methods require high computing capacities. Furthermore, ML methods require a sufficiently large data set to enable training [4]. In perspective of frequently changing gait behavior, such as for humanoid robots, we see a high effort to generate the training data sets.

In contrast to the approaches mentioned above, we use a rule-based method because it offers both the necessary real-time capability and the flexibility to adapt to the varying use cases. Based on defined transition conditions, we detect the gait phases using a state machine. We estimate the inclination angle of the surface during the stance phase through the orientation of the IMU.

The contribution of our method is that we can determine the inclination of the walking surface. This allows adapting the planned trajectory of the feet of humanoid robots for walking on inclined surfaces. For this purpose, the planned orientation and the leg length are adjusted depending on the gait phase. With the help of this information, the actuated orthosis can allow disabled people to walk on inclined surfaces by actively adjusting the foot pitch angle to the angle of inclination.

3   Method

Some previous publications [8,10] divide the human gait cycle into four different phases. We use the same four gait phases in our work. These can be detected with a single IMU. For the detection of more finely differentiated gait phases, we would need additional sensors, such as additional IMUs or force sensors, as shown in [14].

For this classification of the gait phases, the contact phase begins with the initial contact with the ground surface. When the foot is entirely touched down, the stance phase is initiated. As soon as the heel lifts off, the pre-swing phase begins. After the foot is entirely lifted off the surface, the swing phase begins. With the following contact of the heel with the surface, the current gait cycle ends. Likewise, this contact represents the starting point of the next gait cycle.

For the transition between the gait phases, we define the transition conditions T1 to T4, shown in Fig. 2. The transition conditions are described by logical operators. Here, three measurements in a row and/or their time derivatives must fulfill these conditions to switch the state. We calculate the derivatives by a backward difference. The quantitative threshold values are determined and set using recorded IMU data from the orthosis and the simulation. The transition options depending on the current gait phase are defined by the state machine.

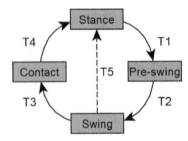

Fig. 2. Model of the ruled-based algorithm implemented. The detectable gait phases are visualized in gray boxes. The transition conditions are labeled with the abbreviation T1 to T5, where T5 is only used for the robot.

For humanoid robots, we add the dashed transition condition T5 in Fig. 2. This allows the transition from the swing to the stance phase. We need this to reinitialize the gait phase detection in case of a deviating gait pattern due to external perturbations or instabilities.

We use the sensor signals of the IMU *Grove IMU 9DOF* and the rotation angle sensor *AS5048* as the input signals. The IMU provides the spatial orientation, of which we use the Cardan angle about the y-axis (γ_y) and the linear acceleration vector a of a body segment in the transition conditions. Furthermore, we use the time derivative of the ankle joint angle β_y, the acceleration coefficients a_x and a_z, as well as that of the rotation angle γ_y in the transition conditions. We use the time derivative to qualitatively identify the trend of the curve development. We do not use their quantitative values. The sensors are mounted on the orthosis at the height of the upper ankle joint and on the foot segment of *Sweaty*.

Generally, the data from the IMU is overlaid with white noise. We filter the input signals by a moving average. We record the sensor data from the orthosis with a sampling frequency of 300 Hz and we choose a filter length of the 10th order. Physical and simulated locomotion data of the robot *Sweaty* are also used for the application of the algorithm. For the simulation, *Webots* by *Cyberbotics Ltd.* is used (cf. Fig. 3). This simulator was used for the virtual *RoboCup 2021*.

The field bus frequency of the robot *Sweaty* is 125 Hz. Similarly, we simulate the robot with the fixed step size of 8 ms. We choose a filter length of the 4th order. Furthermore, we subtract the acceleration due to gravity as a function of IMU rotation from the measured acceleration vector a. We verify the algorithm by the detection precision and the time delay.

The rule-based algorithm was tested with custom made orthosis on three human participants, two healthy and one with foot drop syndrome, and on the robot *Sweaty*. The orthosis is adapted to the individual anatomy of the lower extremity and is not applicable to other participants. Due to this the count of human participants was limited to three. For the evaluation of the algorithm, video data of the human subjects walking on a treadmill was compared with the

Fig. 3. *Sweaty* walks on a ramp with a slope of 7.9° in the simulation environment *Webots*.

synchronized data of the IMU. We also used this procedure in the evaluation of the real data and simulation results of *Sweaty*. In the following we show the data and results from one healthy participant and the robot *Sweaty*.

4   Results

We evaluate the gait phase algorithm on the level and inclined surface using sensor data from the orthosis and sensor data as well as simulation results from *Sweaty*. Since the robot *Sweaty* can currently only walk on level surfaces, we use just the simulation results for the evaluation of the gait phase detection on the inclined surfaces. For the evaluation, we determine the transition conditions as shown in Table 1.

Figure 4 (a) and (c) show curves of the acceleration coefficients a_x and a_z for one gait cycle. Thereby, Fig. 4 (a) contains the sensor data of the orthosis and Fig. 4 (c) the simulated data from the robot *Sweaty*. These curves show similarity in their behavior.

The gait cycle begins and ends with the contact phase. This phase is shown in both Fig. 4 (a) and (c) characterised by a local acceleration minimum for a_x. The stance phase is recognisable by acceleration values $a_x \approx 0$ and $a_z \approx 0$. The pre-swing is characterized by a positive deflection of a_z. Thereby, the deflection is more pronounced in human motion than in the gait of the robot *Sweaty*. The swing phase is defined by the most prominent curve section of a_x. There is a positive acceleration peak followed by a negative peak in both figures. The zero intersection of a_x ais approximately at the midpoint of the swing phase. The described negative acceleration peak of a_x marks both the endpoint of the gait cycle and the starting point of the periodically repeating gait cycle. In contrast, the duration of a gait cycle shows a more distinct difference. Humans need 1.40 s, the robot *Sweaty* needs 0.44 s for a complete gait cycle.

Figure 4 (b) and (d) shows the curves of the ankle angle β_y and the IMU angle γ_y. For an easier interpretation, we have mirrored the curves. Therefore, for example, the dorsal extension of the foot is now assigned with a positive value. This convention applies to all the following figures that contain angle curves.

The curves in Fig. 4 (b) and (d), however, are different in their progression. In the contact phase in Fig. 4 (b) the ankle joint is almost in the neutral zero position. The IMU rotation is negative because the leg is in front of the body and the toe points upwards in this phase. Compared to Fig. 4 (b), (d) shows a curve with significantly lower amplitudes for γ_y. While the humans have a range of approx. $-31°$ to $40°$, the robot has only a range of $-23°$ to $3°$. The ankle joint angle β_y also has a much steadier progression than that of humans and is at a much lower baseline. During the stance phase, the ankle joint β_y is close to zero in humans and *Sweaty*'s averages approx. $-18°$. The subjects' amplitudes for the ankle joint angle β_y range between approx. $-12°$ and $9°$. The robot *Sweaty* shows a smaller range of approx. $-27°$ to $-17°$.

4.1   Detection Precision and Time Delay

To determine the detection precision and the overall time delay, we randomly selected 300 gait cycles from a data set. The ground truth for evaluating our algorithms applied to the orthosis are synchronized video, force plate and IMU data. For *Sweaty*, we use the synchronized 6-axis force/torque and IMU sensor signals and simulation data.

For the participant wearing our orthosis, we correctly detect 92 % of the gait phases. When applying the algorithm to *Sweaty*, we detect the stance, pre-swing and swing phase with an accuracy of 98 % and 95 % for the contact phase.

The overall time delay $\tau_{overall}$ is caused, among other factors, by the moving average filter used. Here, we can calculate this time delay with the equation $\tau = \frac{n-1}{2} \cdot \Delta t$. The order of the filter is n. Δt is the difference between two measured values in time.

Considering that the sensor measurements and/or their time derivatives have to fulfill the transition conditions three times in a row. We have to add two additional Δt to τ. This leads to an overall time delay $\tau_{overall}$ of 21.67 ms for the gait of the participant. The time delay for the detection of *Sweaty* is 28 ms.

4.2   Determination of the Slope

Figure 5 shows γ_y of the IMU for walking on level and inclined surface of a subject with orthosis (a) and of *Sweaty* (b). Figure 5 (a) shows similar developments between the two curves. The curves significantly differ in the amplitude of the peaks. For walking on inclined planes, the positive angular deflections are less prominent. In addition, in the stance phases the inclination angle of the surface can be recognized by the significant, almost constant curve progression. In the stance phases of the displayed gait cycles in Fig. 5 (a) and (b), the orthosis

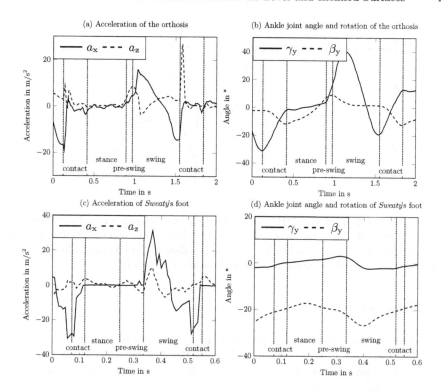

Fig. 4. Plot of accelerations a, ankle joint angle β_y and the orientation of the IMU γ_y for bipedal locomotion of a human being and the robot *Sweaty*. The recognized gait phases are delineated and labeled with vertical lines.

measures inclination angles of $-0.4°$ and $-0.3°$ for the flat ground. For the inclined surface it measures inclination angles of $6.6°$ and $7.2°$.

The curves from Fig. 4 (b) are also similar in shape. A characteristic difference is that the curve of $\gamma_{y,\text{sloped}}$ is shifted in a positive direction. The IMU in *Sweaty*'s foot determines $-0.2°$ and $0.1°$ for the level ground. For the inclined ground it gives $7.4°$ and $7.8°$. The reference ramp has an inclination angle of $7.9°$.

4.3 Validation with Physical Data

Figure 6 (a) and (b) shows recorded data of the physical robot *Sweaty* for walking on the flat surface. In Fig. 6 (a) and (b), we see also the characteristic peaks of the acceleration coefficients a_x and a_z compared to the simulation. Differences are in the amplitudes of the local extremes. Equally, the gradients of the curves during the swing phase are smaller in contrast to Fig. 4 (c). On the other hand, the ankle joint angles β_y and the rotation of the IMU γ_y are almost identical to the values of the simulation in Fig. 4 (d).

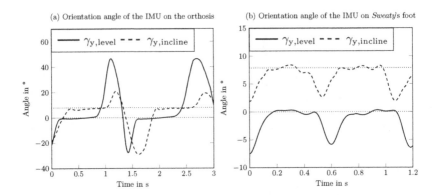

Fig. 5. The rotation of the IMU γ_y displays the participant with the orthosis (a) and the robot *Sweaty* (b) for level and inclined surfaces. The solid curves illustrate the rotation of the IMU $\gamma_{y,level}$ on leveled and $\gamma_{y,incline}$ dashed lines on the inclined ground. The horizontal dotted line indicates the inclination of the surface.

5 Discussion

From Figs. 4, 5 and 6 it can be seen that different curve progressions are determined for each application scenario. The selection of the rule-based algorithm proved to be practicable. Due to the flexibility, we assume the advantages of ML-based algorithms here.

The occurring differences in the curves can be explained, for instance, by the fact that *Sweaty* adjusts the leg length difference during walking in the double and single support phases by knee and hip flexion. In the current gait patterns of *Sweaty*, there is only a minimal rolling motion implemented. On the one hand, this results in a negative shifted curve for the ankle joint angle. On the other hand, there is much less movement of the feet compared to humans. The differences in the amplitudes of the acceleration curves between physical and simulated sensor data of the robot *Sweaty* can also attribute to the elastic and shock-absorbing sole on the physical robot feet.

Table 1. The transition conditions applied for gait phase detection for a participant and the humanoid robot *Sweaty*.

	Human conditions	Robot conditions
T1	$a_z > 3\,\mathrm{m/s^2} \wedge \beta_y > 5°$	$a_x < 0\,\mathrm{m/s^2} \wedge a_z > 0\,\mathrm{m/s^2}$
	$\dot{a}_z > 0\,\mathrm{m/s^3} \wedge \dot{\beta}_y > 0°/s$	$\gamma_y > 0° \wedge \dot{a}_z > 0\,\mathrm{m/s^3}$
T2	$\dot{a}_z < 0\,\mathrm{m/s^3} \wedge \dot{\beta}_y < 0°/s$	$\dot{a}_x > 0\,\mathrm{m/s^3} \wedge \dot{a}_z > 0\,\mathrm{m/s^3} \wedge \dot{\gamma}_y < 5°/s$
T3	$a_x < 0\,\mathrm{m/s^2} \wedge \gamma_y < 0° \wedge \dot{\gamma}_y > 0°/s$	$a_z > 0\,\mathrm{m/s^2} \wedge \gamma_y < 0° \wedge \dot{a}_x > 0\,\mathrm{m/s^3}$
T4	$\dot{\beta}_y > 0°/s$	$a_x \approx 0\,\mathrm{m/s^2}$
T5	NA	$a_x \approx 0\,\mathrm{m/s^2}$

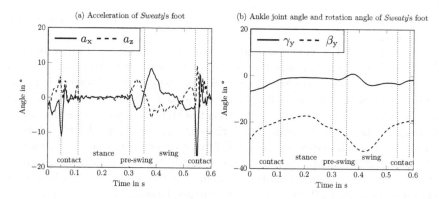

Fig. 6. The plots show under physical conditions the accelerations a in (a), the ankle joint angle β_y and the orientation of the IMU γ_y in (b) for the robot *Sweaty*. The recognized gait phases are delineated and labeled with vertical lines.

Although there were significant differences in the curves and the motion, the algorithm was able with adapted transition conditions to correctly identify the gait phases for all three scenarios. Compared to the algorithm described in the review by Vu et al. [14], we achieve similar detection precision values. For human with orthosis, we detect 92 % and for the robot *Sweaty* 95 % correctly. Also in case of the time delay we lie with 21.67 ms for humans and 28 ms for robots in the average compared to the reviewed approaches. Consequently, our algorithm seems suitable for online applications.

We were able to show that the algorithm is able to deal with high variance of gait patterns. We can detect the gait phases of them after adapting the declared parameters to the specific characteristics of the gait patterns. Therefore, we assume our approach is applicable for other participants and robots.

For the reference inclination of 7.9°, we detected a mean value of approx. 6.9° with the orthosis and 7.6° with the robot in the simulation. With this adjustment of the β_y, it was now possible for the test person with foot-drop symptoms to walk on a ramp.

The difference in the referenced inclination angle can be explained by the positioning inaccuracy of the IMU and the elasticity of the soles on the feet. Similar behavior is evident for the robot *Sweaty*. By applying the algorithm, it can walk on the inclined surface in the simulation. Due to the rigid-body modeling of the segments in the simulation, the deviation for the determined inclination angle is smaller compared to the determinations of the orthosis.

6 Conclusion and Future Work

With this work, we demonstrate that with our approach for a gait phase detection in combination with the self-designed orthosis, the test person was able to walk on inclined surfaces. By similarity, the method transfered to the robot *Sweaty* in the simulation. As a result, the robot *Sweaty* was able to move stable on the inclined plane.

The selection of a rule-based algorithm in combination with a filtering method leads to the requirement of measurement data from the previous time steps. On the one hand, the real-time ability is ensured by the underlying algorithm because no optimization iterations are needed. The flexibility of the rule-based algorithm is an advantage concerning the highly varying curves depending on the application. One the other hand, there is an overall time delay. To further reduce the time delay, specially adapted filter methods could be implemented.

Next, we will validate the algorithm in further investigations using a force plate and an optical motion capture system. For the application on humanoid robots, the algorithm will be extended for curves, lateral and backward walking. In the future, the algorithm will also be used to allow the robot *Sweaty* to walk on inclined surfaces.

Acknowledgment. Special thanks to Seifert Technical Orthopaedics for their support in the field of orthosis and for the possibility to do the trials with the participants. The research work on the orthosis was financed by the Federal Ministry of Economic Affairs and Climate Action of Germany.

References

1. Agostini, V., Ghislieri, M., Rosati, S., Balestra, G., Knaflitz, M.: Surface electromyography applied to gait analysis: how to improve its impact in clinics? Front. Neurol. **11**, 994 (2020). https://doi.org/10.3389/fneur.2020.00994
2. Blaya, J.A., Herr, H.: Adaptive control of a variable-impedance ankle-foot orthosis to assist drop-foot gait. IEEE Trans. Neural Syst. Rehabil. Eng. Publ. IEEE Eng. Med. Biol. Soc. **12**(1), 24–31 (2004). https://doi.org/10.1109/TNSRE.2003.823266
3. Feuvrier, F., Sijobert, B., Azevedo, C., Griffiths, K., Alonso, S., Dupeyron, A., Laffont, I., Froger, J.: Inertial measurement unit compared to an optical motion capturing system in post-stroke individuals with foot-drop syndrome. Ann. Phys. Rehabil. Med. **63**, 195–201 (2019). https://doi.org/10.1016/j.rehab.2019.03.007
4. Huang, L., Zheng, J., Hu, H.: A gait phase detection method in complex environment based on DTW-mean templates. IEEE Sens. J. **21**(13), 15114–15123 (2021). https://doi.org/10.1109/JSEN.2021.3072102
5. Islam, M., Hsiao-Wecksler, E.T.: Detection of gait modes using an artificial neural network during walking with a powered ankle-foot orthosis. J. Biophys. (2016). https://doi.org/10.1155/2016/7984157
6. Kim, S.K., Hong, S., Kim, D.: A walking motion imitation framework of a humanoid robot by human walking recognition from IMU motion data. In: 2009 9th IEEE-RAS International Conference on Humanoid Robots, pp. 343–348. IEEE (07122009–10122009). https://doi.org/10.1109/ICHR.2009.5379552
7. Mannini, A., Genovese, V., Maria Sabatini, A.: Online decoding of hidden Markov models for gait event detection using foot-mounted gyroscopes. IEEE J. Biomed. Health Inform. **18**(4), 1122–1130 (2014). https://doi.org/10.1109/JBHI.2013.2293887
8. Pappas, I.P., Popovic, M.R., Keller, T., Dietz, V., Morari, M.: A reliable gait phase detection system. IEEE Trans. Neural Syst. Rehabil. Eng. Publ. IEEE Eng. Med. Biol. Soc. **9**(2), 113–125 (2001). https://doi.org/10.1109/7333.928571

 9. Pham, M.H., et al.: Validation of a step detection algorithm during straight walking and turning in patients with Parkinson's disease and older adults using an inertial measurement unit at the lower back. Front. Neurol. **8**, 457 (2017). https://doi.org/10.3389/fneur.2017.00457

10. Sánchez Manchola, M.D., Bernal, M.J.P., Munera, M., Cifuentes, C.A.: Gait phase detection for lower-limb exoskeletons using foot motion data from a single inertial measurement unit in hemiparetic individuals. Sensors (Basel, Switzerland) **19**, 2988 (2019). https://doi.org/10.3390/s19132988

11. Sarshar, M., Polturi, S., Schega, L.: Gait phase estimation by using LSTM in IMU-based gait analysis-proof of concept. Sensors (Basel, Switzerland) **21**, 5749 (2021). https://doi.org/10.3390/s21175749

12. Sijobert, B., Feuvrier, F., Froger, J., Guiraud, D., Coste, C.A.: A sensor fusion approach for inertial sensors based 3D kinematics and pathological gait assessments: toward an adaptive control of stimulation in post-stroke subjects. In: EMBC: Engineering in Medicine and Biology (2018). https://doi.org/10.1109/EMBC.2018.8512985

13. Taborri, J., Palermo, E., Rossi, S., Cappa, P.: Gait partitioning methods: a systematic review. Sensors (Basel, Switzerland) **16**, 66 (2016). https://doi.org/10.3390/s16010066

14. Vu, H.T.T., et al.: A review of gait phase detection algorithms for lower limb prostheses. Sensors (Basel, Switzerland) **20** (2020). https://doi.org/10.3390/s20143972

Ultra-Fast Lidar Scene Analysis Using Convolutional Neural Network

Houssem Moussa[1], Valentin Gies[2]([✉]) [iD], and Thierry Soriano[3] [iD]

[1] Université de Toulon, SeaTech, Toulon, France
[2] Université de Toulon, Laboratoire IM2NP - UMR 7334, Toulon, France
`gies@univ-tln.fr`
[3] Université de Toulon, Laboratoire COSMER - EA 7398, Toulon, France
`thierry.soriano@univ-tln.fr`
`https://cosmer.univ-tln.fr/en/`

Abstract. This work introduces a ultra-fast object detection method named FLA-CNN for detecting objects in a scene from a planar LIDAR signal, using convolutional Neural Networks (CNN). Compared with recent methods using CNN on 2D/3D lidar scene representation, detection is done using the raw 1D lidar distance signal instead of its projection on a 2D space, but is still using convolutional neural networks. Algorithm has been successfully tested for RoboCup scene analysis in Middle Size League, detecting goal posts, field boundary corners and other robots. Compared with state of the art techniques based on CNN such as using Yolo-V3 for analysing Lidar maps, FLA-CNN is 2000 times more efficient with a higher Average Precision (AP), leading to a computation time of $0.025\,ms$, allowing it to be implemented in a standard CPU or Digital Signal Processor (DSP) in ultra low-power embedded systems.

Keywords: Lidar processing · Scene analysis · Convolutional neural networks · Low-power

1 Introduction

Mobile autonomous robots in unknown or changing environment need to take decisions based on the surrounding scene analysis. For this task, two types of sensors are mainly used: cameras and LIDAR. This latter is an interesting exteroceptive sensor in robotics providing reliable maps of the surrounding environment, with a better precision in object positioning than using only cameras. This interesting feature greatly helps to ensure a high level of safety in human-robots interactions.

This paper focuses on autonomous robot soccer scene analysis, including robots, humans, goals (posts), and field boundary using a planar Lidar, but with limited computing capabilities, and at least without using a GPU (such as for image processing). Chosen Lidar is a Pepperl+Fuchs R2000 UHD one generating a 1D sequence of 1440 distance measurements at each scan of the scene, 50 times

© The Author(s), under exclusive license to Springer Nature Switzerland AG 2023
A. Eguchi et al. (Eds.): RoboCup 2022, LNAI 13561, pp. 50–61, 2023
https://doi.org/10.1007/978-3-031-28469-4_5

per second, with a maximum range of 60 m. Application of this paper is analysis of RoboCup Middle Size League scenes. RoboCup is an international competition where robot teams play soccer autonomously. Its main objective in the future is to won against a professional, human soccer team by 2050. In the Middle Size League (MSL), teams are composed of five robots playing autonomously on a 22 m by 14 m soccer field with a real soccer ball.

Having limited computing capabilities have a deep impact on the algorithms that can be used in a robot. Most scene analysis algorithms are using 2D or even 3D representations of the scene, leading to huge computation time on limited computing systems. This paper aims at introducing a novel algorithm for a LIDAR scene analysis, using state of the art CNN with end to end learning, but without using a 2D representation of the scene. Instead of that CNN is applied to the raw 1D lidar signal, using the 1440 distance data of each LIDAR rotation as input tensor. This algorithm is called Fast Lidar Analysis using Convolutional Neural Network (FLA-CNN). It achieves an excellent precision while having a very low detection time and a low power consumption, making it usable for mobile robots with CPU or DSP for real time detection, without requiring a GPU.

This work is divided into 3 parts:

- In Sect. 2, an overview of scene analysis techniques using cameras or LIDAR is presented, focusing on their advantages and disadvantages.
- In Sect. 3, Fast Lidar Analysis using Convolutional Neural Network (FLA-CNN) algorithm is introduced, allowing to analyse scenes with low latency and computing power requirements.
- In Sect. 4, application to the RoboCup scene analysis is presented, with a focus on dataset creation and labeling, training process, and a discussion on results and performance.

2   State of the Art in Robot Scene Analysis

Deep neural networks, and particularly convolutional neural networks are considered as state of the art models for feature extraction due to their great ability to learn their features and classifiers from data into an end to end learning process. However, their main drawback is to require high computing capabilities, making them difficult to implement in an embedded computer or microcontroller. Moreover, using these algorithms for controlling fast robots in real time, requires a fast processing time considering the speed of the robots (up to 6 m/s in Middle Size League) corresponding to at least 30 frame per second (FPS). Combining these two aspects, high frame rate and embedded processing, is the key for efficient embedded robot control.

2.1   Object Detection Based on Images

Camera is one of the most used sensor in robotics combined with processing for scene analysis. Among the most efficient ones are detectors based on CNN, having one stage or two stages. Both of them have interesting performances in terms of accuracy, but are relatively slow [1,2]. [1] gives an interesting comparison of these detectors on Coco dataset: using an Intel i7-8700 CPU and an NVIDIA GeForce RTX 2080Ti 12GB GPU on 640×960 images, Average Precision (AP) reaches 32.4 on the most accurate models (*i.e.* Faster RCNN Res2Net101), but at the price of a computation time of 63 ms. Reaching a computation speed of 30 FPS requires to use Mobile Net Models at the price of a loss in accuracy (mAP = 24.3).

An interesting result in [1] is that two stages models [3–8] are more accurate than one stage ones such as YOLO [9–11], even if the main trend is nowadays to use these latter. Figure 1) shows a result of scene analysis on a 360°C image using YOLOv3. This image has been recorded using an omnidirectional camera and transformed to a panorama image because Yolo algorithms are not rotation invariant and cannot be applied directly to omnidirectional images. Models have been proposed to cope with this issue [12,13] but are more resource consuming.

Fig. 1. YOLOv3 for object detection running on a GPU with omnidirectional camera.

In conclusion, cameras are potentially rich sensors but requires a high processing power for extracting segmenting the scene at a high FPS rate, making them difficult to use on embedded CPU or micro-controllers. Considering stereovision cameras would also be interesting, but it would require more computing power making them out of scope on proposed application to MSL robots.

2.2 Object Detection Based on 2D Map Lidar Images

An alternative to the use of cameras, is to use 2D or 3D lidars. An example of 2D map obtained using a Pepperl+Fuchs R2000 lidar is presented in Fig. 2.

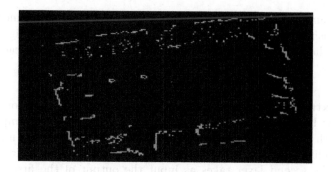

Fig. 2. 2D map image created with 1D lidar signal

This kind of 2D image can be analysed using image processing algorithms for detecting shapes or objects. However, doing that would lead to the same drawbacks as for image processing: important computation time and computer power requirement.

Considering the strong constraints of embedded systems in terms of limitations in computing power and the need for high FPS, another approach is proposed in this paper, using CNN on th raw 1D-Lidar signal.

3 Contribution: Fast Lidar Analysis Using Convolutional Neural Network (FLA-CNN)

In this paper, Fast Lidar Analysis using Convolutional Neural Network (FLA-CNN) is introduced. It aims at analysing 1D raw lidar distance data using a state of the art deep learning model predicting corners of a RoboCup field and posts of a RoboCup goal. FLA-CNN design has been inspired by YOLO: its design is being presented in details.

Notation. We use $P = (P_d, P_\theta, P_{conf}, P_{class}) \in \Re^4$ to denote a ground-truth points, where P_d, P_θ are the distance and its corresponding angle in robot referential, $P_d \in [0, 60]$ and $P_\theta \in [0, 2\pi]$. P_{conf} is the confidence score of an existing object and P_{class} is the index of the corresponding class name to the detected object. Similarly $\hat{P} = (\hat{P}_d, \hat{P}_\theta, \hat{P}_{conf}, \hat{P}_{class}) \in \Re^4$ denotes a predicted object.

3.1 CNN Architecture

FLA-CNN network can be divided into two parts: feature extractor neural network (FNN) (Eq. 1) and object regression network (ORN) (Eq. 2).

$$F = FNN(D) \tag{1}$$

$$T = ORN(F) \tag{2}$$

where $D \in \mathbb{R}^{1440}$ is the input tensor featuring the measured distance for each angle during a full rotation. F denotes a feature vector and T denotes a list of predicted points that represents the objects in 1D signal in transformed notation (the relationship between T and P will be defined soon - Eq. 11). Figure 5 shows the overall the FLA-CNN architecture, while below we describe each stage in some depth.

Feature Extractor Network FNN: The feature extractor network, takes an input vector D and outputs a list of features T. This stage is composed with 2 layers, each layer apply a convolution 1D, batch normalisation, max pooling and leaky relu as an activation function. The first layer takes as input the 1D signal(D), the second layer takes as input the output of the first layer. This network extract features from signal, only one layer is not capable to identify the signal for that reason we added the second layer that will complete the recognition stage (Fig. 3).

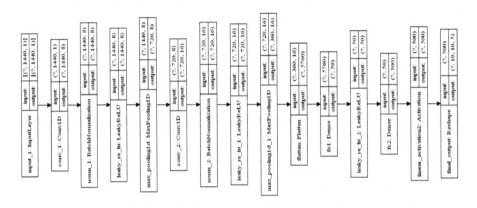

Fig. 3. Model network for object detection using 1D lidar signal detection

The convolution 1D is faster than 2D one, decreasing drastically computational complexity, making it suitable for running on mobile devices, embedded systems, and on some microcontrollers and DSP.

In each convolutional layer, the forward propagation is expressed as follows:

$$x_k^l = b_k^l + \sum_{i=1}^{N_{l-1}} conv1D(w_{ik}^{l-1}, s_i^{l-1})$$ (3)

where x_k^l is defined as the input, b_k^l is defined as the bias of the kth neuron at layer 1, s_i^{l-1} is the output of the ith neuron at layer l-1, w_{ik}^{l-1} is the kernel from the ith neuron at layer l-1 to the kth neuron at layer l.

The output of the convolutional layer is followed by a batch normalisation layer for fixing means and variances. Learning is done using stochastic optimisation due to the memory limits, for reducing over-fitting and training time. Normalisation process is expressed as follows, where B is a mini-batch of size m of the whole training dataset. The mean and variance of B could be expressed as:

$$\mu_B = \frac{1}{m} \sum_{i=1}^{m} x_i \tag{4}$$

$$\sigma_{B^2} = \frac{1}{m} \sum_{i=1}^{m} (x_i - \mu B)^2 \tag{5}$$

For a layer of the network with d-dimensional input, $x = (x^{(1)}, ..., x^{(d)})$, each dimension of its input is then normalized (i.e. re-centered and re-scaled) separately,

$$\hat{x}_i^k = \frac{x_i^k - \mu_B^k}{\sqrt{\sigma_{B^2}^{k^2} + \epsilon}} \tag{6}$$

where $k \in [1, 1440]$ and $i \in [1, m]$; $\mu_B^{(k)}$ and $\sigma_B^{(k)^2}$ are the per-dimension mean and variance, respectively.

ϵ is added in the denominator for numerical stability and is an arbitrarily small constant we used $\epsilon = 1e^{-6}$. The resulting normalized activation $\hat{x}^{(k)}$ have zero mean and unit variance, if ϵ is not taken into account. To restore the representation power of the network, a transformation step then follows as

$$y_i^k = \gamma^k x_i^k + \beta^k \tag{7}$$

Object Regression Network. Following the FNN, a separate MLP is applied to each feature vector F to produce a transformed version of object points predictions, noted $T < N_\theta, N_d, 3 + N_{classes} >$, a 3-dimensional matrix where $N_\theta = 10$ is the numbers of angular cells (size of each cell is 0.63 rad). $N_d = 10$ is the numbers of radial cells (size of each cell is 6 m). First and second indexes correspond to the location (θ, d) in the prediction polar grid, and last index corresponds to the intra-cell predicted position and classification of the point in the considered polar grid cell with the following information: $T_{Cell\theta}$ the predicted point angle in the considered cell, T_{Celld} the predicted distance in the considered cell, $T_{CellConf}$ the probability that an object exist in the considered cell and $T_{CellClass_i}$ the probability that the object belongs to the ith class.

Exact coordinates of the object in the signal can be obtained using the following equations:

$$P_d = (d + Sig(T_{Celld}))S_d \tag{8}$$
$$P_\theta = (\theta + Sig(T_{Cell\theta}))S_\theta \tag{9}$$
$$P_{conf} = Sig(T_{CellConf})) \tag{10}$$
$$P_{class_i} = softmax(T_{CellClass_i}) \tag{11}$$

where $Sig(\cdot)$ is the logistic (sigmoid) activation function is defined by:

$$sig(x) = \frac{1}{1 + e^{-x}} \tag{12}$$

where Softmax() is the normalized exponential function is defined by:

$$softmax(z)_i = \frac{e^{z_i}}{\sum_{j=1}^{k} e^{z_j}} \text{ for i=1...k and } z = (z_1...z_k) \in \Re^k \qquad (13)$$

where S_d is the radial grid size $S_d = \frac{max(d)}{N_d}$
and where S_θ is the angular grid size $S_\theta = \frac{2\pi}{N_\theta}$

A flatten layer has been added between the FNN and OPN, and all the feature vectors obtained with the CNN 2 layers in the FNN are transformed to one vector that will be the input for the final multi layer perceptron. The number of neurons in the last MLP layer must be equal to $(N_d * N_\theta * (N_c + 3))$, where N_θ is the number of grid bellowing to x, N_d is the number of grid bellowing to y axis and 5 represent the predictions coordinates ($\hat{T} = (\hat{T}_{Celld}, \hat{T}_{Cell\theta}, \hat{T}_{CellConf}, \hat{T}_{CellClass_i})$ $\in \Re^5$) where N_c is the class numbers.

4 Application to the RoboCup Scene Analysis

4.1 Data Set Creation and Labelling

Data uses in this work have been recorded using Robot Club Toulon (RCT) robots participating to RoboCup Middle Size League, using a Pepperl+Fuchs 1D lidar delivering 1440 distance measurements per rotation (angular resolution is 0.25°), 50 times per second with a maximum range of 60m. Its precision is approximately ±1 cm. This work aims at detecting field boundary corners and goal posts. These information are sufficient for computing position of our robot in the field using distance and angles of each detected post, and developing strategies for playing soccer and shooting with precision.

Signal labelling application has been designed for labelling 1D lidar signal easily, with labels having the same format as the model output discussed in (3.1). This labelling tool uses data files containing timestamped Lidar data that is decoded and extracted to create one lidar file for every lidar sample, each file name containing the timestamp of the lidar acquisition. Each file contains 1440 lines and 2 columns, first one for the distance and second one for the lidar angle in robot referential.

In a LIDAR scene recorded in our RoboCup field, four points can be considered as field boundary corners, and two of them can be considered as posts (our dataset has been recorded with only one goal in the scene). Creating labels directly on the 1D signal is difficult and would lead to many faults in the dataset: using a 2D representation, only for labeling 1D signal, is a better way to easily find and label field corners and posts in the scene (Fig. 4).

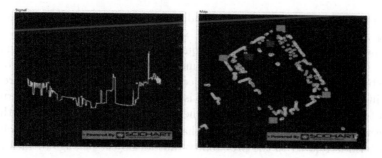

Fig. 4. LIDAR data signal labeling: 1D signal (left) and 2D map (right). (Color figure online)

For transforming each 1D lidar sample to a 2D map, following transformation is used, where d is the measured distance, θ the measurement angle, and (x, y) the cartesian position of the obstacle in the robot referential:

$$x = \cos(\theta) * d \qquad (14)$$
$$y = \sin(\theta) * d \qquad (15)$$

Once the map is created, object class is selected manually, and cartesian map coordinates are transformed back to polar coordinates for labeling a specific point in the 1D signal. As shown in (Fig. 4), 4 points have been labelled in pink for the boundary corners, and 2 points have been labelled in blue for the posts. Inverse transformation (from cartesian to polar point) is expressed as follows:

$$d = \sqrt{x^2 + y^2} \qquad (16)$$
$$\theta = 2\arctan(\frac{y}{x + d}) \qquad (17)$$

This process is repeated for every lidar Sample, after each annotation for each sample an xml file is created that contain distance, angle and the class name of each object which is represented by a point on the 2D Map. Once the labeling process is finished, a new dataset with a unique *.txt* file containing lidar 1D data and its corresponding *.xml* file for the labels is generated.

4.2   Training

FLA-CNN has been trained from scratch on our dataset, as model has been fully customised for our application. During training, model prediction has been optimised by minimising the loss function (Eq. 18). Output of our model is a tensor of dimansion $< N_d = 10 \times N_\theta = 10 \times (3 + N_{Classes}) = 5 >$, since we have 2 objects classes (goals and corners) to detect, so that every grid may contain 1 or in some cases 2 objects. Anchors boxes for multiple detection in the same grid cell are not used, since object bounding boxes have same size always in the 2D map.

Proposed loss function is a sum of squared errors between the different components of the predictions and the ground truth. First term is related to angle and distance only when there is an object. 1_i^{obj} is equal to 1 when there is an object in the grid cell i, and to 0 if there is no object. Second and third terms are related to prediction confidence when there is an object 1_i^{obj} and when there is no object 1_i^{noobj}. $\lambda_{pred_o bj} = 3$ is higher than $\lambda_{pred_n oobj} = 2.5$ to focus on objects. Fourth term is related to the probability that an object belongs to each one of the classes.

$$
\begin{aligned}
loss = \lambda_{cord} \sum_{i=0}^{S^2} 1_i^{obj} [(P_\theta^i - \hat{P}_\theta^{\,i})^2 + (P_d^i - \hat{P}_d^{\,i})^2] \\
+ \lambda_{pred\_obj} \sum_{i=0}^{S^2} 1_i^{obj} (P_{conf}^i - \hat{P}_{conf}^i)^2 \\
+ \lambda_{pred\_noobj} \sum_{i=0}^{S^2} 1_i^{noobj} (P_{conf}^i - \hat{P}_{conf}^i)^2 \\
+ \sum_{i=0}^{S^2} \sum_{j=1}^{k} 1_i^{obj} (P\_class_j^i - \hat{P}\_class_j^i)^2
\end{aligned}
\tag{18}
$$

Network has been trained on 336 samples of 1440 elements. 200 epochs have been iterated, with a batch size of 8. Optimiser used is adam, with a learning rate $= 1e - 4$, decay$= 0.005$ and beta $= 0.99$. To avoid overfitting issues, training has been stopped when loss function was increasing for 10 epochs. Only the best weights of the last 10 epochs will be saved.

5   Results

The dataset that we created in Sect. 4 is devided in 2 parts, 80% for training and 20% for the evaluation, we used the AUC of the ROC curve to evaluate our detection results, ROC curve is a graph showing the performance of a detection model at all detection thresholds. This curve plots two parameters: True Positive Rate (TPR) (Eq. 19) and False Positive Rate (FPR) (Eq. 20)

$$
TPR = \frac{TP}{TP + FN}
\tag{19}
$$

$$
FPR = \frac{FP}{FP + TN}
\tag{20}
$$

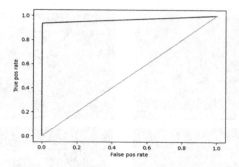

Fig. 5. Model evaluation with ROC curve

The ROC curve (Fig. 5) that we obtain on the evaluation dataset for different threshold starting with 0 ending with 1 with a step of 0.1, for our model the AUC = 0.94 which is an excellent result in term of precision.

We used other metric to evaluate the detection results which is the mean absolute error between the predictions of our model and the ground truth of the evaluation dataset (Fig. 5), the maximum error of the distance is 0.17 cm, and the maximum angular error is 0.05 rad (2°C).

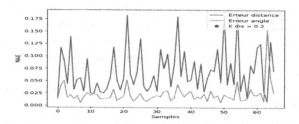

The detection results are shown above, the model predicts the distance and angle of each object that below to pink and blue on the signal and using the transformation discussed in (3) we are showing the correspondent object position in the 2D MAP (Fig. 6).

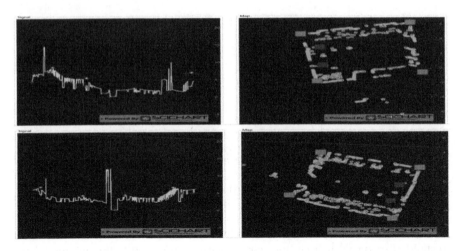

Fig. 6. Model predictions for different samples : : 1D signal (left) and 2D map (right).

6   Conclusion

This work introduces a ultra-fast object detection method named FLA-CNN for detecting objects in a scene from a planar LIDAR signal, using convolutional Neural Networks (CNN). Compared with recent methods using CNN on 2D/3D lidar scene representation, detection is done using the raw 1D lidar distance signal instead of its projection on a 2D space, but is still using convolutional neural networks.

Algorithm has been successfully tested for RoboCup scene analysis in Middle Size League, detecting goal posts, field boundary corners and other robots. Prediction inferences are computed in 25 μs with 94% of mAP. Thanks to the reduced size of the proposed CNN processing directly 1D lidar signal instead of converting it to a 2D image, implementing it in an embedded system is possible and doesn't require a high computational power such as a GPU, but can be achieved in a DSP or a microcontroller for real time detection in mobile robots.

In term of precision and speed our model achieve the better results compared to the start of art methods that we tested ourselves as explained as follows:

Method	Input size	mAP	Time(ms)	Computing system
SSD	$321 \times 321 \times 3$	45.4	61	GPU (GTX 1080)
SSD	$513 \times 513 \times 3$	50.4	125	GPU (GTX 1080)
YOLO_v3	$416 \times 416 \times 3$	55.3	29	GPU (GTX 1080)
YOLO_v3	$608 \times 608 \times 3$	57.9	51	GPU (GTX 1080)
Faster_RCNN	$1000 \times 600 \times 3$	73.2	142	GPU (GTX 1080)
Ours (FLA-CNN)	1440×1	94	0.025	CPU (i5-9500)

Even if proposed results are preliminary and require a large database with many objects and labels to detect for a full validation of the proposed algorithm, first results are promising and clearly show that converting a 1D lidar data into a 2D map leads to dramatically increase computation power requirement.

References

1. Carranza-Garcia, M., Torres-Mateo, J., Lara-Benitez, P., Garcia-Gutierrez, J.: On the performance of one-stage and two-stage object detectors in autonomous vehicles using camera data. Remote Sens. **13**(1), 89 (2021)
2. Hu, Z., et al.: Fast image recognition of transmission tower based on big data. Prot. Control Modern Power Syst. **3**(1), 1–10 (2018). https://doi.org/10.1186/s41601-018-0088-y
3. Girshick, R., Donahue, J., Darrell, T., Malik, J.: Rich feature hierarchies for accurate object detection and semantic segmentation. In: Proceedings of the IEEE Conference on Computer Vision and Pattern Recognition, pp. 580–587 (2014)
4. Redmon, J., Farhadi, A.: Yolov3: an incremental improvement (2018)
5. Van de Sande, K.E., Uijlings, J.R., Gevers, T., Smeulders, A.W.: Segmentation as selective search for object recognition. In:2011 International Conference on Computer Vision, pp.1879–1886 (2011)
6. Sultana, F., Sufian, A., Dutta, P.: A review of object detection models based on convolutional neural network. Adv. Intell. Syst. Comput. 1–16. Springer, Singapore (2020)
7. Ren, S., He, K., Girshick, R., Sun, J.: Faster R-CNN: towards real-time object detection with region proposal networks. CoRR, abs/1506.01497 (2015)
8. Ren, S., He, K., Girshick, R., Sun, J.: Faster R-CNN: towards real-time object detection with region proposal networks. In: Advances in Neural Information Processing Systems, vol. 28 (2015)
9. Redmon, J., Divvala, S.K., Girshick, R.B., Farhadi, A.: You only look once: unified, real-time object detection. CoRR, abs/1506.02640 (2015)
10. Redmon, J., Farhadi, A.: YOLO9000: better, faster, stronger. CoRR, abs/1612.08242 (2016)
11. Bochkovskiy, A., Wang, C.Y., Liao, H.Y.M.: Yolov4: optimal speed and accuracy of object detection. CoRR, abs/2004.10934 (2020)
12. Duan, Z., Tezcan, M.O., Nakamura, H., Ishwar, P., Konrad, J.: Rapid: rotation-aware people detection in overhead fisheye images. CoRR, abs/2005.11623 (2020)
13. Xia, G.S., et al.: DOTA: a large-scale dataset for object detection in aerial images. CoRR, abs/1711.10398 (2017)

Towards a Real-Time, Low-Resource, End-to-End Object Detection Pipeline for Robot Soccer

Sai Kiran Narayanaswami[1](✉), Mauricio Tec[1], Ishan Durugkar[1], Siddharth Desai[1], Bharath Masetty[1], Sanmit Narvekar[1], and Peter Stone[1,2]

[1] Learning Agents Research Group, The University of Texas at Austin, Austin, USA
saikirann94@gmail.com
[2] Sony AI, Austin, Texas, USA

Abstract. This work presents a study for building a Deep Vision pipeline suitable for the Robocup Standard Platform League, a humanoid robot soccer tournament. Specifically, we focus on end-to-end trainable object detection for effective perception using Aldebaran NAO v6 robots. The implementation of such a detector poses two major challenges, those of speed, and resource-effectiveness with respect to memory and computational power. We benchmark architectures using the YOLO and SSD detection paradigms, and identify variants that are able to achieve good detection performance for ball detection, while being able to perform rapid inference. To add to the training data for these networks, we also create a dataset from logs collected by the UT Austin Villa team during previous competitions, and set up an annotation pipeline for training. We utilize the above results and training pipeline to realize a practical, multi-class object detector that enables the robot's vision system to run 35 Hz while maintaining good detection performance.

1 Introduction

Object detection [12,25,31] is one of the paramount challenges of computer vision and a key component of robotic perception. This paper develops an effective perception module for robot soccer under the stringent hardware constraints of the Robocup Standard Platform League (SPL) [2,5,19], in which the vision system needs to identify various objects and landmarks in real time, such as the ball or other robots.

Neural network-based detectors have progressed tremendously over the past decade. However, many network architectures require computational power that limits their applicability in low-resource real-time scenarios such as the SPL [6]. On the other hand, the necessity for fast, resource-friendly solutions for mobile computing and the IoT has led to increasing attention on specialized hardware and network architectures [6,10,16,18]. In this project, we aim to leverage these advances to develop an object detection pipeline suitable for use in the SPL. The implementation uses a software stack based on TensorFlow Lite [1] (TFLite).

The paper starts with an investigation of candidate object detection architectures by attending to reliable detection in the field and testing speed with the robot hardware. Next, it describes the methods used for data collection, labeling, and data augmentation. Finally, the performance of the detector is evaluated, and it is demonstrated to be effective at detecting objects under the computational constraints for real-time operation.

© The Author(s), under exclusive license to Springer Nature Switzerland AG 2023
A. Eguchi et al. (Eds.): RoboCup 2022, LNAI 13561, pp. 62–74, 2023
https://doi.org/10.1007/978-3-031-28469-4_6

2   Background

This section goes over some preliminaries, a description of SPL, the hardware and compute setup, and the object detection challenge this paper seeks to solve.

2.1   Robocup Standard Platform League

The Robocup SPL is a humanoid soccer tournament in which our team competes as the *UT Austin Villa*. Soccer games are played between teams of five Aldebaran NAO v6 robots each, on a scaled version of a soccer field. The competition tests four main systems: vision, localization, motion and coordination. Localization relies on the vision results, such as by using detected markers to estimate the robot's position. Subsequently, the localization information is used to direct coordination. Thus reliable and fast object detection is important to achieve effective gameplay. No assistance from external hardware, including remote, is allowed (i.e., the robot's hardware is the Standard Platform).

2.2   Hardware Setup

Camera. There are two identical cameras available on the top and bottom of the NAOs (located above and below the eyes). The top camera is meant to give a full view of the field, while the bottom camera provides a close-up view of the ground immediately in front of the robot. Both cameras are, in principle, able to capture video at a resolution of 1280×960 pixels at 30 frames per second, with automatic exposure adjustment capabilities [27]. However, to save computation in various steps of the vision pipeline, and since the bottom camera does not view anything distant, we use a bottom camera resolution of 320×240 resolution. Example views are shown in Fig. 2.

Compute. The NAO has an Intel® Atom E3845 CPU, along with 4 GB of RAM [39]. The CPU has 4 cores running at 1.91 GHz, and supports some advanced SIMD instruction sets such as SSE4, enabling parallel computation. This CPU is optimized for low-power applications, and its performance compares to today's mobile processors.

An integrated GPU (Intel® HD Graphics for Intel Atom® Processor Z3700 Series) is present in the CPU, clocked at 542 MHz. It is capable of compute acceleration through OpenCL, enabled using a custom compiled Linux Kernel and drivers. This GPU is comparable in clock speed to today's mobile GPUs as well. However, experiments in Sect. 7.1 will show that it is not able to perform significantly faster than the CPU. Nevertheless, it remains viable as an additional source of compute power.

2.3   Object Detection Challenges

Object detection needs to run alongside the other components involved. To guarantee the functionality of all the components, the perception loop needs to maintain a processing rate 30 Hz, which comes to 33.3 ms. Within this time, both the top and bottom camera images need to be processed, and any other stages of the vision pipeline also need to be completed. These stringent timing requirements, in conjunction with the limited compute capabilities described above, challenge the implementation of a neural-network based object detector. Even real-time desktop applications such as YoloV3-Tiny [3] prove to be too expensive for these purposes.

3   Related Work

This section presents an overview of previous work related to various aspects of low-resource real-time object detection. Additional discussion of the merits and demerits of relevant approaches is handled later in the context of our design choices.

General Object Detection. There are three main classes of methods for detection. One, a two-stage approach which proposes regions of interest, followed by the actual detection. Regions with CNN features (R-CNN) [12] and related methods such as Faster-RCNN [34], or Mask-RCNN [13] for segmentation fall into this class. Two, single stage detectors with *You Only Look Once* (YOLO) [7,31,32] being one of the most well-known paradigms that performs direct bounding box regression. Related approaches have also been developed based on corner point detection [20]. Three, detection at multiple resolutions and scales simultaneously, with methods like Single Shot MultiBox Detector (SSD) [25] and Feature Pyramid Networks [23], using existing architectures as backbones [22,25].

Low-Resource Object Detection. MobileNets [16,17,37] are a series of architectures developed for inference on mobile devices that use depthwise separable convolutional layers to achieve fast performance. Other lightweight architectures include Squeezenet [18] and NASNet [44], though these are mainly optimized for parameter count rather than speed as sought here. OFA Net [9] is an approach for training full scale networks, and subsequently deploying distilled versions with greatly reduced sizes. One class of approaches such as XNOR-Net [29] use binarized or heavily quantized versions of larger networks to reduce the computational cost for a given network size. Although these architectures are much less resource intensive than real-time detectors for desktop hardware [3], they are still significantly slower than what SPL requires. One major reason for this slowness is that these works target a large number of object classes, whereas there are many fewer classes in the Robocup setting.

Existing Approaches for Humanoid Soccer. The SPL introduced the current NAO v6 robots in 2017. Until then, NAO v5 robots were used, which had an older, single core CPU that almost entirely prohibited deep learning pipelines. As a result, Deep Vision detectors in this competition are still in nascent stages of development, and even the winner of the 2018 competition did not use them [8]. UT Austin Villa has also been relying on more classical techniques such as SVM or feedforward classification on features extracted from region proposals [26]. The deep approaches that have been proposed [11,28,36] have largely been based on RoI proposals. To the best of our knowledge, these RoI-based approaches detect one object at a time. [11] use XNOR-Nets for ball detection, while [35] use an encoder-decoder architecture for ball localization on pre-processed RoI images. [28] propose Jet-Nets based on MobileNet for robot detection.

xYOLO [6] is a lightweight detector proposed for the Robocup Humanoid League that is based on Tiny-YOLO [3]. It achieves good ball-detection and speed using XNOR operations and a reduced number of layers and filters. We note however, that this architecture was developed for a different league with differing robots and slightly more permissive compute constraints, including access to more powerful modern hardware such as the Raspberry Pi 3 [6]. A more recent work proposes YOLO-Lite [43], which uses lightweight inference libraries (as we also do here) and was tested on the NAO v6 robots. However, their proposed networks are not nearly fast enough to meet our target frame-rates, and they do not test the NAO's GPU.

4 Data Curation and Data Augmentation

We used three sources of data used to train our Deep Vision system. These datasets provide a good amount of variety in lighting conditions, decor, and clutter:

- **xYOLO:** This dataset contains 1400 images collected using a webcam in the Robocup Humanoid League for training the xYOLO network [6]. Annotations are provided for the ball and goalposts in the form of rectangular boxes, and we manually label other objects.
- **UTAV:** We created this internal dataset from our team logs collected during calibration and practice sessions in the SPL 2019 competition. There are about 1000 images from the top and bottom cameras, which were annotated manually for balls, robots and crosses. The images are recorded after the autoexposure adjustment.
- **NaoDevils:** This dataset contains 10 thousand images from the NaoDevils team's matches at SPL 2019 and the German Open 2019 [4]. The images contain segmentation labels from six classes: line, ball, robot, center circle, goal, and penalty cross. We transformed the segmentation masks to rectangular bounding boxes. Approximately 10% of the images are high-quality manually annotated images, the rest are lower-quality automatically segmented used for pre-training.

Manual Annotation. Manual annotation of existing and future data is a time-consuming process. To expedite this process, we utilize general purpose, full scale detectors to provide a seed set of labels, which can then be refined by human annotators. This idea is tested using Efficient-Det D7 [40], which is among the state-of-the-art detectors at present. We use a detector that is pretrained on the MS-COCO dataset [24]. Efficient-Det D7 was able to detect the ball and robots in many of the data images.

Thus, we use these annotations as a starting point in order to save time, as well as to allow us to use data not directly from the robot's camera for more robust detection. To this end, we used the CV Annotation Tool (CVAT) [38] by Intel, which provides a convenient user interface to execute detectors and correct their output, as well as enabling users to upload data in many data formats. Using CVAT, we annotated the ball, robots and the penalty crosses in the xYOLO and UTAV dataset images. Although we focus most of our study on ball detection with a single-class detection architecture, we also train and deploy a multi-class detector that also detects robots and penalty crosses.

Data Augmentation. Multiple data augmentation techniques are used based on the Pytorch implementations available in the YOLOv5 codebase [42]. One kind of augmentations we use are kinematic augmentations which include rotations, translations and scaling. These serve to improve detections at image locations where objects are not frequently present in dataset images, and also in sizes that are not present frequently. Mosaic augmentation [7] is also used to combine multiple images, which is useful when multiple objects are present. The other class of augmentations is photometric in nature. We use transformations in the HSV color space in order to bring about invariance to lighting conditions, material color and texture, and camera exposure fluctuations.

5   Analyzing Computational Constraints

The computational requirements are critical considering that the amount of compute resources available is non-negotiable in the competition. This restriction has meant that a significant portion of our efforts so far have been directed towards understanding these constraints and studying what can be accomplished within them.

This section first describes the motivations behind the choice of software stack used. It then presents a benchmark study involving this stack to give an idea of the scale of the networks that can be executed in real time on the NAO.

5.1   Software Stack

There are several low-resource libraries available for performing inference on Machine Learning models, both general purpose [1,30] as well as ones made by other teams competing in Robocup [14,41]. Models are first trained in PyTorch with an adapted version of the YOLOv5 repository [42]. The network weights are then transformed to TensorFlow Lite (TFLite). We use TFLite [1] for several reasons:

- Advanced NN Layer support: TFLite has support for a much wider range of activation functions and layer types than the other libraries (e.g. batch normalization).
- Multi-threaded operation: The special purpose libraries mentioned above do not allow for the use of more than one thread, whereas TFLite does.
- Optimizations for NN inference using the XNNPack [21] library, including the usage of advanced instruction sets (mainly SSE) as many other libraries also do.
- GPU Support: TFLite is the only lightweight library to have OpenCL support, thus enabling us to use the GPU as additional hardware.

5.2   Computational Benchmarks

We conduct a set of first principle benchmarks to study the computational cost as a function of network size:

We generate CNNs with random weights, with varying layer sizes and numbers of filters in each layer. Each convolutional layer has n filters, except for the last one, which has $2n$. They convert to a dense layer through a softmax of 10 units, representative of the number of objects we will eventually have to detect. The convolutional layers all use ReLU activations, and each is followed by a (2,2) Max Pooling layer. This template is based on an architecture that has been tested by the team B-Human [36]. Weights for the network were initialized at random (Gaussian with $\sigma = 0.01$). Four threads on the CPU were used to perform inference.

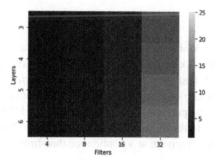

Fig. 1. Runtime benchmark results for synthetically generated CNN models. Times are presented in milliseconds.

The results are presented in Fig. 1. For a target inference rate of 60 frames per second for both cameras together, we see that the network with 16 filters per layer, with 6 layers is comfortably feasible. The main impact on inference time appears to be due to the number of filters rather than the layers. This suggests that narrow, but deep architectures with skip connections might be the best computational choice.

For the above, quantization of weights is not performed. When testing with MobileNet-v2, we observed that the quantized models performed slower (64 ms vs. 58 ms).

6 Detector Design

As mentioned earlier, in the past, region proposal methods have been used in the SPL for detecting a single class of objects. However, this design aims to be scalable and to detect the full set of objects on the field. Region proposals tend to be more expensive with more objects to detect. Thus, this study focuses on YOLO and SSD setups.

Although SSD is more expensive than YOLO, it has potential advantages in terms of detecting objects at multiple scales. A suitable backbone architecture needs to be provided on top of which the SSD layers are implemented. We use Mobilenet-v2 [37] due to the fast runtimes while maintaining good performance in an SSD setup. SSD requires designing prior boxes for each of the feature maps from the SSD layers. However, such a design is not much more expensive compared to the design of anchor boxes for YOLO-9000 [32], which is necessary to get good performance with YOLO setups.

xYOLO [6] uses a greatly reduced version of YOLO in terms of layers and number of filters. It gets good performance on the accompanying dataset while 10 Hz detection rate on a Raspberry Pi 4, which is hardware similar to the NAOs. Taking into account the simplicity of the YOLO paradigm, we consider xYOLO a good starting point for creating reduced models. xYOLO also uses XNOR layers instead of regular convolution. They are not used in this study since XNOR layers can be detrimental to accuracy [6], and the TFLite library was not able to leverage quantization for significant speed improvements on the NAO's hardware. We also incorporate the improvements up to YOLOv3 [33], such as multiple class labels and anchor boxes.

7   Experiments

We create variants of the Mobilenet-v2+SSD and xYOLO architectures, and evaluate their performance on the task of ball detection, with the goal of identifying suitable architectures that are fast, and also maintain reasonable performance. The models are trained on both the xYOLO as well as the UTAV datasets, holding out 10% of the data as validation data. Test time performance is reported on the UTAV (test) dataset alone, as these images are closer to the expected playing conditions. Performance is measured in terms of Mean Average Precision at a confidence threshold of 0.5 (mAP@0.5).

7.1   MobileNet and xYOLO Results

The following variants of the architectures are tested:

- SSD-Mobilenet-v2: This is the same architecture as in [37].
- SSD-Mobilenet-v2-6: This is a reduced version of the architecture with only 6 layers total. This reduction is achieved by not repeating layers and using a stride of 2 at each layer.
- SSD-Mobilenet-v2-8n: This is also a similarly reduced version, but it has 8 layers and half as many filters.
- xYOLO: This is the architecture from the xYOLO paper [6], but as mentioned earlier, we do not use XNOR layers in place of regular convolutional ones.
- xYOLO-nano: We drop the two widest layers from xYOLO, in addition to xNOR.
- xYOLO-pico: We replace the maxpool layers in xYOLO-nano with strided convolutions. We also drop the extra layers in the detection/classification head.

We use 224×224 inputs in batches of 4 for training all the above models. Mosaic loading is used, in conjunction with standard data augmentations such as cropping and resizing. Training iterations are performed for 100 epochs of the combined datasets. Training is done using the Pytorch framework, and the saved models are exported to TFLite format for runtime benchmarks. We use 4 threads on the NAO's CPU. 150 inferences are performed to estimate the inference times for each model. We use existing codebases for both MobileNet+SSD[1] and xYOLO [42].

The results are presented in Table 1 (left). We see that the xYOLO models are all able to achieve good mAP, while the MobileNet models struggle, particularly the reduced variants. The xYOLO-pico variant appears to have the best trade-off between speed and mAP. However, the unreduced xYOLO is also able to easily meet the 16.7 ms bound. Thus, it would remain a viable option when data becomes available for a larger number of objects.

7.2   Reduced Models for Bottom Camera

While it would be feasible to use the above models identically for both the top and bottom cameras, the bottom camera needs to detect only the ball and field markers like the cross. These also always appear at a fixed scale, and so we can reduce the detector

---

[1] https://github.com/qfgaohao/pytorch-ssd.

Table 1. Left: Performance of Mobilenet and xYOLO variants with runtimes on the CPU and GPU. Runtimes are given in ms. Right: Performance of extra xYOLO variants for only the bottom camera images.

Architecture	mAP@0.5(%)	CPU	GPU
xYOLO (no XNOR)	**95.6**	7.7	11.0
xYOLO-Nano	95.1	7.1	10.4
xYOLO-Pico	89.5	2.7	4.7
Mobilenet-v2	78.7	58.1	88.1
Mobilenet-v2-6	46.7	6.9	7.2
Mobilenet-v2-8	30.8	**1.1**	3.4

Architecture	mAP(%)
xYOLO-femto	85.0
xYOLO-atto	75.0
xYOLO-zepto	2.5

size and the number of anchor boxes greatly. The input image size can also be reduced, leading to significant computational savings. Thus, we also benchmark the following variants of xYOLO:

- xYOLO-femto: Operates on 224×224 input with 6 layers.
- xYOLO-atto: Operates on 80×80 input, with only 4 layers.
- xYOLO-zepto: Operates on 40×40, with only 3 layers.

We also pretrain these networks on the same data as above, and fine tune on only the bottom camera images. The results are presented in Table 1 (right). We see that xYOLO-atto is able to attain decent performance comparable to xYOLO-femto, despite operating on lower 80×80 input. Thus, this size range is a viable option for the bottom camera since very high accuracy is not necessary in this case. xYOLO-zepto is unable to learn due to the drastically reduced network and input size.

8 Practical Deployment

Although we use the above results as the basis for designing the network used in the 2021 competition, we did not base our final networks on the above architectures, and adaptations needed to be made. We describe the chosen architectures and the major adaptations in this section, followed by benchmarks and results.

Architectures. We settled on particular architectures for the top and bottom camera networks as summarized in Fig. 2c. The inputs are YUV-422 format images that are transformed to RGB using a (non-learnable) convolution. (See next section for additional discussion.) The bottom camera architecture is similar to the top camera but with the fist backbone layer removed. The overall architecture is based on YOLOv3 tiny [3], but aggressively shrunk in the number of channels and layers to increase speed. For the top camera, we trained and deployed the network both for single ball detection (`micro-256`) and a multi-class variant that also detects robots and crosses (`microx-256`). These two variants only differ on the last detection layer. For the bottom camera, the input size is 128×96 (`micro-128`).

Color Conversion and Subsampling. The cameras are set to capture images in YUV422 format due to the hardware characteristics favoring this format, as well as the requirements of other existing vision components. This means that the images need to be converted to RGB, the format they were trained on. We express this conversion as a pointwise convolution, and add another layer in TF-Lite to our models to perform this convolution. As mentioned earlier, the cameras capture at a higher resolution than what the model uses. As the scale factor is an integer (5), we subsample the image by dropping 4 pixels for every pixel kept. This method avoids the need to perform expensive interpolation for resizing the image.

Results. Table 2 reports the individual runtimes and performances of the architectures described above. Also reported are the vision system frame-rates after the detection has been fully integrated into the vision system with the above features in place. Note that motion and localization algorithms are also running concurrently. A visualization of the results is in Fig. 2. We see that the networks exhibit high detection performance and the entire vision system is able to run at 30 Hz for multi-class detection, which goes up 35 Hz when the top and bottom detectors are run simultaneously in a multithreaded setup. Higher framerates enable more reliable estimates of object positions, as well as better localization.

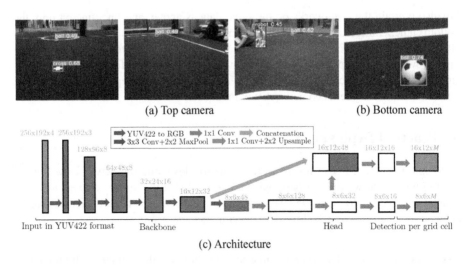

(a) Top camera   (b) Bottom camera

(c) Architecture

Fig. 2. Detection examples for the network used in the 2021 competition. (c) shows the architecture for the top camera.

Table 2. Left: Performances and runtimes (ms) of deployment architectures. Right: vision system framerates (Hz) of different combinations (bottom network is always micro-128). "Simultaneous" indicates top and bottom networks run simultaneously on CPU (3 threads) and GPU respectively. Otherwise, they are run one after the other.

Network	mAP@0.5	CPU(3 threads)	GPU
micro-256	98.7	9.9	15.4
microx-256	97.2	14.6	19.1
micro-128	98.3	2.9	5.3

Top Network	Framerate
micro-256	33
microx-256	30.5
microx-256 (simultaneous)	35

9 Conclusions

Through this project, we have gained understanding of the challenges involved in object detection for the SPL. We implemented a data collection pipeline to enable training our object detectors, and data augmentation strategies for robustness of detections. We identified network architectures and ways to improve on runtime while maintaining good detection performance. We put these to the test, and realize a practical, multi-class object detector that is robust to lighting conditions, and is able to work 35 Hz for both cameras. There are several avenues to pursue in the future:

- There has been work based on Stochastic Scene Generation specific to Robosoccer [15] using the Unreal Engine to render realistic Robosoccer game scenes, which provides a promising avenue for data augmentation in our proposed training pipeline, and we are currently investigating the feasibility of using such an approach. In conjunction with GANs, such a setup could be used to perform high quality data augmentation for increased robustness.
- Depthwise Separable Convolutional are used in MobileNets, leading to significant computational savings, without degradation in detection performance. We conducted initial tests in our setting obtaining promising results. Thus, these layers open avenues for using larger networks for better performance and speed.

Acknowledgements. The authors thank Juhyun Lee, Terry Heo and others at Google for their invaluable help with setting up and understanding the best practices in using TensorFlow Lite. This work has taken place in the Learning Agents Research Group (LARG) at the Department of Computer Science, The University of Texas at Austin. LARG research is supported in part by the National Science Foundation (CPS-1739964, IIS-1724157, FAIN-2019844), the Office of Naval Research (N00014-18-2243), Army Research Office (W911NF-19-2-0333), DARPA, Lockheed Martin, General Motors, Bosch, and Good Systems, a research grand challenge at the University of Texas at Austin. The views and conclusions contained in this document are those of the authors alone. Peter Stone serves as the Executive Director of Sony AI America and receives financial compensation for this work. The terms of this arrangement have been reviewed and approved by the University of Texas at Austin in accordance with its policy on objectivity in research.

References

1. Abadi, M., et al.: TensorFlow: large-scale machine learning on heterogeneous systems (2015). Software available from https://www.tensorflow.org/
2. Achim, S., Stone, P., Veloso, M.: Building a dedicated robotic soccer system. In: Proceedings of the IROS-96 Workshop on RoboCup, pp. 41–48 (1996)
3. Adarsh, P., Rathi, P., Kumar, M.: YOLO v3-tiny: object detection and recognition using one stage improved model. In: 2020 6th International Conference on Advanced Computing and Communication Systems (ICACCS), pp. 687–694 (2020)
4. Broemmel, P., et al.: RoboCup SPL instance segmentation dataset (2019). https://www.kaggle.com/pietbroemmel/naodevils-segmentation-upper-camera
5. Alami, R., Biswas, J., Cakmak, M., Obst, O. (eds.): RoboCup 2021: Robot World Cup XXIV. Springer, Cham (2022). ISBN 978-3-030-98681-0
6. Barry, D., Shah, M., Keijsers, M., Khan, H., Hopman, B.: XYOLO: a model for real-time object detection in humanoid soccer on low-end hardware. In: 2019 International Conference on Image and Vision Computing New Zealand (IVCNZ), pp. 1–6 (2019)
7. Bochkovskiy, A., Wang, C.Y., Liao, H.Y.M.: YOLOv4: optimal speed and accuracy of object detection. arXiv preprint arXiv:2004.10934 (2020)
8. Brameld, K., et al.: RoboCup SPL 2018 rUNSWift team paper (2019)
9. Cai, H., Gan, C., Han, S.: Once for all: train one network and specialize it for efficient deployment. CoRR abs/1908.09791 (2019)
10. Cai, H., Gan, C., Wang, T., Zhang, Z., Han, S.: Once-for-all: train one network and specialize it for efficient deployment (2019)
11. Cruz, N., Lobos-Tsunekawa, K., Ruiz-del-Solar, J.: Using convolutional neural networks in robots with limited computational resources: detecting NAO robots while playing soccer. CoRR abs/1706.06702 (2017)
12. Girshick, R., Donahue, J., Darrell, T., Malik, J.: Rich feature hierarchies for accurate object detection and semantic segmentation. In: 2014 IEEE Conference on Computer Vision and Pattern Recognition, pp. 580–587 (2014)
13. He, K., Gkioxari, G., Dollar, P., Girshick, R.: Mask R-CNN. In: 2017 IEEE International Conference on Computer Vision (ICCV), pp. 2980–2988 (2017)
14. Hermann, T.: Frugally-deep: header-only library for using Keras models in C++. RoboCup (2019)
15. Hess, T., Mundt, M., Weis, T., Ramesh, V.: Large-scale stochastic scene generation and semantic annotation for deep convolutional neural network training in the RoboCup SPL. In: Akiyama, H., Obst, O., Sammut, C., Tonidandel, F. (eds.) RoboCup 2017. LNCS (LNAI), vol. 11175, pp. 33–44. Springer, Cham (2018). https://doi.org/10.1007/978-3-030-00308-1_3 ISBN 978-3-030-00308-1
16. Howard, A., et al.: Searching for MobileNetV3. In: 2019 IEEE/CVF International Conference on Computer Vision (ICCV), pp. 1314–1324 (2019)
17. Howard, A.G., et al.: MobileNets: efficient convolutional neural networks for mobile vision applications. CoRR abs/1704.04861 (2017)
18. Iandola, F.N., Han, S., Moskewicz, M.W., Ashraf, K., Dally, W.J., Keutzer, K.: SqueezeNet: AlexNet-level accuracy with 50x fewer parameters and <0.5 mb model size (2016)
19. Kitano, H., Asada, M., Kuniyoshi, Y., Noda, I., Osawa, E.: RoboCup: the robot world cup initiative. In: Proceedings of the First International Conference on Autonomous Agents, AGENTS 1997, pp. 340–347. Association for Computing Machinery, New York (1997). https://doi.org/10.1145/267658.267738. ISBN 0897918770
20. Law, H., Deng, J.: CornerNet: detecting objects as paired keypoints. In: Proceedings of the European Conference on Computer Vision (ECCV) (2018)

21. Lee, J., et al.: XNNPACK (2019). https://github.com/google/XNNPACK
22. Lin, T., Goyal, P., Girshick, R.B., He, K., Dollár, P.: Focal loss for dense object detection. CoRR abs/1708.02002 (2017)
23. Lin, T.Y., Dollar, P., Girshick, R., He, K., Hariharan, B., Belongie, S.: Feature pyramid networks for object detection. In: Proceedings of the IEEE Conference on Computer Vision and Pattern Recognition (CVPR) (2017)
24. Lin, T.-Y., et al.: Microsoft COCO: common objects in context. In: Fleet, D., Pajdla, T., Schiele, B., Tuytelaars, T. (eds.) ECCV 2014. LNCS, vol. 8693, pp. 740–755. Springer, Cham (2014). https://doi.org/10.1007/978-3-319-10602-1_48 ISBN 978-3-319-10602-1
25. Liu, W., et al.: SSD: single shot multibox detector. CoRR abs/1512.02325 (2015)
26. Menashe, J., et al.: Fast and precise black and white ball detection for RoboCup soccer. In: Akiyama, H., Obst, O., Sammut, C., Tonidandel, F. (eds.) RoboCup 2017. LNCS (LNAI), vol. 11175, pp. 45–58. Springer, Cham (2018). https://doi.org/10.1007/978-3-030-00308-1_4
27. OmniVision Technologies: Ov5640: color CMOS QSXGA (5 megapixel) image sensor with OMNIBSI(TM) technology (2011)
28. Poppinga, B., Laue, T.: JET-Net: real-time object detection for mobile robots. In: Chalup, S., Niemueller, T., Suthakorn, J., Williams, M.-A. (eds.) RoboCup 2019. LNCS (LNAI), vol. 11531, pp. 227–240. Springer, Cham (2019). https://doi.org/10.1007/978-3-030-35699-6_18
29. Rastegari, M., Ordonez, V., Redmon, J., Farhadi, A.: XNOR-Net: imagenet classification using binary convolutional neural networks. CoRR abs/1603.05279 (2016)
30. Redmon, J.: Darknet: open source neural networks in C (2013–2016). http://pjreddie.com/darknet/
31. Redmon, J., Divvala, S., Girshick, R., Farhadi, A.: You only look once: unified, real-time object detection. In: 2016 IEEE Conference on Computer Vision and Pattern Recognition (CVPR), pp. 779–788 (2016)
32. Redmon, J., Farhadi, A.: Yolo9000: better, faster, stronger. In: 2017 IEEE Conference on Computer Vision and Pattern Recognition (CVPR), pp. 6517–6525 (2017)
33. Redmon, J., Farhadi, A.: YOLOv3: an incremental improvement. CoRR abs/1804.02767 (2018)
34. Ren, S., He, K., Girshick, R., Sun, J.: Faster R-CNN: towards real-time object detection with region proposal networks. In: Cortes, C., Lawrence, N., Lee, D., Sugiyama, M., Garnett, R. (eds.) Advances in Neural Information Processing Systems, vol. 28, pp. 91–99. Curran Associates, Inc. (2015)
35. Rofer, T., Laue, T., Baude, A.: Team report and code release 2019 (2019)
36. Röfer, T., et al.: B-Human team report and code release 2018 (2018). http://www.b-human.de/downloads/publications/2018/CodeRelease2018.pdf
37. Sandler, M., Howard, A.G., Zhu, M., Zhmoginov, A., Chen, L.: Inverted residuals and linear bottlenecks: mobile networks for classification, detection and segmentation. CoRR abs/1801.04381 (2018)
38. Sekachev, B., Zhavoronkov, A., Manovich, N.: Computer vision annotation tool: a universal approach to data annotation (2019). https://cvat.org
39. Softbank Robotics: Nao 6 datasheet (2018). https://www.generationrobots.com/media/Specifications_NAO6.pdf
40. Tan, M., Pang, R., Le, Q.V.: EfficientDet: scalable and efficient object detection. CoRR abs/1911.09070 (2019)
41. Thielke, F., Hasselbring, A.: A JIT compiler for neural network inference. In: Chalup, S., Niemueller, T., Suthakorn, J., Williams, M.-A. (eds.) RoboCup 2019. LNCS (LNAI), vol. 11531, pp. 448–456. Springer, Cham (2019). https://doi.org/10.1007/978-3-030-35699-6_36 ISBN 978-3-030-35699-6

42. Ultralytics: YOLOv5 rocket in PyTorch (2020). https://zenodo.org/badge/latestdoi/264818686
43. Yao, Z.B., Douglas, W., O'Keeffe, S., Villing, R.: Faster YOLO-LITE: faster object detection on robot and edge devices. In: Alami, R., Biswas, J., Cakmak, M., Obst, O. (eds.) RoboCup 2021. LNCS (LNAI), vol. 13132, pp. 226–237. Springer, Cham (2022). https://doi.org/10.1007/978-3-030-98682-7_19 ISBN 978-3-030-98682-7
44. Zoph, B., Vasudevan, V., Shlens, J., Le, Q.V.: Learning transferable architectures for scalable image recognition. CoRR abs/1707.07012 (2017)

Object Tracking for the Rotating Table Test

Vincent Scharf$^{(\boxtimes)}$ ⓘ, Ibrahim Shakir Syed$^{(\boxtimes)}$ ⓘ, Michał Stolarz$^{(\boxtimes)}$ ⓘ,
Mihir Mehta$^{(\boxtimes)}$ ⓘ, and Sebastian Houben$^{(\boxtimes)}$ ⓘ

Hochschule Bonn-Rhein-Sieg, Sankt Augustin, Germany
{vincent.scharf,ibrahim.syed,michal.stolarz,
mihir.mehta}@smail.inf.h-brs.de, sebastian.houben@h-brs.de

Abstract. In the RoboCup@Work competition, the Rotating Table
Test problem refers to the task of automatically grasping an object
from a circular table, rotating at constant angular velocity. This task
requires the robot to track the target object's position and grasp it.
In this work, we propose a camera-based online tracking system which
works in real-time. Our approach is based on the YOLOv5 detection
backbone and uses a novel, modified version of the SORT tracker. The
tracker is trained solely on a pre-existing detection dataset containing
annotated static images, thanks to which the collection of additional
situation-specific video data is not required. We evaluate and compare
SORT with YOLOv5 and SqueezeDet backbones and demonstrate the
improvement in tracking performance when using the former. The evaluation dataset and corresponding annotations are made available for use
in the community.

Keywords: Rotating Table Test · RoboCup · Tracking by detection ·
YOLO · SORT

1 Introduction

RoboCup@Work is a competition focused on the use of mobile manipulators
and their integration with automation equipment to perform relevant industrial
tasks [11]. From the several tasks in the competition, this work focuses on the
Rotating Table Test (RTT). In this, as depicted in Fig. 1, several objects are
placed on a circular table of 1 m diameter, which rotates at an angular velocity
of 0.5 rad s^{-1}. The robot is then instructed to pick up one moving object of a
particular class (with a velocity between 5 cm s^{-1} and 20 cm s^{-1} [1]). A low-
latency, robust object detection and prediction is essential to solve this test.

In this work, we leverage tracking-by-detection, which has emerged as the
preferred paradigm to solve multi-object tracking problems like the RTT task
described above, in which the robot needs to perceive the live feed from the rotating table and identify the objects on it. Given a target object, the robot should
track the object and grasp it when it moves into the manipulator's workspace.
Ultimately, successful completion of the grasping task not only requires detecting
and tracking the objects but also predicting their trajectories. However, predicting

ⓒ The Author(s), under exclusive license to Springer Nature Switzerland AG 2023
A. Eguchi et al. (Eds.): RoboCup 2022, LNAI 13561, pp. 75–86, 2023.
https://doi.org/10.1007/978-3-031-28469-4_7

the trajectory is not within the scope of this work. Our contribution is the extension of the SORT [5] tracker and the experimental evaluation deploying two different backbones[1]. We investigate the tracking quality as well as robustness against lower video frame rates, which is a crucial aspect of the implementation on the real robot with limited computational resources. We use a youBot robot with an Intel NUC Core i7 and a RealSense D435 cam-

Fig. 1. Overview of the RTT setup

era (the camera view can be seen in Fig. 1 and Fig. 3c). We make available our dataset[2] that features 1,200 images and annotations from the RTT task.

The remainder of this report is structured as follows: Sect. 2 introduces important concepts and recent work done from the field of object tracking. Section 3 presents the applied detection algorithm and datasets used for training and evaluation. In Sect. 4, we discuss the approach we use to address the tracking problem of RTT. Section 5 describes the dataset created for evaluation and discusses the evaluation results. Finally, in Sect. 6, we present conclusions and future work.

2   Related Work

In the domain of object tracking, most state-of-the-art approaches follow the *tracking by detection* paradigm [8], depicted in Fig. 2. Such methods involve two independent steps: (i) object detection on all individual frames and (ii) tracking or association of those detections across frames [13]. This heavily relies on the performance of the object detector.

Neural Network based detectors have become the current state-of-the-art, particularly since the proposal of detectors such as Faster-RCNN [19] and SDP [26]. Most state-of-the-art methods for *data association* in the scope of object tracking follow a formalization of the problem as a graph, in which each detection is a node and every edge indicates a possible link. The data association can then be formulated as a *maximum flow* [2] or equivalently *minimum cost* [9] problem. Such delineations as optimization problems suffer from high complexity and are infeasible for online applications [3]. Since the task at hand has a critical real-time constraint, a simpler frame-by-frame approach with an underlying assumption of a *small change* in position from one frame to another is chosen. Bergmann et al. [3] proposed Tracktor, a tracking algorithm that firstly initializes the track from the detection done by Faster-RCNN and secondly performs the tracking. The latter is a process based on bounding box regression, namely the process of applying ROI pooling on the features obtained from the new frame

---

[1] https://github.com/VincentSch4rf/rtt_tracking
[2] https://github.com/VincentSch4rf/RoboCup-RTT-Dataset

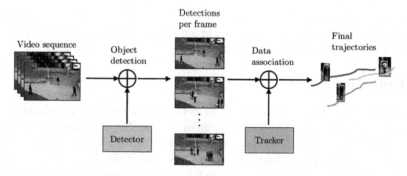

Fig. 2. Tracking-by-detection paradigm. Firstly, an independent detector is applied to all image frames to obtain likely pedestrian detections. Secondly, a tracker is run on the set of detections to perform data association, i.e., link the detections to obtain full trajectories [13].

and bounding box coordinates from the previous frame. Although the proposed approach is an online approach, it cannot run at high frequency, which is a strict requirement for our application. Moreover, it is based on the assumption of only little motion between frames, which cannot always be fulfilled.

Another approach, named SORT, has been presented by Bewley et al. [5] which uses a Kalman filter to estimate the position of the objects in the next frame. An extension of this approach, named DeepSORT, was published by Wojke et al. [23]. Both of these methods focus on leveraging a detector with minimal additions to enable real-time tracking. Naturally, this focus also leaves them more vulnerable to errors. SORT calculates affinities based on the overlap of the estimated bounding box generated using the Kalman filter and the bounding box generated by the detector for the next frame, and matches these with previous detections using the Hungarian algorithm [12]. DeepSORT builds upon this framework, and additionally incorporates appearance feature vectors extracted with a CNN architecture. The affinity is then calculated using the cosine distance between two feature vectors. Lastly, information from the motion features and the appearance features is combined (using the Hungarian algorithm) in order to associate objects. This method has shown improvement over the previous approach, primarily in reducing the number of ID switches (IDSWs) [23]. One major contrast of our work when compared to the original work is the dataset. While the original work showed a significant improvement in reducing IDSWs by using re-ID features, the objects belonging to the same class, in our case, do not have significantly discernible features. Therefore, the information from the feature vectors doesn't add significant value. Moreover, the addition of a feature extracting CNN also increases the execution time. For this reason, our work focuses on utilizing the SORT tracker for the RTT.

Commonly, *re-identification*, i.e., associating an object that has not been visible for some time to an existent track, is of concern in tracking tasks [21], but we do not consider it. As the majority of the rotating table is visible from the camera's perspective, we assume that there is sufficient time to accurately estimate the position of the object and grasp it before it leaves the frame. Thus, we deem re-identification not critical for successful execution of RTT.

3   Detection

Tracking-by-detection, as introduced in Sect. 2, is a tracking paradigm which relies on object detection to identify an unknown number of individual objects of interest in each frame [13]. A typical challenge for tracking-by-detection systems, especially when applied online, has always been the limited performance of the underlying detector, which may produce false positive and missed detections [6]. This problem only becomes more prominent in the context of deployment on a mobile manipulator, as we are not only dealing with an online scenario but also face limited compute capabilities. These in turn limit the complexity of the detector and therefore add to the performance problem.

Performing object detection online with fast inference while maintaining a base level of accuracy was the declared goal of the "You Only Look Once" (YOLO) architecture [16]. YOLO combines the problem of localization and classification in one end-to-end differentiable network by interpreting it as a regression problem of spatially separating bounding boxes and associating class probabilities [16]. While a lot faster, the initial architecture suffered from lower recall and larger localization errors [17], compared to two-stage detectors, like Faster-RCNN [19], which perform localization and classification in two separate steps. With the following YOLOv2 [17], YOLOv3 [18] and particularly YOLOv4 [7], the architecture and training procedure were incrementally improved, rendering it one state-of-the-art real-time object detector [7].

One of their main contributions, which allows for efficient training of a YOLO detector when used in a tracking framework, is Mosaic [7]. It is a data augmentation method introduced by Glenn Jocher for YOLOv3, which mixes four training images and thereby allows for detection of objects outside their normal context. An example of this can be seen in Fig. 3b. It reduces the problem of YOLO relying on the scenery an object appears in, to perform correct classification. This problem is significantly more pronounced for YOLO architectures than in two-stage detectors, as the one-stage detection architectures enable the model to consider the entire image for classification rather than only the region of interest. Therefore, training a model on a general detection dataset unspecific to the RTT scenario is more feasible and the labor-intensive step of producing perfectly tailored datasets for a tracking task at hand is no longer required. In addition, batch normalization calculates activation statistics from four different images on each layer. This significantly reduces the need for a large mini-batch size and thus reduces the memory footprint of the model during training.

In this work, we therefore use the most recent iteration of the YOLO detector, YOLOv5 [10], which is a PyTorch implementation of the YOLOv4 architecture with a CSPNet [22] backbone, PANet [14] neck and anchor-based YOLOv3 [18] head with three levels of detection granularity. The model is trained on an existing object detection dataset [15], consisting of 18 distinct objects, 13 workshop items and five additional helper objects used in the RoboCup@Work competition [1]. The dataset contains 1,150 images, captured under different conditions, in various surroundings and ensembles. In order to assess the model's generalization capabilities with respect to the RTT scenario that this work is concerned with, we use a 300-sample subset of the tracking dataset, which we created in the context of this publication (see Subsect. 5.1). The final results are reported on 902 held-out samples of the same. Example images from the two datasets are presented in Fig. 3.

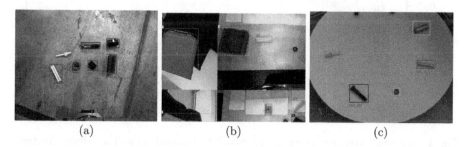

(a)   (b)   (c)

Fig. 3. (a) Sample from the training dataset, (b) training sample after mosaic augmentation, consisting of four images in four tiles of random ratio (c) sample from the tracking sequence. As one can see from this comparison, the detection dataset used for training, does not resemble the RTT scenario of the tracking dataset.

We use the hyperparameter tuning provided with the YOLOv5 implementation. The model is then trained for 1,500 epochs with a batch size of 376 using those optimized hyperparameters. Validation is performed after each epoch. As we can see from Fig. 4, which depicts the learning curve of the model, together with the respective validation metrics, the model converges to a maximum of around 0.85 validation $mAP_{0.5:0.95}$ around the 400th epoch. The iteration of the model with the best validation performance is saved and evaluated on the held-out test set. It has to be noted that the test dataset only contains eight of the 18 object classes, on which the detector was trained. Usually, this would make the evaluation insufficient. However, we are only interested in the generalization capabilities of the detector with respect to the task we are deploying it for. Hence, the evaluation on the tracking dataset reflecting only a subset of the object classes is acceptable. The results are summarized in Table 1. The large difference between the 0.5 Intersection-over-Union (IOU) and 0.95 IOU mean average precision (mAP) of some classes can be explained by imperfect annotations. Since the M20 and Bearing objects are a lot smaller compared to the other objects present in the test set, imprecise annotations have a much larger impact on the IOU of the predicted bounding box. Hence, the object is not correctly detected at higher IOU thresholds and, thus, the mAP score is lower. Apart from this, the S40_40_B class shows a much lower recall than all other object classes. Investigating the missed detections, we found, that the S40_40_B, once

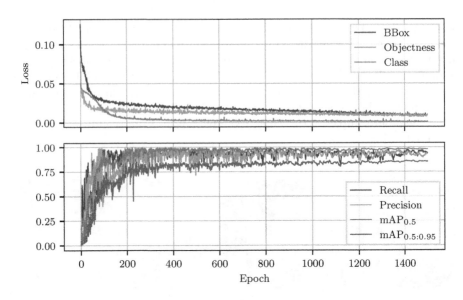

Fig. 4. Learning curve, depicting the losses on the training, as well as relevant metrics on the validation set.

it is positioned directly underneath the camera, does not feature any recognizable perspective or contours any more. It becomes a black rectangle without any depth perception to it. This probably makes it not only hard for a human to infer the correct object class, but also for the YOLO detector, as the lack of texture and perspective, only leaves the pure two-dimensional shape of the object as a remaining criterion for discrimination. Unfortunately, this two-dimensional shape of the S40_40_B is very similar to the R20's, which, however, exhibits a cylindrical form considering all three dimensions. Another problem with the evaluation, as such, is that the dataset follows the convention of annotating objects only once they are between 50 and 70% visible. However, the YOLO detector does detect a lot of the objects even after they are more than 70% outside the frame. This produces a large portion of incorrect false positives.

However, even despite those difficulties, the detector generalizes very well to the RTT scenario, even though it was trained on data whose scenery could be considered out-of-distribution for the particular problem. The near perfect precision and recall (with the exception of the discussed S40_40_B), combined with the good average precision, even at high IOUs, leads us to conclude that the detector is sufficiently trained and can produce reliable and robust detections under most circumstances.

Table 1. Evaluation results of YOLO detector

Class	Images	Labels	Precision	Recall	mAP$_{0.5}$	mAP$_{0.5:0.95}$
All	902	4889	0.972	0.983	0.993	0.784
Axis	902	378	0.998	1.000	0.995	0.911
Bearing	902	817	0.997	0.999	0.993	0.681
F20_20_G	902	779	0.946	0.996	0.994	0.776
M20	902	1122	0.932	0.994	0.990	0.556
M20_100	902	351	0.997	1.000	0.995	0.946
Motor	902	340	0.958	1.000	0.995	0.938
S40_40_B	902	349	1.000	0.875	0.995	0.639
S40_40_G	902	753	0.952	1.000	0.985	0.827

4 Tracking

Once the detections have been generated by the detector, we use a tracker to assign, associate and keep track of the objects. In this work, we primarily focus on the SORT tracker [5] and utilize the implementation available online.

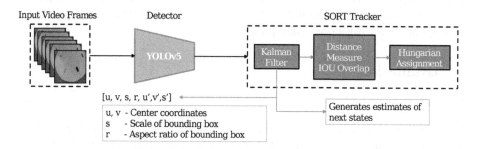

Fig. 5. SORT Architecture

The SORT tracker [5] initializes tracks based on the received input detections. It then generates estimates of the next state of each track using a Kalman filter. The detections for the next frame are then compared with the estimates generated by the Kalman filter. All bounding boxes which have a significant overlap are considered as potential candidates to be matched as the same objects. Finally, tracks and new detections are associated using Hungarian matching. The procedure has been depicted in Fig. 5. It provides a lightweight, yet efficient framework for tracking objects on the rotating table. It should be noted that the original SORT implementation does not take the object class into consideration that is returned by the detector. We extended the code such that when a track is initialized, the tracker stores the object class. This is essential as we are interested in knowing which object is being tracked so that it can be grasped.

4.1   Updated Track Handling

Several modifications were made to the original SORT implementation which significantly change the results. In the original implementation, in order for a track to be initialized, it must be associated with the detections for T_{min} frames. A track then needs consecutive detections in order to remain active. If a track is temporarily lost, i.e., no detections are associated with it for more than 1 frame, it requires T_{min} detections to be matched with it to be reactivated. In case there are no detections for T_{lost} frames, the corresponding track is discarded. It should be noted that for both active and inactive tracks, the tracker keeps estimating the bounding boxes. However, it only returns outputs for active tracks.

For our case, some of these conditions are problematic, as we require the tracker to remain active even in the absence of detections. We modify the track handling method in two major ways. Firstly, we remove the need for consecutive detections of a track to remain active, i.e., once a track is initialized, it returns outputs for the next T_{lost} frames even without detections. Secondly, we modify the track re-initialization scheme such that even if a single detection is associated with a track, the T_{lost} counter is set to zero, i.e., it no longer requires T_{min} detections to be re-initialized. Not only do these changes allow us to deal with missing detections but also to use intermittent detections for tracking. Furthermore, this change enables us to return predicted positions of tracked objects several frames into the future. This is particularly useful in deployments with limited computational resources, where running the detection head multiple times isn't practical, as it is the case with mobile manipulators.

5   Evaluation

5.1   Dataset

The aforementioned (in Sect. 3) additional dataset was created as part of this work. It contains not only annotated objects but also their trajectories, thus it can also be used for evaluating the tracking performance. We use part of the test split of this dataset (which is basically a short video) to evaluate the chosen tracking approach. The video was recorded with a frequency of 30 frames per second, consists of 451 frames and ten unique tracks. It captures one full table rotation with a RealSense D435 camera. The experiment setup reflects the setup of the real competition, where the mobile manipulator is positioned next to the table. Each frame has a size of 640 × 480. There are eight different object types in the video (all of them are described in [1]): Bearing, F20_20_G, S40_40_G, S40_40_B, M20, axis, M20_100 and Motor.

5.2   Metrics

Finally, to evaluate the performance of the model, we use the following metrics. (↑) for a metric denotes that higher scores are preferable and (↓) denotes that lower scores are better. As it is difficult to quantify the performance of the tracker by a single metric, we use multiple commonly used metrics from Classical metrics [25], CLEAR MOT metrics [4] and ID scores [20]:

- MOTA($\uparrow$): Multi Object Tracking Accuracy [4]
- MOTP($\downarrow$): Multi Object Tracking Precision [4]
- IDF1($\uparrow$): Ability of tracker to track the same object with same ID for the longest duration possible [20]
- TP($\uparrow$): Number of correct detections
- FP($\downarrow$): Number of incorrect detections
- FN($\downarrow$): Number of missed detections
- IDSW($\downarrow$): Number of times IDs are switched
- MT($\uparrow$): Tracks overlapping with ground truth for at least 80% of the sequence
- PT: Tracks overlapping with ground truth from 20% to 80% of the sequence
- ML($\downarrow$): Tracks overlapping with ground truth for at most 20% of the sequence
- t($\downarrow$): Average detector inference time in for a single frame on the youBot

5.3   Results

We compare the performance of the SORT tracker with different detection backbones, i.e., SqueezeDet [24] and YOLOv5 [10]. The results have been tabulated in Table 2. It should be noted that for the following tests, we have set $T_{min} = 3$ and $T_{lost} = 11$. Our trained YOLO detector performs better in detection, having higher MOTA and MOTP scores, which also improves the IDF1 score. Furthermore, we should also add that even though the IDF1 score of SqueezeDet is significantly less than YOLO, on inspection of the outputs, a single switch of a partially visible object causes this difference. For the purposes of the RTT task, this difference is insignificant. This is also evidenced by a less significant difference in the MOTA score of the two detectors. Nevertheless, YOLO tracker performs slightly better than its SqueezeDet counterpart. Hence, we use this as the default backbone detector. It should be noted that because the total number of tracks in the test dataset is 10, the number of partially tracked objects can then be estimated based on this information as PT = 10 − MT − ML.

Table 2. Comparison of different tracker backbones

Detector	MOTA$\uparrow$	MOTP$\uparrow$	IDF1$\uparrow$	TP$\uparrow$	FP$\downarrow$	FN$\downarrow$	IDSW$\downarrow$	MT$\uparrow$	ML$\downarrow$	t[ms]$\downarrow$
YOLO	**88.7%**	**79.4%**	**0.89**	**2260**	252	**0**	3	**10**	**0**	114
SqueezeDet	86.3%	72.2%	0.86	2164	**211**	94	**5**	9	**0**	**93**

As the computational resources are limited on deployable systems such as robots, we evaluate the effect of low input frame rate on tracking. We simulate lower frame rates by skipping frames from the original video recorded at 30 Hz. The results have been tabulated in Table 3. We observe that although there is a slight and steady decline in the MOTA, MOTP and IDF1 scores, the effect

Table 3. Effect of video frame rate on tracking performance (YOLO backbone)

FPS (Hz)	Frames	MOTA↑	MOTP↑	IDF1↑	TP↑	FP↓	FN↓	IDSW↓	MT↑	ML↓
30	451	88.7%	79.4%	0.89	2260	252	0	3	10	0
15	226	87.5%	78.7%	0.89	1132	139	2	1	10	0
10	151	86.7%	76.7%	0.88	755	97	4	0	10	0
5	76	78.9%	66.9%	0.82	373	71	9	1	9	0
3	46	19.8%	67.0%	0.36	102	56	122	8	5	4

of low frame rates is relatively insignificant even if the frame rate is reduced to 7.5 Hz (FPS). Additionally, the performance at 5 Hz remains satisfactory. The performance only starts to dip when the input frame rate is reduced below 5 Hz. This shows that the system is able to function properly even at frame rates as low as 5 Hz. This has also been illustrated in Fig. 6.

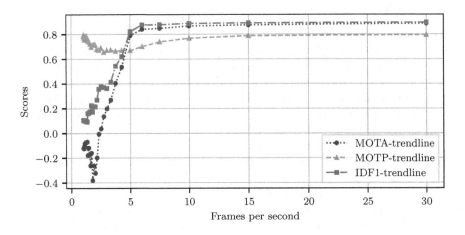

Fig. 6. Effect of frame rate on tracking performance (YOLO backbone)

6   Conclusion

In this study, we evaluated the SORT tracker deploying two different detection backbones, namely: SqueezeDet and YOLOv5. The results are reported in terms of established tracking performance measures. We evaluated the performance of the tracker at lower frame rates. Based on our findings, we propose a tracking system, a combination of a YOLOv5 detector backbone and a modified SORT tracker, capable of tracking objects in real-time which still functions satisfactorily with low computational resources. We provide a test dataset which can be used to evaluate the performance of the tracker.

As an extension of our work, we plan to extend 2D tracking with 3D sensor data. Based on this, we can estimate the entire three-dimensional object trajectory, such that we can forecast the object position to grasp it at a proper time. Since the objects circle around the same center, we will investigate how simultaneously tracking multiple object trajectories can improve the accuracy of an individual trajectory.

Real-world end-to-end experiments investigating the targeted grasping mechanism's speed, accuracy, and success rate will allow for a rigorous, quantitative evaluation of our proposed systems and all its individual components.

Acknowledgments. We gratefully acknowledge the continued support by the b-it Bonn-Aachen International Center for Information Technology and Hochschule Bonn-Rhein-Sieg.

References

1. Bartsch, L., et al.: RoboCup@Work 2022 - Rulebook (2022). https://atwork. robocup.org/rules/
2. Berclaz, J., Fleuret, F., Turetken, E., Fua, P.: Multiple object tracking using k-shortest paths optimization. In: IEEE Trans. Pattern Anal. Mach. Intell. **33**(9), 1806–1819 (2011)
3. Bergmann, P., Meinhardt, T., Leal-Taixe, L.: Tracking without bells and whistles. In: International Conference on Computer Vision, pp. 941–951 (2019)
4. Bernardin, K., Stiefelhagen, R.: Evaluating multiple object tracking performance: the clear MOT metrics. EURASIP J. Image Video Process. **2008**, 1–10 (2008)
5. Bewley, A., Ge, Z., Ott, L., Ramos, F., Upcroft, B.: Simple online and realtime tracking. In: International Conference on Image Processing, pp. 3464–3468. IEEE (2016)
6. Bochinski, E., Eiselein, V., Sikora, T.: High-speed tracking-by-detection without using image information. In: International Conference on Advanced Video and Signal Based Surveillance (AVSS), pp. 1–6. IEEE (2017)
7. Bochkovskiy, A., Wang, C.Y., Liao, H.Y.M.: YOLOv4: optimal speed and accuracy of object detection. arXiv preprint arXiv: 2004.10934 (2020)
8. Ciaparrone, G., Sánchez, F.L., Tabik, S., Troiano, L., Tagliaferri, R., Herrera, F.: Deep learning in video multi-object tracking: a survey. Neurocomputing **381**, 61–88 (2020)
9. Jiang, H., Fels, S., Little, J.J.: A linear programming approach for multiple object tracking. In: Conference on Computer Vision and Pattern Recognition, pp. 1–8 (2007)
10. Jocher, G., et al.: ultralytics/yolov5: v6.1 - TensorRT, TensorFlow Edge TPU and OpenVINO Export and Inference (2022)
11. Kraetzschmar, G.K., et al.: RoboCup@Work: competing for the factory of the future. In: Bianchi, R.A.C., Akin, H.L., Ramamoorthy, S., Sugiura, K. (eds.) RoboCup 2014. LNCS (LNAI), vol. 8992, pp. 171–182. Springer, Cham (2015). https://doi.org/10.1007/978-3-319-18615-3_14
12. Kuhn, H.W.: The Hungarian method for the assignment problem. Naval Res. Logistics Q. **2**(1–2), 83–97 (1955)
13. Leal-Taixé, L.: Multiple object tracking with context awareness. arXiv preprint arXiv:1411.7935 (2014)

14. Liu, S., Qi, L., Qin, H., Shi, J., Jia, J.: Path aggregation network for instance segmentation. In: Conference on Computer Vision and Pattern Recognition, pp. 8759–8768 (2018)
15. Padalkar, A., et al.: b-it-bots: our approach for autonomous robotics in industrial environments. In: Chalup, S., Niemueller, T., Suthakorn, J., Williams, M.-A. (eds.) RoboCup 2019. LNCS (LNAI), vol. 11531, pp. 591–602. Springer, Cham (2019). https://doi.org/10.1007/978-3-030-35699-6_48
16. Redmon, J., Divvala, S., Girshick, R., Farhadi, A.: You only look once: unified, real-time object detection. In: Conference on Computer Vision and Pattern Recognition, pp. 779–788 (2016)
17. Redmon, J., Farhadi, A.: YOLO9000: better, faster, stronger. In: Conference on Computer Vision and Pattern Recognition, pp. 7263–7271 (2017)
18. Redmon, J., Farhadi, A.: YOLOv3: an incremental improvement. arXiv preprint arXiv:1804.02767 (2018)
19. Ren, S., He, K., Girshick, R., Sun, J.: Faster R-CNN: towards real-time object detection with region proposal networks. In: Advances in Neural Information Processing Systems, vol. 28 (2016)
20. Ristani, E., Solera, F., Zou, R., Cucchiara, R., Tomasi, C.: Performance measures and a data set for multi-target, multi-camera tracking. In: Hua, G., Jégou, H. (eds.) ECCV 2016. LNCS, vol. 9914, pp. 17–35. Springer, Cham (2016). https://doi.org/10.1007/978-3-319-48881-3_2
21. Ristani, E., Tomasi, C.: Features for multi-target multi-camera tracking and re-identification. In: Conference on Computer Vision and Pattern Recognition, pp. 6036–6046 (2018)
22. Wang, C.Y., Liao, H.Y.M., Wu, Y.H., Chen, P.Y., Hsieh, J.W., Yeh, I.H.: Cspnet: a new backbone that can enhance learning capability of CNN. In: Conference on Computer vision and Pattern Recognition Workshops, pp. 390–391 (2020)
23. Wojke, N., Bewley, A., Paulus, D.: Simple online and realtime tracking with a deep association metric. In: International Conference on Image Processing, pp. 3645–3649. IEEE (2017)
24. Wu, B., Forrest, I., Peter H, J., Keutzer, K.: SqueezeDet: unified, small, low power fully convolutional neural networks for real-time object detection for autonomous driving. In: Conference on Computer Vision and Pattern Recognition Workshops, pp. 129–137 (2017)
25. Wu, B., Nevatia, R.: Tracking of multiple, partially occluded humans based on static body part detection. In: Conference on Computer Vision and Pattern Recognition, pp. 951–958. IEEE (2006)
26. Yang, F., Choi, W., Lin, Y.: Exploit all the layers: fast and accurate CNN object detector with scale dependent pooling and cascaded rejection classifiers. In: Conference on Computer Vision and Pattern Recognition, pp. 2129–2137 (2016)

Evaluating Action-Based Temporal Planners Performance in the RoboCup Logistics League

Marco De Bortoli[✉] and Gerald Steinbauer-Wagner

Graz University of Technology, Graz, Austria
mbortoli@ist.tugraz.at

Abstract. Due to increased demands related to flexible product configurations, frequent order changes, and tight delivery windows, there is a need for flexible production using AI methods. A way of addressing this is the use of temporal planning as it provides the ability to generate plans for complex goals while considering temporal aspects such as deadlines, concurrency, and durations. A drawback in applying such methods in dynamic environments is their high and unpredictable planning time. In this paper, we present an evaluation of the current state-of-the-art temporal planners within the RoboCup Logistics League. Among the many factors that impact automated planners applicability, the level of abstraction of the planning model is paramount. We center our study on the effect that modeling choices have on the performance of the assessed planners. Our experimental results suggest that seeking for the right level of abstraction of planning domain models allows for compromising solutions between plan quality and plan solving time.

1 Introduction

Temporal planning is a technology allowing to deal with the changes introduced by Industry 4.0. It allows for time and resource optimization, without violation of constraints which may change on-the-fly. Drawbacks in applying such methods in dynamic environments (e.g., autonomous mobile robots in a warehouse fulfilling on-the-fly orders) are their limited ability to react to unexpected difficulties in the plan execution as well as their high and unpredictable planning time (e.g., robots need to come up with decisions in short time). Moreover, replanning may be necessary frequently, for example to handle a new order or to tackle a failing execution in the physical world. To better react to external changes, the replanning process has to be conducted fast. In this paper we focus on these issues, analyzing the performance of state-of-the-art temporal action-based planners and their applicability to the RoboCup Logistics League (RCLL) [15]. Moreover, different ways of encoding the domain are also object of this evaluation. The RoboCup Logistics League was founded with the goal of providing a benchmark for dynamic intra-logistics domains. The main challenge posed by this competition is building a full software stack, comprising elements from both

© The Author(s), under exclusive license to Springer Nature Switzerland AG 2023
A. Eguchi et al. (Eds.): RoboCup 2022, LNAI 13561, pp. 87–99, 2023
https://doi.org/10.1007/978-3-031-28469-4_8

AI and Robotics. Subcomponents of the robots, like sensing and object manipulation, together with the general management and task assignments strategies have to be developed and connected together. We propose different modelings of the problem, featuring various levels of abstraction. Abstracting away some details of the domain represents a way to mitigate the planning complexity at the cost of plan quality. For each modeling, a set of instances with increasing complexity (in terms of size and number of orders) is tested with each planner. The metric for this evaluation is formed by the combination of planning time and plan makespan. In fact, since in dynamic domains there is no separation between planning and execution time and goals are not known in advance, the sum of the two values represents the expected end of the plan. Execution is out of the scope of this paper, yet the ability of computing a good plan quickly provides advantages to the execution phase as well. A fast planning process implies more time for dispatching the plan. Moreover, the plan dispatcher can better rely on replanning to tackle unexepected events, further optimizing the final outcome w.r.t. a greedy approach like plan repairing.

The remainder of the paper is organized as follows. In the next section, the RoboCup Logistics League is briefly described. In Sect. 3, we discuss about the actual research involving scheduling and execution in the RCLL. In Sect. 4, the planners used to solve the planning domains for the RCLL are presented. In the next section we present the different encodings, highlighting the featured strategies and level of abstractions. A comprehensive evaluation is presented in Sect. 6. Finally, in Sect. 7 we draw the conclusions and discuss future work.

2   The RoboCup Logistic League

The league aims to stimulate the development of Robotics and AI using robotics competitions. A fleet of three autonomous mobile robots cooperate to assemble a set of products, by interacting with production stations in a real world environment. Orders (product configuration and delivery time) to accomplish are randomly generated during the game on an incremental basis. A product is mimicked by stacks of one base, from zero up to three rings, and a single cap. The different intermediate production steps are provided specific stations. The amount of rings determines the complexity of the product (C0 to C3). Moreover, the mounting of some rings asks for a payment to the corresponding station by providing extra pieces. In general, several refining steps of intermediate products by different machines are needed to assemble a product. Depending on the complexity of a delivered product, the corresponding amount of points is awarded.

3   Research and Strategies in the RoboCup Logistics League

Most of the research in the area of scheduling and execution within the RCLL of the last years was conducted mainly by two teams participating in the league, namely the Carologistics team from Aachen University and the GRIPS team from

Graz University of Technology. In this section we will discuss their work and the adopted strategies to tackle this domain in more details. Focus is mainly on planning and scheduling. However, some approaches adopted by the teams interleave both scheduling and execution, making it difficult to draw a clear separation. For example, the Carologistics team uses a greedy decentralized approach based on Goal Trees [16]. Such system ensures the flexibility to integrate scheduling, goal reasoning, and execution needed for a dynamic domain such as RCLL. Preconditions and a priority are assigned to every task. Then, every time a robot is idle, it selects and executes task with the highest priority among the ones whose preconditions are satisfied. This strategy has the advantage to keep the robots busy without requiring heavy computational load, coordinating tasks, and allocating resources to the robots on-the-fly, but without considering long-term scheduling and optimization. The GRIPS team was adopting a similar strategy [13]. Recently the team adapted a long-term strategy based on temporal planning. The Carologistics team did an attempt with a similar paradigm, by using ASP [18] to calculate a plan for a short time window. The use of temporal planning for a dynamic domain such as the RCLL poses some issues, related to the heavy computational load typical for solving techniques based on enumeration. In order to make this technology applicable for dynamic domains one has to apply measures to mitigate the complexity of solving. For instance, one will not plan for all the available orders at the same time, in order to avoid increasing the problem size. As a consequence, a Goal Reasoning strategy is used to heuristically select different sets of goals where it is likely the planner finds a solution in time. Then, a separate planning process is performed over each set, at the same time, in order to exploit the multi-threads capabilities of modern CPUs. The best plan within a fixed amount of time is selected and executed. However, further modifications are needed to shorten the planning time. Abstracting away some details of the domain may significantly reduce the size of the problems. This abstraction comes often at the cost of plan quality (which, in our case, is related to the makespan), since the planner is provided with less information. This encoding is discussed in details in Sect. 5.

The Freiburg team developed a mixed approach, making use of both long-term planning and on-the-fly task assignment to the robots. In [10], the strategy is generalized to all multi-agent domains. The plan is generated abstracting away the agents, and the actions are auctioned off to robots during the execution.

4   Planner Candidates

In this section a brief overview of the evaluated planner is given. The paper is focused on a comparison of action-based temporal planners supporting the standardized PDDL language [7]. This comparison involves the following planners: (1) POPF [3], (2) Optic [1], (3) Temporal Fast-Downward (TFD) [6], (4) C4PT [8], (5) ITSAT [17] and (6) YAHSP3 [21]. All of them participated in the temporal track of the International Planning Competition (IPC) 2014 [20] and/or 2018 [12]. A few others has been excluded, like the PDDL temporal planners tBurton [22] and DAEYAHSP [5], because public implementations were not available. The evaluation including also Timeline-Based planners [9] or hierarchical temporal planners is more difficult, since there are no standardized languages widely adopted by all of them and a specific encoding for each planner is needed. Although the support for the HDDL language [11] increased in the last years thanks to the IPC, many hierarchical planners from the past still use their own language, like FAPE [2] with ANML [19]. However, a similar evaluation of the state-of-the-art planners using these approaches will be part of future work. All the tested planners support the PDDL language. More specifically they support the 2.1 version of the modelling language, which includes durative actions. Durative actions allows for concurrent execution and time management and are therefore necessary to encode temporal domains. Although temporal constraints and concurrency could be handled by classical planners expressive enough with some post-processing [14], temporal planners has been built to manage specifically such features. POPF and Optic are two planners developed by the planning group at King's College London. Optic is built over POPF, and it augments to support PDDL feature such as preferences and time-dependent goal-collection costs, making it possible to encode soft constraints. POPF, which is itself an evolution of the previous planner Colin [4], applies forward-chaining state-based search strategy to partial-order planning, in combination with a late-commitment approach and linear programming to handle continuous linear numeric change. The TFD planner is a forward-chained planner that performs a heuristic search in the space of time-stamped states, where the two types of search steps are the insertion of a durative action at the current time point and the advancement of the current time by a certain increment. C4PT is a porfolio planner, which tries to solve the problem with different solving algorithm given a priority order, after a compilation of the temporal domain into a classic one. If a planner fails within a certain amount of time, the next one is selected. ITSAT is a planner based on satisfiability. It applies two preprocessing methods for mutex relation extraction and action compression, compiling the planning problem into a SAT formula. Violation of temporal constraints are detected through a Simple Temporal Network, and solved by adding the corresponding formula to the problem, preventing that inconsistency. YAHSP3 is a forward state-space heuristic search planner that embeds a lookahead policy based on an analysis of relaxed plans.

5   Domain Encodings and Abstraction

In this section we present five PDDL modelings of the challenge of the RoboCup Logistics League with different levels of abstraction. Abstraction allows to speed up the search process, usually at the cost of plan makespan. Since RCLL is a dynamic domain, we are interested in the sum of the solving time and the makespan. In fact, in a dynamic domain such value corresponds to the real end of the plan execution. We refer to this value as *total-time*. The least and the most abstracted domains have been developed by the Carologistics and GRIPS team respectively. We call these modeling CARO[1] and GRIPS[2]. The other three

---

[1] https://github.com/timn/ros-rcll_ros.

[2] https://tinyurl.com/2dkbasft.

domains lay in the middle, encoding an intermediate level of abstraction. Two of them are derived from the GRIPS encoding. Each one reintegrates the explicit modeling of an aspect which was abstracted away in the original GRIPS domain. We refer to them as GRIPS-MOVE and GRIPS-MPS, that reintegrate the move actions and the station processing respectively. The last domain has been developed by the Freiburg team (FREIBURG)[3].

5.1   The GRIPS and CARO Encodings

We start by introducing the main difference between the CARO and GRIPS PDDL encodings. The former is very accurate and models every detail of the domain. It has been compared to other modelling and planning approaches in [18]. The latter abstracts away the following details: (1) the sending of preparation messages to production stations and the corresponding processing task, (2) the move action of agents and (3) the representation of every base workpiece present in the environment. In GRIPS, all of these aspects are considered implicitly and not modeled as standalone actions or objects. This modeling is currently used by the GRIPS team in the RCLL competition. The performance of this approach can be seen in the result of RoboCup Asia Pacific 2021[4].

Machine Processing and Preparation Messages: One can identify two types of preparation messages. The ones required to interact with a station, which needs to be sent before performing some kind of delivering or retrieving task and the ones used to start the processing task on a ring or a cap station. A processing task consists of performing an activity on a piece placed on the input side of a station, mounting a ring or a cap on it, and making it available on the output side for the retrieval. In the GRIPS domain, both type of messages are abstracted away. The plan dispatcher will handle the preparation messages implicitly during the execution. In the first case, the message is sent when the robot starts moving towards the station, and there is no advantage in anticipating it. This means that the makespan is not increased by this abstraction. However, the same does not apply for the second type of preparation messages. Such messages are sent automatically by the GRIPS plan dispatcher as soon as a piece is delivered to the input side of a machine. As a consequence, both the input and output sides of the station's conveyor need to be empty during the delivery. Being able to postpone the processing w.r.t to the delivery action, possible in CARO, means that a piece can be delivered to the input even if the output is occupied. Such freedom increases the options for action concurrency and may shorten the makespan.

Move Actions: In GRIPS movements of the robots are not represented as atomic actions. Every movement is integrated into an interaction of the robot with a machine. If a robot needs to use a station, the time needed to travel between its actual position and the station is added to the duration of the

---

[3] https://github.com/GKIFreiburg/rcll-sim-freiburg.
[4] https://tinyurl.com/bdzk8v6v.

action representing the interaction with the machine. This results in potentially suboptimal solutions in terms of concurrency, but the solving time improves drastically because an explicit move action can be applied almost anytime and increases the search space a lot. In fact, removing the possibility for a robot to move in advance towards a position may cut out plans with better makespans. For instance, a robot is allowed to retrieve a piece from the output side of a station only after the processing of the piece was performed by the station. The corresponding PDDL action is then scheduled after the delivery of the piece to the input side. However, in the Carologistics modeling a different robot can move to the output side beforehand, waiting there until the piece is ready to be retrieved. In the GRIPS modeling, instead, the absence of a standalone move action means that the robot can not anticipate the movement to the output side of a station w.r.t. the retrieving task. It will start moving towards it only after the processing. Listing 1.1, which depicts the plans obtained using the GRIPS and Carologistics encoding for the interaction with a ring station, shows this behaviour. The Carologistics model can finish 15 s earlier, namely the time needed by a robot to move to the right position.

Workpiece Representation: Another important difference between the two modelings is that CARO explicitly models every base with a dedicated object. Every base can either be linked to an actual order or used as a resource to be provided to the ring station. In the former case, its features (like the presence and colors of rings and/or a cap mounted on top of the workpiece) are added as soon as the workpiece is processed by other stations. The bases situated on the shelf of the cap station, which are used as cap carriers and have to be delivered to the input side to provide a cap, are also individually represented with an object. In the GRIPS encoding the workpieces are not directly modeled. For each order, a dedicated predicate to keep track of the progressing is used. Stations' resources which can not be refilled by robots and need human intervention, namely the bases used as cap carriers or stored inside the Base Station, are supposed to be infinite. However, the usage of such resources by robots is modeled, to avoid the derivation of invalid plans. We can conclude that adopting a more general modeling which supposes infinite resources does not increase the makespan, as long as the status of the intermediate product and the usage of resources are properly modeled.

Dealing with the Abstracted Aspects During the Plan Execution. In the GRIPS overall software architecture, which includes also other aspects like plan execution and monitoring, all the details abstracted away during the planning phase are managed by the plan dispatcher. For instance, the GRIPS *get-BaseFromBaseStation* action represents a robot retrieving a base piece from the base station. This PDDL action is actually split into three different subtasks by the plan dispatcher: the (1) the sending of a preparation message to the station, which dispenses a base piece onto the conveyor; (2) the move task of the robot from its actual position to the base station, together with the alignment to the right side; (3) the actual grasping of the base piece by the robot.

Listing 1.1. GRIPS and Carologistics plans to retrieve a base. We assume 15 seconds for robot movements, 30 seconds for interaction with the station and 1 second for machine processing. The action descriptions of GRIPS and Carologistics models have been simplified and uniformed for better readability. Numbers between square brackets represent the time point of the start or the end of the corresponding action.

```
GRIPS:
[30.000]   deliverProductToRS1_start(r1,p1)
[76.000]   deliverProductToRS1_end(r1,p1)
[76.001]   getProductFromRS_start(r2,p1)
[121.001]  getProductFromRS_end(r2,p1)
Carologistics:
[30.000]   moveTo_start  (r1,rs1_input)
[30.000]   moveTo_start  (r2,rs1_output)
[30.000]   prepare−rs_start  (rs1)
[30.001]   prepare−rs_end  (rs1)
[45.000]   moveTo_end  (r1,rs1_input)
[45.000]   moveTo_end  (r2,rs1_output)
[45.001]   deliverProductToRS1_start(r1,p1)
[75.001]   deliverProductToRS1_end(r1,p1)
[75.002]   rs−mount−ring1_start  (rs1)
[76.002]   rs−mount−ring1_end  (rs1)
[76.003]   getProductFromRS_start(r2,p1)
[106.003]  getProductFromRS_end(r2,p1)
```

5.2   GRIPS-MOVE and GRIPS-MPS Encodings

The GRIPS-MOVE and GRIPS-MPS modelings are derived from the GRIPS encoding, by adding some of the aspects which have been abstracted away. GRIPS-MOVE features the standalone move actions. Robots are free to travel between locations and the move action is no more integrated into an interaction task.

The GRIPS-MPS encodings adds the handling of processing tasks by production stations. Simple preparation messages are still abstracted away, since they do not provide any benefit in terms of plan quality. The sending of a message to trigger the processing of a workpiece by a station is encoded as a standalone action. As a result, a delivery task on a station can be performed even if the output side is occupied by another piece. As soon as the output side is freed, the processing action can be performed.

5.3   FREIBURG Encoding

The FREIBURG encoding, we are evaluating in this paper, is similar to GRIPS-MPS in terms of abstraction level. Simple preparation messages are not modeled, while processing messages are present. Move actions are partially abstracted away, but in a different way as in GRIPS-MPS. The key differences between GRIPS-MPS and FREIBURG are the following:

Modelling of Processing Tasks: Like in GRIPS-MPS, the processing tasks of production stations are explicitly modeled. However, there is an important difference. In FREIBURG, it is not allowed to deliver an object to the input side as long as an workpiece is present on the conveyor of the station, even if it is on the output side. As a result, the modeling of processing tasks does not provide any advantage in terms of makespan w.r.t. the original GRIPS domain. GRIPS-MPS is instead able to find plans with better degrees of concurrency.

Modelling of Move Actions: In FREIBURG the move actions are also abstracted away, but in a different way w.r.t. GRIPS. In GRIPS, every move action is integrated into one interaction action. Differently, the FREIBURG domain include the encoding of a *transport* action, which combines a move action with two interaction actions. More specifically, a transport action includes the (1) retrieve task of a workpiece from a station by a robot, (2) the move of that robot to another station and (3) the delivery of the workpiece to that station. However, a standalone move action is still necessary, to place the robot in the right place before the execution of a transport action. In other words, a robot can freely move around only if he is not carrying anything. Otherwise, the movement is combined with two interaction tasks.

6   Evaluation

In this section detailed results of the evaluation are presented, discussing the planners and the encodings separately.

Metrics: In the evaluation we are considering different kind of metrics: (1) the *total-time*, (2) the makespan, (3) the solving time and (4) the number of planning problems instances the planner is able to solve in a given time. Regarding the evaluation of planners, the main metric is the number of solved instance. For the evaluation of encodings we focus on the average *total-time* of the best plan in each solved instance. The *total-time* of a plan is the sum of the plan makespan and the solving time needed to derive that plan. In this case, with *best* plan we do not mean the plan with the best makespan, but the one with the best *total-time*. We consider a time limit of 17 min for the solving time, as 17 min is the duration of a RCLL game. Since we are testing a dynamic domain, and the solving phase needs to be integrated into the game, we are interested into minimizing the sum of the solving time and the makespan. All the tested planners iteratively finds better and better plans during the solving process. The best plan (in terms of makespan) can not be executed if we spent all the 17 min for planning. On the opposite, the first found plan may have a very long makespan w.r.t. a plan found after a few more seconds of solving. To estimate when to stop the planning process and commit to the best plan found that far is not possible. As a consequence, the idea is to commit to the best plan (in terms of *total-time*) found so far. If a better plan is found, the old plan is discarded and the new one is executed. As long as the *total-time* is smaller, we know that we are able to finish its execution before the old one. The saved time can then be used to derive and dispatch a plan for other products. The same reasoning does not apply if we consider the makespan alone. If a plan with a better makespan is found, we may have no time to dispatch it anymore. A possible issue is that this strategy only works together with the hypotesis that we are able to immediately dispatch a new found plan. If we were dispatching a different plan, we have to adapt the new plan to the actual situation. However, since both the old and the new plan involve the same products, this task can be done without increasing the makespan. On the contrary, since in the old plan we may have already solved some of the steps needed to assemble the same products, the real makespan may decrease. To summarise, a new found plan can be easily merged with the current one and the *total-time* may further decrease. By using *total-time*, we are considering both the solving time and the makespan, allowing to choose to which plan commit on-the-fly during the execution.

Setup: For each planner and each encoding we test 5 different product configurations in terms of complexity. For each product configuration we use 10 different game configuration. A game configuration determines the position of the stations on the field, the features of the product (color of caps, bases and rings), and the requested payment for each ring color. The game configurations have been randomly generated using the Referee Box, the official software referee of the competition. The product configuration represents the complexity of the orders, in terms of number and size of the products: p1 (1 C0 product), p2 (1 C1 product), p3 (1 C2 product), p4 (1 C3 product), p5 (1 C0, 1 C2, 1 C3 products). The total number of runs are 1500, decreasing to 1350 due to incompatibility issues between some encodings and some planners (e.g., C4PT with FREIBURG, C4PT and TFD with CARO). The evaluation has been executed on a laptop with 16 GB of RAM and an Intel i5 8500u CPU, featuring Ubuntu 16.04.

Table 1. Number of solved instances by different planners in relation to the complexity of the problem instance. Number of tested instance per planner can vary due to unsupported features like numeric fluents or math operators.

	Optic	POPF	TFD	C4PT	ItSAT	YAHSP3
P1	40/50	40/50	40/40	30/30	40/50	42/50
P2	41/50	40/50	31/40	30/30	40/50	41/50
P3	40/50	38/50	29/40	30/30	4/50	38/50
P4	34/50	35/50	30/40	30/30	0/50	31/50
P5	25/50	26/50	25/40	30/30	0/50	31/50
Overall	180/250	179/250	155/200	150/150	84/250	183/250

Results: For the evaluation of planners and domain encodings in the context of RCLL we performed an extensive evaluation on various combinations of planners, encoding, and problem instances. While we present the results of this evaluation in a condensed form here, all detailed results can be found on the

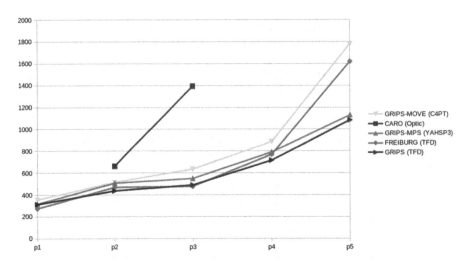

Fig. 1. Comparison of different problem encodings for different complex problems. The *total-time* is given in seconds for the best performing planner per encoding. Failure of such planner to find a solution within 17 min results into missing data points.

web[5]. In Table 1 the comparison between the planner is shown. Due to incompatibility issues, C4PT has not been tested on CARO and FREIBURG, while TFD has not been tested on CARO. Optic, POPF, ItSAT and YAHSP3 run flawless with all the domains. Between them, YAHSP3 is the clear winner, since it was able to solve most of the instances. Optic and POPF are close, while ItSAT shows the overall worse performance in the comparison. Overall, C4PT is the only planner able to solve 100% of the tested instances. However, we have to highlight that almost no instance of CARO, the most complex encoding, has been solved by any planner. The fact that C4PT has not been tested on CARO has had a significant impact on the score. Nevertheless, by looking at the set of common tested instances between C4PT and Optic, or C4PT and YAHSP3, the portfolio planner beats both of them. TFD manages to solve 155 instances out of 200. Considering that it has not been tested on CARO, it shows worse performance w.r.t to all the other planners, except for ItSAT. The comparison between the modelings is depicted in Fig. 1. For space reason, and to ensure a better readability, for each modeling we show the results of the planner which performs better, depending on the number of solved instances for that particular modeling. If two or more planners solved the same number of instances, we select the one with the best average *total-time*. In the plot, for each encoding and each product configuration the average *total-time* of the corresponding 10 game configurations is shown. From the plot, the contribution of the various levels of abstraction is well depicted. First, it is clear that a detailed and deep representation of the domain, like CARO, is too complex for an dynamic domain. Only a few instances have been solved, more specifically two p1, four p2 and two p3.

---

[5] https://tinyurl.com/2p86r4hj.

The reason why some planners manage to solve p2 and p3 instances but not p1 is unclear. However, finding a solution for them required a huge amount of time, as can be seen on the plot. In this case, most of the contribution to the *total-time* comes from the solving time. Even considering a perfect solution in terms of makespan, the quantity of time invested in planning is too high for a dynamic domain. While this is conditional to the available compute resources, it is hard to imagine that such combinatorial explosion may be compensated by a more powerful hardware. The best performing modeling is GRIPS, namely the one with the higher degree of abstraction. From the results, it is clear how the drawback of abstraction, namely a potential worse makespan, become meaningless if compared with the gain obtained in terms of better solving time. The trade-off between makespan and solving time favors the usage of abstraction. By analyzing the other encodings, we can weight the impact of each abstracted aspect. Abstracting away the move actions is the aspect affecting the *total-time* the most. In fact, in the two domains where move actions are not (GRIPS-MOVE) or only partially (FREIBURG) abstracted away, the *total-time* of the most complex instance is significantly higher w.r.t. GRIPS and GRIPS-MPS, where move actions are completely abstracted away. This behaviour is also confirmed by the other metrics we tested. For example, CARO and GRIPS-MOVE have worse makespans than expected. This can be explained by the high complexity of the two domains. In fact, the 17 min limit for the planning time prevents the planners to find the real optimal solutions for these encodings. For some planners, the suboptimal plans found within the time limit are worse than the ones found with more abstracted encodings. This means that, due to the short time window w.r.t. to the size of the problem, abstracting away the move actions does not even results into an higher makespan. We obtain both a faster computation and a better plan. Finally, abstracting away the processing task of production stations has also a positive impact, how can be seen by the comparison of GRIPS and GRIPS-MPS, although way smaller than the movement abstraction.

7   Conclusion and Future Work

In this paper we performed a comparison between PDDL temporal planners in the context of the RoboCup Logistics League. Moreover, we evaluated different domain encodings in order to investigate the impact of abstraction. Abstracting away some aspects allows to decrease the problem complexity, but it may results in worse plans. Since RCLL is a dynamic domain and solving time is crucial, we exploited the trade-off between planning time and makespan to verify if abstraction leads to an overall benefit, considering the sum of the two times. Looking at the final results, we can conclude that it is the case. As part of future work, we want to investigate the performance of other type of planner which support temporal domain, like some Hierarchical Task Network (HTN) planners and Timeline-based planners. Different modelings will be considered as well. For instance, we plan to exploit the task hierarchy property of HTN planners, to verify that using knowledge of the structure of a solution (e.g., order of production steps) available in RCLL can speed up planning significantly.

References

1. Benton, J., Coles, A., Coles, A.: Temporal planning with preferences and time-dependent continuous costs, vol. 22 (2012)
2. Bit-Monnot, A., Ghallab, M., Ingrand, F., Smith, D.E.: FAPE: a constraint-based planner for generative and hierarchical temporal planning. arXiv preprint arXiv:2010.13121 (2020)
3. Coles, A., Coles, A., Fox, M., Long, D.: Forward-chaining partial-order planning, pp. 42–49 (2010)
4. Coles, A.J., Coles, A.I., Fox, M., Long, D.: Colin: planning with continuous linear numeric change. J. Artif. Intell. Res. **44**, 1–96 (2012)
5. Dréo, J., Savéant, P., Schoenauer, M., Vidal, V.: Divide-and-evolve: the marriage of descartes and darwin. In: Proceedings of the 7th International Planning Competition (IPC). Freiburg, Germany, vol. 91, p. 155 (2011)
6. Eyerich, P., Mattmüller, R., Röger, G.: Using the context-enhanced additive heuristic for temporal and numeric planning. In: ICAPS (2009)
7. Fox, M., Long, D.: Pddl2.1: An extension to PDDL for expressing temporal planning domains. J. Artif. Intell. Res. (JAIR) **20**, 61–124 (2003)
8. Furelos-Blanco, D., Jonsson, A.: Cp4tp: A classical planning for temporal planning portfolio (2018)
9. Gigante, N.: Timeline-based planning: Expressiveness and complexity (2019)
10. Hertle, A., Nebel, B.: Efficient auction based coordination for distributed multi-agent planning in temporal domains using resource abstraction. In: Trollmann, F., Turhan, A.Y. (eds.) KI 2018: Advances in Artificial Intelligence, pp. 86–98 (2018)
11. Höller, D., et al.: Hddl - a language to describe hierarchical planning problems
12. IPC: International Planning Competition 2018 (2021). https://ipc2018.bitbucket. io/#. Accessed 27 Apr 2022
13. Kohout, P., De Bortoli, M., Ludwiger, J., Ulz, T., Steinbauer, G.: A multi-robot architecture for the RoboCup logistics league. Elektrotechnik und Informationstechnik **137**(6), 291–296 (2020)
14. Nau, D.S., et al.: Shop2: an HTN planning system. J. Artif. Intell. Res. **20**, 379–404 (2003)
15. Niemueller, T., Ewert, D., Reuter, S., Ferrein, A., Jeschke, S., Lakemeyer, G.: RoboCup Logistics League Sponsored by Festo: A Competitive Factory Automation Testbed, pp. 605–618 (2016)
16. Niemueller, T., Hofmann, T., Lakemeyer, G.: Goal reasoning in the clips executive for integrated planning and execution. Proc. Int. Conf. Autom. Plan. Sched. **29**(1), 754–763 (2021)
17. Rankooh, M.F., Ghassem-Sani, G.: ITSAT: an efficient sat-based temporal planner. J. Artif. Intell. Res. **53**, 541–632 (2015)
18. Schäpers, B., Niemueller, T., Lakemeyer, G., Gebser, M., Schaub, T.: Asp-based time-bounded planning for logistics robots (2018)
19. Smith, D.E., Frank, J., Cushing, W.: The ANML language. In: The ICAPS-08 Workshop on Knowledge Engineering for Planning and Scheduling (KEPS), vol. 31 (2008)

20. Vallati, M., Chrpa, L., Grześ, M., McCluskey, T.L., Roberts, M., Sanner, S., et al.: The 2014 international planning competition: Progress and trends (2015)
21. Vidal, V.: YAHSP3 and YAHSP3-MT in the 8th international planning competition. In: Proceedings of the 8th International Planning Competition (IPC-2014), pp. 64–65 (2014)
22. Wang, D., Williams, B.: tBurton: a divide and conquer temporal planner. In: Proceedings of the AAAI Conference on Artificial Intelligence, vol. 29, no. 1 (2015)

An Embedded Monocular Vision Approach for Ground-Aware Objects Detection and Position Estimation

João G. Melo[(⊠)] and Edna Barros

Centro de Informática, Universidade Federal de Pernambuco, Av. Prof. Moraes Rego, 1235 - Cidade Universitária, Recife, Pernambuco, Brazil
{jgocm,ensb}@cin.ufpe.br

Abstract. In the RoboCup Small Size League (SSL), teams are encouraged to propose solutions for executing basic soccer tasks inside the SSL field using only embedded sensing information. Thus, this work proposes an embedded monocular vision approach for detecting objects and estimating relative positions inside the soccer field. Prior knowledge from the environment is exploited by assuming objects lay on the ground, and the onboard camera has its position fixed on the robot. We implemented the proposed method on an NVIDIA Jetson Nano and employed SSD MobileNet v2 for 2D Object Detection with TensorRT optimization, detecting balls, robots, and goals with distances up to 3.5 m. Ball localization evaluation shows that the proposed solution overcomes the currently used SSL vision system for positions closer than 1 m to the onboard camera with a Root Mean Square Error of 14.37 mm. In addition, the proposed method achieves real-time performance with an average processing speed of 30 frames per second.

Keywords: Autonomous navigation · Position estimation · Object detection

1 Introduction

At the RoboCup Small Size League (SSL) robot soccer competition, games occur between two teams of omnidirectional mobile robots with eight players for division A and six for division B. Frames from cameras placed above the field are processed by a dedicated computer, which runs SSL Vision: a standard vision system for detecting and tracking elements such as robots, goals, balls and field lines [24]. Off-field computers, one for each team, receive the position information and referee commands and perform most of the computation and exchange of information with robots using Radio Frequency (RF) communication with minimal bandwidth.

In recent RoboCup editions, the League has proposed a new technical competition in which teams are only allowed to use embedded sensing information

Supported by Centro de Informática (CIn - UFPE), Fundação de Amparo a Ciência e Tecnologia do Estado de Pernambuco (FACEPE), and RobôCIn Robotics Team.

© The Author(s), under exclusive license to Springer Nature Switzerland AG 2023
A. Eguchi et al. (Eds.): RoboCup 2022, LNAI 13561, pp. 100–111, 2023
https://doi.org/10.1007/978-3-031-28469-4_9

for executing basic soccer tasks inside the field [15]. The Vision Blackout challenge encourages teams to propose autonomous navigation solutions for the SSL environment. In 2021 the teams were assigned three tasks for the competition: grabbing a stationary ball somewhere on the field (1), scoring with the ball on an empty goal (2), and scoring with the ball on a statically defended goal (3). Therefore, detecting and locating balls, robots, and goals from embedded devices is needed.

SSL robots are constrained to a 180mm diameter limit and achieve up to 3.7 m/s velocities, requiring low-power, small-size, high-throughput solutions. For example, SSL ball detection using scan lines and color segmentation has been previously proposed [17]. However, even though this approach has achieved accurate results in recent competitions, it cannot detect other SSL objects. Moreover, it lacks robustness concerning local illumination or field changes.

With the advances in Deep Neural Networks (DNN) architectures and parallel processing technologies, the use of Convolutional Neural Networks (CNN) for object detection has grown considerably in embedded applications [2]. This approach presents significant advantages compared to traditional computer vision techniques, especially in robustness to environmental conditions and adaptability to new object classes. For that, an open-source SSL dataset[1] containing 2D bounding boxes for detecting balls, goals, and robots on images is available [6].

Computing position and orientation (pose) from detected objects is also essential for autonomous navigation. Considering the robot's resource-constraints, solutions employing a single monocular camera for object detection and position estimation are preferred. Exploiting prior knowledge from the environment, Inverse Perspective Transformation can be applied for computing three dimensional positions from camera frames [3].

This research proposes a monocular vision solution for detecting and estimating the relative positions of SSL objects, using bounding boxes' 2D coordinates from a CNN-based Object Detection model and previously calibrated camera parameters for back-projecting objects positions on the soccer field. The proposed solution is tested on an NVIDIA Jetson Nano Developer Kit [7] employing a Logitech C922 camera, both mounted on the top of a SSL robot, and we evaluate accuracy results for ball localization, achieving a 14.37 mm Root Mean Square Error (RMSE), overcoming position accuracy from the standard SSL Vision system. The implemented system runs at 30 frames per second, with an average 10.8 W power consumption, and respects all League's restrictions, showing it can be applied for the desired challenge, and the main contributions of this work are:

- A complete architecture for detecting and locating SSL objects using CNN-based object detection.
- Presenting procedures for calibrating onboard camera intrinsic and extrinsic parameters.
- A detailed pipeline and configurations for training SSL Object Detection and deploying to NVIDIA Jetson Nano.

---

[1] https://github.com/bebetocf/ssl-dataset.

2   Related Work

The SSL Vision Blackout Challenge was introduced in the League in 2019, with Tigers Mannheim achieving the best results [17]. Their work aims to detect the ball and estimate its relative position to the robot and their research reports that CNNs may achieve robust results [13], but the team has discarded its use due to the lack of embedded hardware accelerators by the time; blob detection methods are too slow and highly sensitive to lighting changes; an edge detection method took too much processing time as well and identified moving balls as ellipsoids. Therefore, Tigers comes up with a novel approach: an algorithm that searches for ball candidates by applying a sharp edge detection kernel along logarithmically distributed horizontal lines. A vertical scan is performed when candidates are found, and the ball's radius, confidence, and position are estimated. The team also shares a deployment infrastructure for the on-bot vision software [21] and a full architecture for autonomous robot-ball interaction [10], on which ball distances to the robot are estimated from the object size on the image.

At other RoboCup soccer leagues, Deep Learning methods are proposed for detecting elements such as balls, robots, goalposts, and field lines. At the Standard Platform League, for instance, CNNs are used for detecting field boundaries [9], robots and balls, while also estimating ball's positions [14]. Since detecting goals and field lines are essential for self-localization on the soccer field, another approach defines four object classes for the network to detect: balls, line crossings, robots, and goalposts [19]. Exploiting prior knowledge about the ball characteristics, another research presents an algorithm for searching ball regions proposal to accelerate inference time by applying the object detection CNN to a smaller region of the image [20].

An open-source dataset containing 2D bounding boxes labels for SSL robots, balls and goals was introduced in 2021's RoboCup edition [6]. The research also benchmarks Deep Learning state-of-the-art object detection models evaluating their accuracy and inference speed on a Google Coral TPU accelerator, reporting SSD MobileNets V1 and V2 to achieve the best overall performances.

The main bottleneck for using CNNs on SSL was the lack of hardware-accelerated embedded platforms for running DNNs in real-time [10]. With Deep Learning empowering several IoT applications, the concept of Edge Computing rapidly gains huge attention, urging for low power devices capable of running complex DNNs [18]. Edge Devices comparisons show that NVIDIA Jetson Nano and Google Coral Developer Board achieve the best overall results when running object detection models, concerning the accuracy, inference time and power consumption [2].

As minor performance improvements are extremely relevant in real-time applications, numerous CNN architectures are proposed in the object detection domain, being mostly divided into two categories: two-stage and one-stage, such as single shot detectors (SSD). For mobile applications, single-shot ones gain most of the attention and previous comparisons between state-of-the-art models report SSD MobileNet V2 [11,16] to achieve great speed-accuracy trade-offs, especially under hardware constrained scenarios [6,8,12].

As 2D object detection models are only capable of computing two-dimensional bounding box positions on images, we also search for three-dimensional position estimation solutions. A DNN combining human semantic segmentation and depth prediction achieves remarkable accuracy with high-speed inferences on Jetson Nano [1]. However, the network must be trained with a depth-labeled dataset, which is not available for SSL objects. At the RoboCup@Home League, OpenPose CNN architecture [5] is used for detecting key points, and object poses are calculated from 2D-3D correspondences between detected and ground-truth key points using PnP-RANSAC [22]. However, the method presents limitations when dealing with symmetric objects. Also, works on vision-based self-localization show that relative positions from points on images can be retrieved by Inverse Perspective Transformation if one of its world coordinates is known [3].

In this work, we solve the Perspective-n-Point (PnP) problem for 2D-3D correspondences between a set of hand-marked points on the field for estimating a camera's relative pose to the field coordinates. Also, we employ SSD MobileNet v2 for regressing object's 2D bounding boxes and a linear regression model for calculating their bottom-centers projection on camera frames. Extrinsic and intrinsic parameters are used for solving the Inverse Perspective Transformation problem for points on the ground and object's relative positions are estimated.

3 Proposed Approach

During soccer matches and especially for the Vision Blackout challenge, SSL objects mostly lay on the soccer field, and we exploit this prior knowledge for proposing a monocular vision solution for detecting and estimating their relative positions to the robot. For that, the camera is fixed to the robot and its intrinsic and extrinsic parameters are obtained using calibration and pose computation techniques from the Open Computer Vision Library (OpenCV) [4]. A state-of-the-art CNN-based object detection model, SSD MobileNet v2 [11,16], is used for detecting objects on camera frames. After labeling, linear regression is applied to the bounding box's coordinates, assigning a point on the field that corresponds to the object's bottom center, which has its relative position to the camera, and, therefore, to the robot estimated using pre-calibrated camera parameters. Position information can be delivered to decision-making and path planning algorithms to execute autonomous navigation. Figure 1 illustrates a scheme for the proposed method and all software is open-source[2].

In the following subsections, in-depth explanations for each of the steps from the proposed pipeline for object localization are presented. Firstly, we depict the camera pinhole model and calibration procedures. Then, in sequence, since the proposed approach for estimating objects' positions is based on ground points localization, a method for computing field points' relative positions is presented. Then, CNN model conversion and training details are given, followed by an explanation of the procedure adopted for fitting a linear regression model for estimating the object's ground pixel.

---

[2] https://github.com/jgocm/ssl-detector.

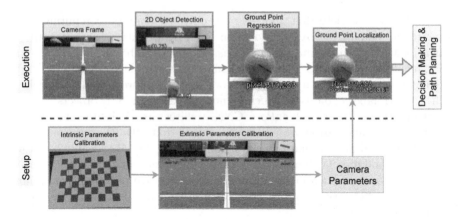

Fig. 1. Designed architecture for detecting and locating SSL objects with a single monocular camera.

3.1 Camera Calibration

Camera parameters are divided into intrinsic and extrinsic, and the process for projecting three-dimensional points to the image plane can be described in three steps: converting 3D world position to the camera coordinates system using the extrinsic parameters (1); projecting points to the image plane using intrinsic parameters (2); and re-scaling pixels using a scale parameter (3). The pinhole camera model describes the mathematical representation for this process [10], which is given in detail by Eq. 1 and simplified on Eq. 2.

$$
s \begin{bmatrix} u \\ v \\ 1 \end{bmatrix} = \begin{bmatrix} \alpha_x & \gamma & u_0 \\ 0 & \alpha_y & v_0 \\ 0 & 0 & 1 \end{bmatrix} \begin{bmatrix} r_{11} & r_{12} & r_{13} & t_1 \\ r_{21} & r_{22} & r_{23} & t_2 \\ r_{31} & r_{32} & r_{33} & t_3 \end{bmatrix} \begin{bmatrix} x_w \\ y_w \\ z_w \\ 1 \end{bmatrix} \tag{1}
$$

$$
s\, p_c = K\, [R \mid t]\, p_w \tag{2}
$$

In Eq. 2: p_w represents a 3D point in the world coordinates system; $[R \mid t]$ describes the camera axis rotation and translation concerning the world coordinates, which are called extrinsic parameters; K is the intrinsic parameters matrix, consisting of α_x and α_y scale factors, u_0 and v_0 coordinates for the principal point and the γ skew factor; p_c is the pixel position on screen; and s is a depth scale factor. Thus, for back-projecting image pixels to three-dimensional world coordinates, camera parameters must be calculated beforehand.

The chosen procedure for estimating intrinsic camera parameters requires multiple images of a planar pattern from different viewpoints, making 2D-3D correspondences between pattern points from different images [23]. For instance, a chessboard pattern was used and OpenCV implementations for camera calibration and chessboard corner detection for pattern recognition were applied, as illustrated in Fig. 2.

Fig. 2. Corner detection and chessboard pattern recognition result. The method is applied to multiple images and, by the end, the 2D points and its 3D correspondences are used for estimating the camera's intrinsic parameters and lens distortions [23].

For extrinsic parameters estimation, 2D-3D correspondences between SSL field points can be used to find the camera rotation and translation regarding field axis by solving a Perspective-n-Point (PnP) problem, which consists of finding a camera's pose based on a set of pixels and its corresponding three-dimensional world coordinates. Figure 3 describes the calibration procedure.

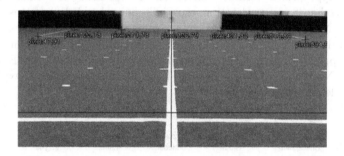

Fig. 3. With the camera mounted onto the robot, pixels corresponding to field points with known positions are hand-marked, for instance, penalty area corners, field corners, goal corners, or the goal center can be used. A PnP problem for the 2D points and their three-dimensional correspondences is solved by minimizing re-projection error. The employed algorithm is an implementation from OpenCV library and computes the camera's rotation and translation vectors with respect to field coordinates.

3.2 Ground Point Localization

By using the calibrated extrinsic and intrinsic parameters, given a pixel on the screen, the camera model can be rewritten as Eq. 3. From that, if rotation matrix R, translation vector t, and intrinsic parameters matrix K are computed, any position p_w can be retrieved if one of its coordinates is known. For example, points on the ground, which in case correspond to $z_w = 0$, can have their x_w and y_w coordinates estimated.

$$\begin{bmatrix} x_w \\ y_w \\ z_w \end{bmatrix} = s\, R^{-1}K^{-1} \begin{bmatrix} u \\ v \\ 1 \end{bmatrix} - R^{-1}t \tag{3}$$

3.3 Object Detection Model

Object detection tasks require high computing power, being the most time-consuming step of our proposed architecture. Thus, the SSD MobileNet v2 model was selected for its better trade-off between accuracy and inference time, among other state-of-the-art models [8].

With pre-trained weights on the COCO dataset, a model from the Tensorflow framework was retrained for $57,565$ steps on a Google Colab Notebook, running with a Tesla K80 GPU. The proposed dataset for SSL Object Detection contains 931 images with up to 4182 instances of balls, robots, and goals labeled in Pascal VOC and YOLO formats [6]. For the training, images were randomly partitioned into train and test sets with an 80/20 proportion, and batch size was configured to 24.

Using appropriate format conversion is essential for enabling high-performance hardware acceleration, especially on embedded platforms. For Jetson Nano, NVIDIA's TensorRT deep learning framework, which is built on CUDA parallel programming model, delivers low latency and high throughput for inference applications while also supporting models generated from Tensorflow, Pytorch, ONNX and other frameworks. Figure 4 gives an in-depth explanation for the model conversion procedure.

Fig. 4. The inference graph was exported from the Tensorflow retrained model checkpoint. TF Lite model was extracted from the graph and converted to ONNX format. NonMaxSupression (NMS) post-processing operation was replaced due to incompatibility issues, and the TensorRT Inference Engine was successfully generated from the ONNX file.

3.4 Ground Point Linear Regression

In order to compute three-dimensional positions with the proposed method, we fit linear regression weights for predicting a pixel that corresponds to the point where the object touches the ground, that is $z_w = 0$. The inputs for the model are bounding box coordinates generated from the 2D object detection inference. Figure 5 illustrates the procedure for finding the regression weights for the ball.

Fig. 5. With landmarks positioned with a 250 mm grid on the field, the ball was placed on each of the markers and its bounding boxes were regressed from the object detection model. Then, removing the ball, the pixels that correspond to the center of each landmark on the screen were annotated. Thirty markers were used in the procedure.

4   Evaluation

The SSL uses standard golf orange balls, with an average 42.7 mm diameter, for the soccer matches, and its localization was tested to evaluate the proposed method. Landmarks were positioned on the SSL field with a 250 mm grid and used as the ground truth during the experiments. We placed the robot at a fixed position, setting camera XY coordinates to 0 and -500. After calibrating camera intrinsic and extrinsic parameters, we estimated object relative positions on a set of images from the ball placed in 30 marked coordinates and compared the results to the SSL Vision system.

For qualitative comparison and behavior analysis, ball XY coordinates were plotted on Fig. 6. As for quantitative measurements, Root Mean Square Error (RMSE) was employed for the set of points taking landmark positions as the ground truth. Angles between the objects and the robot are essential for navigating in the SSL environment and were also computed for evaluation. Coordinates measurements are in millimeters, while angles are in degrees.

4.1   Camera Calibration Results

Intrinsic parameters were estimated from 20 chessboard pictures taken in 640×480 resolution with a Logitech C922 camera, as presented in Sect. 3.1. From the previously presented approach for extrinsic parameters calibration, 5 points were hand-marked on the screen: the bottom left and right goal corners, the lower left and right penalty area corners, and the goal bottom center. Estimating rotation and translation vectors from the PnP solution, camera to field axis relative pose can be computed. Calibration results are exhibited on Eq. 4.

$$K = \begin{bmatrix} 642.41 & 0 & 322.80 \\ 0 & 642.54 & 239.76 \\ 0 & 0 & 1 \end{bmatrix}, \begin{bmatrix} X \\ Y \\ Z \end{bmatrix} = \begin{bmatrix} -5.38 \\ -509.79 \\ 171.40 \end{bmatrix}, \begin{bmatrix} \omega \\ \phi \\ \kappa \end{bmatrix} = \begin{bmatrix} 106.94° \\ -0.43° \\ -0.38° \end{bmatrix}. \quad (4)$$

4.2 Objects Detection Performance

Table 1 and Table 2 show accuracy results for the retrained weights on the test set. After deploying to Jetson Nano, the model runs at an average processing time of 24 ms, equivalent to 41.67 frames per second, and SSL objects can be detected for up to 3.5 m distances using the onboard camera.

Table 1. Average Precision (AP) evaluation on the test set. Results are measured by Intersection over Union (IoU) threshold, where 50 and 75 indexes indicate 0.5 and 0.75 IoU, while AP_S, AP_M and AP_L represent small, medium and large objects.

Model	AP	AP_{50}	AP_{75}	AP_S	AP_M	AP_L
SSD MobileNet v2	62.2%	93.4%	68.2%	35.0%	81.9%	91.1%

Table 2. Average Recall (AR) evaluation on the test set. 1 and 10 indexes indicate the maximum number of objects per image, while AR_S, AR_M and AR_L represent small, medium and large objects.

Model	AR_1	AR_{10}	AR_S	AR_M	AR_L
SSD MobileNet v2	47.3%	68.8%	48.5%	85.6%	93.3%

4.3 Ball Localization

For evaluating the proposed object localization approach, we chose 30 different field positions for the ball. Since errors from points closer to the robot have a higher impact than further ones for autonomous navigation, points were split into subsets according to their distance. The estimated locations from the onboard vision system and RMSE can be seen in Fig. 6.

From the given plot and RMSE measurements, errors increase with the distance to the robot. This is due to objects on screen getting smaller for further distances, resulting in less accurate bounding boxes and position estimation being more sensitive to pixel differences for distant points. Thus, we present an in-depth analysis from the four nearest positions in Table 3, showing that the proposed solution is capable of overcoming SSL Vision accuracy for locations near the robot. Angles are measured by the tangent arc of relative XY positions.

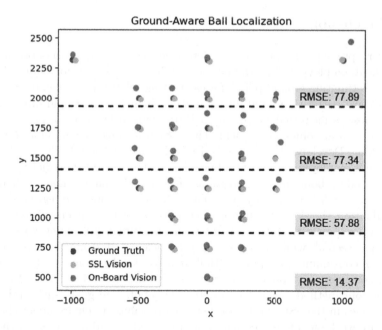

Fig. 6. Ball relative positions were estimated from the on-board detection system (green), being compared to SSL Vision (orange) and the ground truth (blue). Subsets of points were selected and the respective RMSE's were computed, reporting a 14.37 mm error for the four nearest points. The dashed lines highlight the divisions between subsets and the RMSE for the whole set of points was 67.32 mm (Color figure online).

Table 3. Comparison between estimated and ground-truth relative positions and rotation. In the last two lines, mean error and standard deviation of our method are compared to SSL Vision for the correspondent set of points. Coordinates are measured in millimeters, while angles are in degrees.

Ground Truth (x, y, θ)			On-Board Vision (x, y, θ)		
0	500	0	−0.03	508.86	0.00°
−250	750	−18.43	−259.17	762.23	−18.78°
0	750	0	−1.08	772.20	−0.08°
250	750	18.43	247.12	753.32	18.16°
Mean Error (Ours)			**−0.03 ± 3.39**	**3.32 ± 8.38**	**0.00° ± 0.15°**
Mean Error (SSL Vision)			15.31 ± 11.04	−11.03 ± 7.00	0.72° ± 1.12°

Accuracy results can be interpreted from comparisons to the robot and ball dimensions: SSL robots have an approximate 100 mm wide front area for grabbing and shooting the ball, which has a fixed diameter of 42.7 mm. Mean errors of 3.32 mm, in the y axis, −0.03 mm, in the x axis, and 0°, in direction, should suffice the accuracy needed for autonomously approaching the ball, for instance.

5   Conclusion

This work presents an approach for detecting and locating objects on the SSL soccer field, employing SSD MobileNet v2 CNN architecture for Object Detection and estimating ground points for calculating relative positions. Images from a chessboard pattern are used for computing intrinsic camera parameters. The camera pose is measured by solving a PnP problem for a set of 2D-3D correspondences from points on the field. The system is implemented on an NVIDIA Jetson Nano Developer Kit, enabling TensorRT acceleration. CNN-based object detection is employed, and we fit a linear regression model for predicting ground points based on bounding box positions. The pinhole camera model equation is solved for $z_w = 0$, and the object's relative XY coordinates are regressed.

Evaluation shows that our approach overcomes the current vision system accuracy for points near the camera with an average processing speed of 30 frames per second, while also respecting the league's 180 mm diameter restriction and consuming low power. Ball relative localization can be used for autonomously moving on its direction or rotating around it, for instance. In addition, our method is capable of detecting robots and goals, differently from existing ones in the SSL, while also being more robust to environment changes.

Future work include reproducing the presented procedure for estimating the robot and goal positions and searching for field lines detection solutions, which can have their relative coordinates calculated by the proposed method as well, enabling online camera pose calibration. Also, performance evaluations in a more dynamic environment, with moving objects and camera, must be done.

References

1. An, S., Zhou, F., Yang, M., Zhu, H., Fu, C., Tsintotas, K.A.: Real-time monocular human depth estimation and segmentation on embedded systems. CoRR abs/2108.10506 (2021). https://arxiv.org/abs/2108.10506
2. Baller, S.P., Jindal, A., Chadha, M., Gerndt, M.: Deepedgebench: Benchmarking deep neural networks on edge devices. CoRR abs/2108.09457 (2021). https://arxiv.org/abs/2108.09457
3. Bonin-Font, F., Burguera, A., Ortiz, A., Oliver, G.: Towards monocular localization using ground points. In: 2010 IEEE 15th Conference on Emerging Technologies Factory Automation (ETFA 2010), pp. 1–4 (2010). https://doi.org/10.1109/ETFA.2010.5641279
4. Bradski, G.: The OpenCV Library. Dr. Dobb's Journal of Software Tools (2000)
5. Cao, Z., Hidalgo, G., Simon, T., Wei, S., Sheikh, Y.: OpenPose: realtime multi-person 2D pose estimation using part affinity fields. CoRR abs/1812.08008 (2018). http://arxiv.org/abs/1812.08008
6. Fernandes, R., Rodrigues, W.M., Barros, E.: Dataset and benchmarking of real-time embedded object detection for RoboCup SSL. In: Alami, R., Biswas, J., Cakmak, M., Obst, O. (eds.) RoboCup 2021: Robot World Cup XXIV, pp. 53–64. Springer International Publishing, Cham (2022)
7. Franklin, D.: Jetson Nano brings AI computing to everyone. https://developer.nvidia.com/blog/jetson-nano-ai-computing/

8. Franklin, D.: Jetson Nano: deep learning inference benchmarks. https://developer. nvidia.com/embedded/jetson-nano-dl-inference-benchmarks
9. Hasselbring, A., Baude, A.S.: Soccer field boundary detection using convolutional neural networks. In: RoboCup (2021)
10. Litzelmann, R., Ratzel, M.: Robust on-board image recognition for autonomous robot-ball interaction. Technical report, Baden-Wuerttemberg Cooperative State University Mannheim (2020)
11. Liu, W., et al.: SSD: single shot multibox detector. CoRR abs/1512.02325 (2015). http://arxiv.org/abs/1512.02325
12. Meneghetti, D.D.R., Homem, T.P.D., de Oliveira, J.H.R., da Silva, I.J., Perico, D.H., da Costa Bianchi, R.A.: Detecting soccer balls with reduced neural networks: a comparison of multiple architectures under constrained hardware scenarios. CoRR abs/2009.13684 (2020). https://arxiv.org/abs/2009.13684
13. O'Keeffe, S., Villing, R.C.: A benchmark data set and evaluation of deep learning architectures for ball detection in the RoboCup SPL. In: RoboCup (2017)
14. Poppinga, B., Laue, T.: JET-Net: real-time object detection for mobile robots. In: Chalup, S., Niemueller, T., Suthakorn, J., Williams, M.-A. (eds.) RoboCup 2019. LNCS (LNAI), vol. 11531, pp. 227–240. Springer, Cham (2019). https://doi.org/ 10.1007/978-3-030-35699-6_18
15. RoboCup: RoboCup 2021 SSL Vision Blackout technical challenge rules. https:// robocup-ssl.github.io/technical-challenge-rules/2021-ssl-vision-blackout-rules.pdf
16. Sandler, M., Howard, A.G., Zhu, M., Zhmoginov, A., Chen, L.: Inverted residuals and linear bottlenecks: Mobile networks for classification, detection and segmentation. CoRR abs/1801.04381 (2018). http://arxiv.org/abs/1801.04381
17. Seel, F., Jut, S.: On-board computer vision for autonomous ball interception. Technical report, Baden-Wuerttemberg Cooperative State University Mannheim (2019)
18. Shi, W., Cao, J., Zhang, Q., Li, Y., Xu, L.: Edge computing: vision and challenges. IEEE Internet Things J. 3(5), 637–646 (2016). https://doi.org/10.1109/JIOT.2016. 2579198
19. Szemenyei, M., Estivill-Castro, V.: ROBO: robust, fully neural object detection for robot soccer. CoRR abs/1910.10949 (2019). http://arxiv.org/abs/1910.10949
20. Teimouri, M., Delavaran, M.H., Rezaei, M.: A real-time ball detection approach using convolutional neural networks. In: Chalup, S., Niemueller, T., Suthakorn, J., Williams, M.-A. (eds.) RoboCup 2019. LNCS (LNAI), vol. 11531, pp. 323–336. Springer, Cham (2019). https://doi.org/10.1007/978-3-030-35699-6_25
21. Weinmann, F.: Creating a development and deployment infrastructure for the TIGERs Mannheim on-bot vision software. Technical report, Baden-Wuerttemberg Cooperative State University Mannheim (2021)
22. Zappel, M., Bultmann, S., Behnke, S.: 6D object pose estimation using keypoints and part affinity fields. In: RoboCup (2021)
23. Zhang, Z.: A flexible new technique for camera calibration. IEEE Trans. Pattern Anal. Mach. Intell. 22(11), 1330–1334 (2000). https://doi.org/10.1109/34.888718
24. Zickler, S., Laue, T., Birbach, O., Wongphati, M., Veloso, M.: SSL-vision: the shared vision system for the RoboCup small size league. In: Baltes, J., Lagoudakis, M.G., Naruse, T., Ghidary, S.S. (eds.) RoboCup 2009. LNCS (LNAI), vol. 5949, pp. 425–436. Springer, Heidelberg (2010). https://doi.org/10.1007/978-3-642-11876-0_37

Adaptive Team Behavior Planning Using Human Coach Commands

Emanuele Musumeci[1]([⊠])[iD], Vincenzo Suriani[1][iD], Emanuele Antonioni[1][iD],
Daniele Nardi[1][iD], and Domenico D. Bloisi[2][iD]

[1] Department of Computer, Control, and Management Engineering,
Sapienza University of Rome, Rome, Italy
{musumeci,suriani,antonioni,nardi}@diag.uniroma1.it
[2] Department of Mathematics, Computer Science, and Economics,
University of Basilicata, Potenza, Italy
domenico.bloisi@unibas.it

Abstract. In its operating life, an agent that needs to act in real environments is required to deal with rules and constraints that humans ask to satisfy. The set of rules specified by the human might influence the role of the agent without changing its goal or its current task. To this end, classical planning methodologies can be enriched with temporal goals and constraints that enforce non-Markovian properties on past traces. This work aims at exploring the application of real-time dynamic generation of policies whose possible trajectories are compliant with a set of *Pure-Past Linear Time Logic* rules, introducing novel human-robot interaction modalities for the high-level control of strategies for multiple agents. For proving the effectiveness of the proposed approach, we have carried out an evaluation on a partially observable, unpredictable, and dynamic scenario: the RoboCup soccer competition. In particular, we exploit human indications to condition the robot's behavior before or during the time of the match, as happens during human soccer matches.

Keywords: Plan conditioning · Multi-agent planning · Robot soccer

1 Introduction

The flexibility of a robotic player's behavior is a key point in a soccer competition like the RoboCup. Changing team strategy in real-time can be hard to achieve given the fact that in most leagues, state-machine behaviors are still predominant [3], and, even when the deployed behavior is learning-based, the resulting policies can suffer in challenging and dynamic environmental conditions. In order to enable a team behavior to receive external conditioning, we propose a planner system capable of accepting constraints in real-time from the external environment. This allows the online adaptation of team behaviors (even during matches) with the potential outcome of modifying the collective strategy of the team, as shown in Fig. 1.

ⓒ The Author(s), under exclusive license to Springer Nature Switzerland AG 2023
A. Eguchi et al. (Eds.): RoboCup 2022, LNAI 13561, pp. 112–123, 2023
https://doi.org/10.1007/978-3-031-28469-4_10

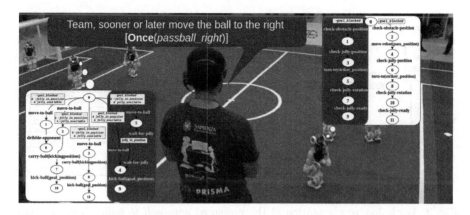

Fig. 1. A use case of the proposed architecture. The human can suggest possible modifications to the team strategy that are grounded over $PLTL_f$ templates.

The application on which we are going to focus is represented by the real-time, in-game interaction with a coach. External guidance is a determinant factor in all human sports. Real-time conditioning may originate from humans that have more information about the current scenario (e.g., a coach or the crowd's reaction to an in-game situation [4]), providing high-level instructions to the team. In human psychology, for instance, we have several examples of how important the coach is in influencing the athletes [13, 17]. In the path to reach the 2050 official goal, it might be beneficial to explore the possibilities of the introduction of adaptive team behaviors using human coach commands. An attempt at involving a human coach in high and low-level control of a team during matches was already made in Simulated 3D League. The possibility of using a coach is not novel even in RoboCup SPL, since a robot coach has been allowed in Standard Platform League (SPL) in the past. The attempt did not achieve resounding success, ruling out the robot coach from the official rulebook. This experience demonstrated that it is extremely difficult to blend past experience and perceptions to condition the actions of an entire team in real-time. To achieve this ambitious goal, human-robot interaction (HRI) can be an intermediate step and an important component worthy of being investigated in such a scenario.

In the adoption of suggestions from human beings, one of the main limitations has been represented by the different planning depths in robots and humans. To this end, performing combined planning, blending the two, can limit the capacities of the usual planner systems. To exploit the human-robot interaction capabilities in RoboCup SPL, we present a novel architecture to condition team behaviors based on the combined use of non-deterministic planning and a set of Pure-Past LTL (PPLTL) rules, also known as $PLTL_f$ rules in literature, which are used to express temporal goals on *finite non-empty traces*. The application allows for real-time generation of non-deterministic policies during a RoboCup SPL match, as shown in Fig. 1. Even though the interaction with the human is

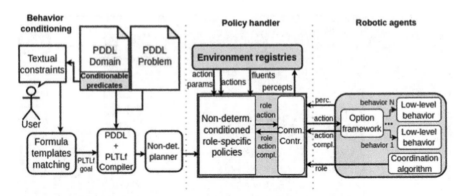

Fig. 2. Work flow representation of the presented approach to online behavior conditioning for robot soccer teams. The user (i.e. a coach) can set vocal constraints to modify the team behavior.

not allowed under the current set of rules, we think that it is worth to be investigated, as high-level commands might give way to new methodologies for team behavior conditioning. In particular, we deployed online conditioning through PLTL$_f$ rules expressing temporally extended goals and non-Markovian properties over traces, natively, using past-temporal operators as shown in Fig. 2. The main contributions of the present work are: the use of PLTL$_f$ over finite past traces to improve over previous planning approaches using LTL$_f$ constraints; the deployment of non-deterministic planning, which is necessary to model the uncertainty of conditions in which actions are performed; the conditioning of the overall strategy by human-understandable commands encoding non-Markovian temporal properties; the release of the presented software as a tool for additional experimentation, built on top of the B-Human framework [16], with additional modules from the SPQR team. The code and the additional material mentioned in this paper is released on https://sites.google.com/diag.uniroma1.it/robocupcoach.

The rest of the paper is organized as follows: in Sect. 2, we survey the current state of the art and compare it to our approach; in Sect. 3, we expose a brief theoretical helpful background to fully understand the concepts expressed in the paper; in Sect. 4 we show more in detail the proposed method; in Sect. 5 we illustrate two use cases of the presented system; in Sect. 6 we discuss the experimental results obtained; finally in Sect. 7 we discuss the conclusion obtained and the possible future developments of this work.

2   Related Work

The first attempt at creating a language to coach a RoboCup team can be found in [15], where *COACH UNILANG* is presented for Simulated 3D League. COACH UNILANG is a standard language for coaching robot soccer teams that enables high-level and low-level coaching, including tactics, formations used in each situation, and giving instructions. An evolution of that language is represented by GOL [14] (Group Organizing Language), a novel language, league independent, that allows designing tactical instructions in robot soccer through a formalization of the Tactical Instruction for RoboCup players. To model those strategies in RoboCup, several approaches have been adopted over the years. Planning-based approaches have been widely deployed in different leagues of the competition. All these approaches cannot easily capture procedural constraints on executions. Linear dynamic logic on finite traces (LDL_f), as an extension of LTL_f by the means of regular expressions, can capture procedural constraints during the execution. In this case, the constraints are typically expressed as regular expressions that must be fulfilled by traces. The logics LTL_f/LDL_f are also used to express non-Markovian rewards/goals in extensions of MDPs [5].

Online generation of plans can be time-consuming. There have been previous approaches handling the generation of behaviors starting from pre-determined plans [2]. To this end, [8] proposes FOND Planning with Linear Temporal Logic over finite traces with temporally extended goals in Fully Observable Non-Deterministic (FOND) domains. To simplify the solving of FOND planning for $LTL_f/PLTL_f$, in [7] $FOND4LTL_f$ is presented, an architecture that compiles FOND planning for LTL_f and $PLTL_f$ goals into standard FOND planning and computes the associated policy.

3   Background

Planning Domain Definition Language (PDDL) is a family of languages for the definition of planning problems. There are now many versions of PDDL available with different levels of expressivity. PDDL1 [1] is the first version of the language that has been released. A predicate logic way of modeling drives it. The model creator defines a set of actions. Each action has a set of preconditions that have to be matched. Actions modify the world's state, the effect of the action on the modeled environment is specified as the effect of the action. Both preconditions and effects are expressed in predicates logic. This logic can be extended to include $\land$, $\lor$, $\neg$, $\implies$, and the other traditional logical operators, allowing to express several complex concepts. The ultimate aim in planning is to achieve some goal state, which is also expressed as a predicate formula. PDDL2 [10] extended PDDL1 by introducing action time durations and numeric fluents while PDDL3 [12] introduced soft constraints to AI planning.

Linear Temporal Logic (LTL) [18] proposes an extension of modal logic in which worlds are organized in an infinite linear structure: each world represents a discrete moment in time. In temporal logic, the evaluation takes place within a set of worlds. Thus, a predicate may be satisfied in some worlds but not in others. How to navigate between the worlds depends on the specified view of time. A temporal accessibility relation between worlds captures the particular model of time. Given a trace $\tau = s_1, ..., s_n$, that is a sequence of states, where s_i at instant i is a propositional interpretation over an alphabet of propositions,

LTL extends the propositional logic by adding temporal operators over states in the trace: X or $\circ$ (*"Next"*), applied to a logical proposition P, is verified in a state s_t only if P is verified in the state s_{t+1}; U (*"Until"*), applied to a couple of propositions P_1 and P_2, is verified if and only if $\exists t_n$ such that $P_2(s_{t_n})$ is true and $P_1(s_i)$ is true $\forall i < t_n$; F or $\diamond$ (*"Eventually"*), applied to a single predicate P, is verified in a state s_t only if there is a state s_{t+i} such that $P(s_{t+i})$ is verified; G or $\square$ (*"Globally"*), applied to a proposition P, is verified in a state s_t only if $P(s_{t+i})$ is verified for all the successors of the state s_t.

Past Time Linear Temporal Logic. Although it is possible to obtain a past-time specification in LTL [11], Pure-Past LTL (PPLTL) describes past-time relationships directly, introducing the operators Y (*"Yesterday"*), S (*"Since"*), O (*"Once"*) and H (*"Historically"*), corresponding to the future operators X, U, F, G, respectively. PPLTL keeps the same expressiveness of LTL although the worst-case complexity of FOND Planning for PPLTL goals is EXPTIME-complete in both the domain and the goal formula, instead of the 2EXPTIME-complete complexity for LTL$_f$ goals [6].

Planning for PPLTL Goals. Following [9] and [8], in classical planning for LTL$_f$ goals, a plan satisfying the LTL goal formula is obtained by first building the deterministic automaton for the planning domain and the nondeterministic automaton for the goal formula, computing their product, and then checking the non-emptiness of the resulting automaton. Following [8], in FOND planning, a DFA game is to be solved on the aforementioned cross-product, to obtain a policy. The result presented in [6] instead converts a *PPLTL* goal ϕ into a set of sub-formulas Σ_ϕ, using only the Y and S past-time operators, such that the evaluation of the PPLTL goal only depends on the propositional interpretation of Σ_ϕ in the current state and the truth value of the other sub-formulas computed up to the previous state. Following this idea, the original planning problem Γ is compiled into a new planning problem Γ', with $\mathcal{D}'$ being a new planning domain where each sub-formula in Σ_ϕ is represented by an additional fluent. This domain is then translated into PDDL, using *derived predicates* to represent the truth value of each sub-formula. The *PPLTL* goal itself is represented as a derived predicate and effects for the already existing domain actions are modified so that they also update predicates associated to propositions in Σ_ϕ. After this compiling, which is shown to have a *polynomial* complexity, any off-the-shelf planner supporting these syntactic devices (such as FastDownward or MyND, which was used in the case at hand) can be used to solve the compiled planning problem Γ'. As shown in [6], the obtained plan or policy is a solution to the original planning problem Γ and requires no further manipulation.

4   Proposed Approach

Behavior engineering is a crucial task when dealing with a robot soccer player. Behaviors are usually hand-crafted and modeled as tree structures, covering all possible cases and encoding team strategies as an emergent property of the multi-agent system itself. This trending habit tends to over-complicate the generation

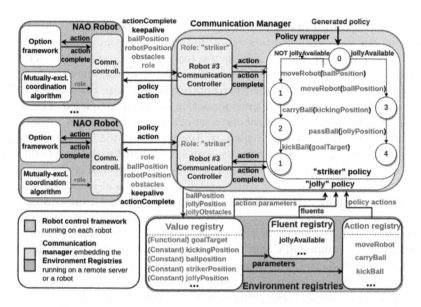

Fig. 3. Overall software architecture. This example, involves policies for Striker and Jolly roles, implementing a collective behavior for passing the ball or scoring.

of context-specific team strategies and task-allocation-based approaches. In the implemented approach, we focus on obtaining dynamic plans through constraints applied using Pure-Past LTL. Given an existing PDDL domain, a new domain embedding the PPLTL constraint is compiled, with the tool presented in [7]. The compiled domain can then be used with any off-the-shelf PDDL planner to generate robot behaviors that are compliant with the PPLTL constraints. In scenarios where fluents from the environment are needed to model conditions that are not known at planning time, policies can be obtained by non-deterministic PDDL domains, where `oneof` constructs are used in the post-conditions of actions to enumerate the set of their possible unpredictable results. Plans generated with non-deterministic planners from such a domain are robust for a set of unpredictable outcomes. We used the MyND planner for our experiments. When the non-deterministic post-conditions contain fluents, i.e., predicates whose value depends on the execution environment, policy execution is directly determined by the agent's own world model and percepts.

4.1 Architecture

The overall architecture is depicted in Fig. 3. The high modularity allows for an easy extension of its components. The architecture features three main parts.

(1) **The robot control framework** instances running on each robot. The framework manages the execution of received actions and announces their completion. Robots communicate at a lower level to ensure synchronization of percepts and to allow role assignment using a mutually exclusive coordination algo-

rithm. **(2) A communication manager** that wraps communication controllers (each one statically assigned to a robot and managing communication with it) and a policy for each robot's role in the team. The current policy action for each role is sent to the robot assigned to that role. Role assignment is performed autonomously by robots, and their percepts are labeled with the robot role. Policies are updated only when the previous action has been completed. The current set of percepts, updated by the robots, a set of selected fluents, and robot actions are stored in the Environment Registries. Fluents, computed from the latest percepts, are stored in the *Fluent Registry*. Percepts, labeled with the role that sent them, are stored in the *Value Registry* and are retrieved during the evaluation of fluents. The *Action registry* stores action templates and instances, mapping them to the available robot skills. **(3) A non-deterministic planning module** with behavior conditioning. Textual or vocal commands are translated to PLTL$_f$ rules by matching them with pre-defined templates, selected according to keywords in the command. Given the expected command structure, conditioned predicates are extracted. The formula is then generated from the retrieved data. For each role, only a chosen set of domain predicates is constrainable and only commands constraining those predicates will be considered valid.

Several constraints can be specified for each role, and the final temporal goal is obtained as their conjunction with the original goal (which is fixed for the case at hand but can be specified in the same way). Given an existing role-specific and context-specific PDDL domain, a new domain that encodes the given conditioning constraint is compiled, following the technique in [6], and a policy is then generated and mapped to the specific role. This process is performed every time a new constraining command is received.

The modularity of this architecture allows running the communication and planning modules directly on each robot. The generated policies will therefore be correctly mapped to the role currently assigned to the local robot.

4.2   Plan Generation and Execution

The proposed solution obtains dynamic behavior generation by harnessing both the power of non-deterministic planning, and the conditioning of behaviors with constraints expressed in Pure-Past Linear Temporal Logic, which encode non-Markovian properties on the trace. As demonstrated in [6], the advantage of using PLTL$_f$ over LTL$_f$ is that the same worst-case computational complexity as in classical planning can be obtained for both deterministic and non-deterministic domains, giving an exponential advantage with respect to LTL$_f$, while keeping the same expressiveness. In our case, policies, generated as in [6], are modeled as graphs where outgoing edges are labeled with fluents.

The main problem with the mentioned approaches is that generated policies are not ready to be executed with temporally-extended actions: PDDL actions are instead considered instantaneous. To solve this problem, percepts sent by the robots are stored in "environment registries". In this way, the current state of a policy is updated only when the completion of the previous action is notified. Fluents for outgoing edges from the current state are evaluated by recovering the necessary percepts, and only one outgoing edge will have verified fluents (by construction). The appropriate edge is selected, and the associated action is sent to the robot. An example of such a policy is shown in Fig. 4, that represents a "striker" which, depending on the non-deterministic fluents that model the presence of an opponent blocking the opponent's goal and the availability of a robot with a "jolly" role (which always tries to receive a pass), decides whether to pass the ball, dribble the opponent or try and kick directly to the goal.

4.3   PLTL$_f$ Temporal Goals over PDDL Domains

Given the role-specific goal g and a mapping between PDDL predicates and their corresponding PLTL$_f$ atoms (such as the PDDL predicate `isat robot waypoint` maps to the atom $isat\_robot\_waypoint$ in the formula), the conditioned goal is obtained as the conjunction of g with the constraining formula. For example, in the temporal goal $g \wedge O(P_1) \wedge H(P_2)$, the trace reaching goal g, is constrained to entail both the requirement that the predicate P_1 happens "*at least once*" and that the predicate P_2 is always verified. Textual command templates featured in our experiments are:

- "['*at least once*' | '*once*' | '*sometimes*' | '*sooner or later*'] $\{P\}$", where P is chosen from a subset of constrainable PDDL predicates, maps to $O(P)$, imposing that the condition expressed by P is verified at least once along the past finite trace.
- "['*at all times*' | '*always*' | '*historically*'] $\{P\}$" maps to $g \wedge H(P)$. The condition expressed by ϕ has to be always verified along the finite trace preceding the goal g.
- "['*never*' | '*avoid*'] $\{P\}$", maps to $g \wedge H(\neg P) \iff g \wedge \neg O(P)$. The condition expressed by P has to never be verified along the finite trace preceding g.

It should be noted that the last two templates are fit to be used as safety rules, stating that some condition has to always (or never) be verified.

5   Examples

To show the versatility of this approach, we propose some use cases.

Single-Agent Examples. Our PDDL domain for a naive striker behavior features three actions: `moverobot`, `kickball` and `carryball`. In our case, `kickball` and `carryball` have the same post-conditions, but they are linked to different low-level skills in the robot (a kick and a dribbling skill). The goal for the striker is to have the ball at the `goaltarget`. To condition the policy, at least one constrainable predicate is needed: in our case, the only conditionable predicate is `isat`, modeling the position of the ball or the robot. The user is allowed to specify a role-specific constraint. Initially, the goal $isat\ ball\ goaltarget$ is unconstrained, resulting in a simple plan requiring the robot to reach the ball and then carry it to the goal. The robot can then be forced to carry the ball to the `kickingposition`

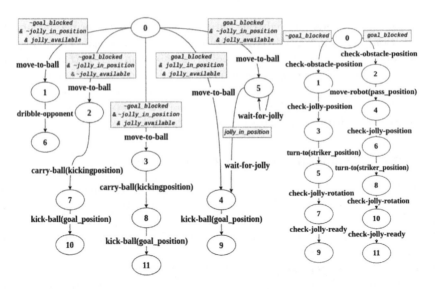

Fig. 4. Non-deterministic striker (left) and jolly (right) policies for the multi-robot example. Fluents are in the gray boxes while actions are the edge labels.

waypoint before carrying it to the goal with the constraint "*at least once isat ball kickingposition*", translated to PLTL$_f$ as $O(isat\_ball\_kickingposition)$ and added as a conjunct to the original goal. All domain objects (such as locations) have to be grounded using the *Environment registries*. Robots perform policy actions, executing their corresponding low-level atomic behaviors implemented in an *option framework* (which is the common approach used in SPL).

A Multi-agent Example. In our multi-agent scenario, the policy for the "striker" and the "jolly" role, which receives a pass, are obtained from non-deterministic domains, both shown in Fig. 4. The jolly turns to the striker if it is already reachable for a pass or reaches its waiting position otherwise. The striker initially reaches the ball, and then, depending on the current situation on the field, it chooses the best policy branch. Without opponents, the striker attempts to kick to goal; if a jolly is not available, it tries to dribble the opponent and bring the ball to the goal; otherwise, it waits for the jolly to be in position, and then it passes the ball. Once the jolly receives the ball, it automatically becomes the striker, according to the role-assignment algorithm running at a lower level on robots, and the policies are reset and reassigned to the respective robots.

6   Experimental Evaluation

The evaluation of the proposed approach has been carried out using real and simulated RoboCup environments. In order to evaluate planning time over an increasing planning depth, a simple simulated RoboCup environment was used, with a single robot starting from one side of a soccer field, tasked with bringing

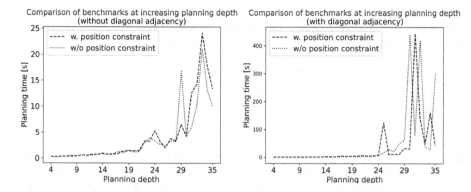

Fig. 5. A comparison between the plot for planning time over an increasing planning depth in a domain where adjacency between waypoints is modeled (on the right, adjacency is considered along diagonals as well). We can notice how including a $PLTL_f$ constraint forcing the robot to pass through a waypoint lying on the optimal path to the goal leads to generally lower planning times.

the ball to the opposite side. Conditioning can be imposed to the whole team, but to evaluate the timing we only analyze the planning performance of the *Striker*, that is the most active role in our architecture. The planning domain is modeled in PDDL as a square grid of waypoints on the field and the robot is only allowed to move the ball between adjacent waypoints, with adjacency modeled by a specific predicate (*"adjacent waypoint1 wapoint2"*). The goal waypoint is placed on the opposite side of the field, such that by increasing the number of grid cells per-side, the length of the shortest path required to reach the goal increases as well and can be used to control the minimal planning depth for a successful plan. Fig. 5 shows how significantly the performance can decrease with a different representation of the environment and how the external guidance can impact the performance of the planner. We evaluated the approach using a simplified adjacency grid (on the left), that does not consider diagonally adjacent waypoints, and a complete one (on the right), considering the whole neighborhood. In both scenarios, planning time was evaluated with and without an additional constraint forcing the robot to pass through a waypoint located approximatively in the middle of the shortest path, expressed using the *"O(is_at ball waypoint)"* constraint. Grid size (therefore the expected minimal planning depth) varies between 4 and 35 in all cases. In the simplified representation, the conditioning does not significantly affect the planning time. Rather, with the complete waypoint adjacency representation, the planning time increases around 20 times, and the improvement given by the conditioning becomes relevant, as can be seen in the plot. Furthermore, the conditioning reduces the search space and improves the planner's performance.

To qualitatively evaluate also the benefit of the conditioning in a Multi-agent scenario, the field was instead modeled such that the robot would try to move the ball between waypoints located respectively at the initial robot position,

Table 1. Results of 10 matches between the human conditioned team and the not conditioned one used as baseline. The human coach conditioned the overall strategy by exploiting the weakness of the opponent team strategy.

	Win %	Average goals per match	Total Goals %
Coach guided behavior	80	2.2	68.75
Baseline	20	1	31.25

the goal target (which has to be reached by the ball as a final goal) and other waypoints regularly arranged along the field. We ran 10 test matches between the guided team and the unconditioned one as a baseline. The baseline behavior consisted in moving the ball between waypoints by kicking it, leaving it slightly unprotected during approach maneuvers. The human coach was able to notice and exploit this vulnerability by forcing the striker robot to always "dribble" the ball to a flank before kicking to goal. The ball is therefore moved more slowly but also more protectively, allowing the robot to exploit the slower opponent approach times, occasionally stealing the ball from the opponent. The results are shown in Table 1. A test with real robots was successfully performed as well, confirming the robustness of this system in a real environment.

7   Conclusions and Future Directions

A key aspect to achieve a successful integration of robots in complex scenarios is to make robots able to perceive and understand the environment around them, conditioning the behavior given indications or rules that can only be acquired during the task. This paper lays the foundation for working on a higher level of abstraction in the decision-making process that can condition the strategies of a robot team through the use of intelligible commands. It uses a modular architecture that is easy to adapt to different purposes and teams as it is based on one of the most popular frameworks in the RoboCup@Soccer SPL competition. Furthermore, the use of hard and soft constraints also allows for adapting the given commands to different areas, such as robot security, allowing to model strategies that can ensure the safety of both the robot and any human operators working in contact with the robot itself. In the future, it would be interesting to extend this work to create a system capable of automatically learning a domain from natural language, for example, dynamically modifying the behaviors of robots from the RoboCup regulation of the current year. In conclusion, this work is a first step towards using and learning new forms of interaction and conditioning between natural language and robot behavior. This allows the creation of new strategies to generalize and deal dynamically with unexpected and complex situations such as those that the RoboCup environment can create.

References

1. Aeronautiques, C., et al.: PDDL—the planning domain definition language (1998)

2. Antonioni, E., Riccio, F., Nardi, D.: Improving sample efficiency in behavior learning by using sub-optimal planners for robots. In: Alami, R., Biswas, J., Cakmak, M., Obst, O. (eds.) RoboCup 2021. LNCS (LNAI), vol. 13132, pp. 103–114. Springer, Cham (2022). https://doi.org/10.1007/978-3-030-98682-7_9
3. Antonioni, E., Suriani, V., Riccio, F., Nardi, D.: Game strategies for physical robot soccer players: a survey. IEEE Trans. Games **13**(4), 342–357 (2021)
4. Antonioni, E., Suriani, V., Solimando, F., Nardi, D., Bloisi, D.D.: Learning from the crowd: improving the decision making process in robot soccer using the audience noise. In: Alami, R., Biswas, J., Cakmak, M., Obst, O. (eds.) RoboCup 2021. LNCS (LNAI), vol. 13132, pp. 153–164. Springer, Cham (2022). https://doi.org/10.1007/978-3-030-98682-7_13
5. Camacho, A., Triantafillou, E., Muise, C., Baier, J.A., McIlraith, S.A.: Nondeterministic planning with temporally extended goals: LTL over finite and infinite traces. In: Thirty-First AAAI Conference on Artificial Intelligence (2017)
6. De Giacomo, G., Favorito, M., Fuggitti, F.: Planning for temporally extended goals in pure-past linear temporal logic: a polynomial reduction to standard planning (2022). https://doi.org/10.48550/ARXIV.2204.09960
7. De Giacomo, G., Fuggitti, F.: FOND4LTL: fond planning for LTL//PLTL/goals as a service (2021)
8. De Giacomo, G., Rubin, S.: Automata-theoretic foundations of fond planning for LTLF and LDLF goals. In: IJCAI, pp. 4729–4735 (2018)
9. De Giacomo, G., Vardi, M.Y.: Linear temporal logic and linear dynamic logic on finite traces. In: Proceedings of the Twenty-Third International Joint Conference on Artificial Intelligence, IJCAI 2013, pp. 854–860. AAAI Press (2013)
10. Fox, M., Long, D.: PDDL2.1: an extension to PDDL for expressing temporal planning domains. J. Artif. Intell. Res. **20**, 61–124 (2003)
11. Gastin, P., Oddoux, D.: LTL with past and two-way very-weak alternating automata. In: Rovan, B., Vojtáš, P. (eds.) MFCS 2003. LNCS, vol. 2747, pp. 439–448. Springer, Heidelberg (2003). https://doi.org/10.1007/978-3-540-45138-9_38
12. Gerevini, A., Long, D.: Preferences and soft constraints in PDDL3. In: ICAPS Workshop on Planning with Preferences and Soft Constraints, pp. 46–53 (2006)
13. Gillet, N., Vallerand, R.J., Amoura, S., Baldes, B.: Influence of coaches' autonomy support on athletes' motivation and sport performance: a test of the hierarchical model of intrinsic and extrinsic motivation. Psychol. Sport Exercise **11**, 155–161 (2010)
14. Hofmann, M., Gürster, F.: GOL-a language to define tactics in robot soccer. In: Proceedings of the 10th Workshop on Humanoid Soccer Robots, in Conjunction with the IEEE-RAS International Conference on Humanoid Robots (HUMANOIDS) (2015)
15. Reis, L.P., Lau, N.: COACH UNILANG - a standard language for coaching a (Robo) soccer team. In: Birk, A., Coradeschi, S., Tadokoro, S. (eds.) RoboCup 2001. LNCS (LNAI), vol. 2377, pp. 183–192. Springer, Heidelberg (2002). https://doi.org/10.1007/3-540-45603-1_19
16. Röfer, T., et al.: B-Human team report and code release 2021 (2021). Only available online http://www.b-human.de/downloads/publications/2021/CodeRelease2021.pdf
17. Sinclair, D.A., Vealey, R.S.: Effects of coaches' expectations and feedback on the self-perceptions of athletes. J. Sport Behav. **12**, 77 (1989)
18. Vardi, M.Y.: An automata-theoretic approach to linear temporal logic. In: Moller, F., Birtwistle, G. (eds.) Logics for Concurrency. LNCS, vol. 1043, pp. 238–266. Springer, Heidelberg (1996). https://doi.org/10.1007/3-540-60915-6_6

Development Track

Omnidirectional Mobile Manipulator LeoBot for Industrial Environments, Developed for Research and Teaching

Martin Sereinig[✉], Peter Manzl, Patrick Hofmann, Rene Neurauter, Michael Pieber, and Johannes Gerstmayr

Department of Mechatronics, University of Innsbruck, Technikerstr. 13, 6020 Innsbruck, Austria
{martin.sereinig,johannes.gerstmayr}@uibk.ac.at

Abstract. This paper presents the approach of the RoboCup@Work team tyrolics of the university of Innsbruck to design, develop and build a mobile manipulator with 10 degrees of freedom. The mobile manipulator LeoBot uses Mecanum wheels to enable omnidirectional movement and includes a Franka Emika Panda serial manipulator. This paper focuses on hardware development and provides information on mechanical, electronic, and mechatronic system components. Basic algorithms developed and used for the competition are briefly described.

Keywords: Robotics · Mobile manipulator · Mobile robot · Mechatronic design · Redundant robot · Real-time control · Franka Emika Panda · Mecanum wheels · Multibody dynamics

1 Introduction

Mobile robotic systems are used in modern flexible industrial solutions. The combination of serial manipulators and mobile robotic systems is called mobile manipulator [1]. Depending on the application there are a lot of different solutions for mobile manipulators regarding load capability, drive technology and dimensions. In 2012 the RoboCup@Work league [2] was introduced to support and increase development of mobile manipulators for industrial related purpose. The challenges have to be completed autonomously and contain detection, manipulation and transport of various objects as well as classical robotics challenges like the peg-in-hole task[1]. A mobile manipulator called LeoBot shown in Fig. 1, to be used in the RoboCup@Work league, was developed and will be discussed in this paper. The field of application for the robot besides the RoboCup@Work competition is research and education.

---

[1] RoboCup@Work rulebook 2022, https://github.com/robocup-at-work/rulebook.

Supported by the University of Innsbruck and the "Foerderkreis 1669 – Wissenschaft Gesellschaft".

© The Author(s), under exclusive license to Springer Nature Switzerland AG 2023
A. Eguchi et al. (Eds.): RoboCup 2022, LNAI 13561, pp. 127–139, 2023
https://doi.org/10.1007/978-3-031-28469-4_11

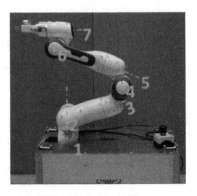

Fig. 1. Mobile manipulator LeoBot including the 7 rotational axis of the serial manipulator (red). (Color figure online)

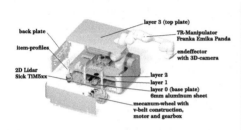

Attribute	Value
overall approx. mass	60 kg
platform length	0.71545m
platform width	0.54495m
platform height	0.2612 m
platform max. velocity	$1.5\,\mathrm{m\,s}^{-1}$
payload platform	5kg
manipulator max. reach	0.855m
manipulator end effector max. velocity.	$2.6\,\mathrm{m\,s}^{-1}$
manipulator joint (1 to 4) max. velocity	$2.6\mathrm{rad\,s}^{-1}$
manipulator joint (5 to 7) max. velocity	$3.14\mathrm{rad\,s}^{-1}$
payload manipulator	3 kg
battery Voltage	24 V
battery Capacity	20A h
mean operating time	2 h

Fig. 2. Design sketch of the mobile manipulator LeoBot with some of its main features and separation of the inner structure with different layers (left); System overview with main parameters (right)

2   LeoBot Hardware Components and Design

LeoBot can be separated into two main parts, the mobile base platform using Mecanum wheels and the mounted Franka Emika Panda serial manipulator. This section describes main parts as well as used hardware components and system properties (see Fig. 2, right). The development of this mechatronic system was done following the so called "V model" for mechatronic developments according to the VDI2206 standard. The requirements on the system where given by the rules of the targeted competition (RoboCup@Work) and additional specifications according human-robot interaction. Some main features of LeoBot are shown in a design sketch in Fig. 2 including the three layers to include electronic components. Additional information can be found in the supplementary material on the teams git-hub page (https://github.com/leobot-UIBK/LeoBotRoboCup).

2.1   Mechanical System and Construction

A laser-cut aluminum plate is used for the mobile base. To investigate the displacement at the tool center point of the dynamic system mobile base with

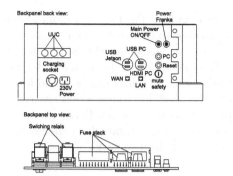
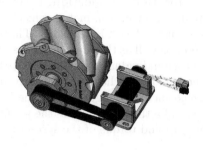

Fig. 3. Backside plate with buttons, switches and ports for easy access and backside plate top view, inside the housing with switching relays and fuse stacks (left). Drive assembly group. Mecanum wheel with timing-belt construction (right).

manipulator, a multibody simulation was done using the multibody simulation framework Exudyn[2]. The base plate was modeled as a flexible body using the floating frame of reference formulation with model order reduction as described in [3]. The deformation of the selected aluminum base plate can be seen in Fig. 5 and is negligible small. For a precise movement of the mobile manipulator it is important that all four Mecanum wheels stay in contact to the floor while driving. With the simulation, further improvements will be done to design the stiffness of the aluminum plate, such that it compensates small unevenness of the ground by small twists. In addition to the base plate, two plates are used to place electronic components onto three levels inside the robot. An additional frame construction made with aluminum profiles is used to set up the used aluminum plates to the right level.

Electronic Components and Payload Placement: The robot is build up of four layers (0 to 3) and a back plate to mount all components. This individual layer can be seen in Fig. 2 and the back plate is shown in Fig. 3. Layer 0 (base plate) contains all motors, gears and timing-belts as well as the four Mecanum wheels the Franka Emika Panda serial manipulator and the laser scanners. Layer 1 is used for the Franka Emika Panda controller and layer 2 contains the majority of electronic devices. Layer 3 (top plate) is used for additional devices and payload.

Mounting Points: The used serial manipulator is located on layer 0 (base plate) to achieve a low center of mass to avoid tilting during movements with high acceleration. Layer 0 (base plate) is not rectangular and includes special mounting points where two laser scanners are mounted (front-left and back-right) on the very outside to achieve a 360° sensor view.

---

[2] Exudyn is a C++ based Python library for efficient simulation of flexible multibody dynamics systems,https://github.com/jgerstmayr/EXUDYN.

Timing-Belt Connection, Bearings and Axles: The four Mecanum wheels are driven by maxon-motors which are connected to the wheels via a timing-belt construction see Fig. 3 (right). This includes the Mecanum wheels the electric motor as well as the timing-belt system. The calculation of the angular velocities for each individual component (e.g.Mecanum wheels, pully) can be found in [4]. The timing-belt has been choosen due to the available construction space and the geometric configuration of the motor-gear assembly group. Furthermore the usage of a timing-belt gives the possibility to reconfigure and adjust the possible velocity and torque of the system by variation of the gear ratio.

Mass Distribution and Influence on Driving Performance: The total mass is about $m_{total} = 60$ kg and the mass for individual components can be seen in Table 1. Due to the fact that it is not possible to equally distribute all components within the robot, the center of mass and the geometric center do not match. During operation, the center of mass (COM), shown in Fig. 4 (left), is also shifted depending on the configuration of the manipulator and payload. In the configurations (Fig. 5) the center of mass is shifted forward. Detailed description of the COM shift and its effect on the movement behavior can be found in [5].

Table 1. Overview of the individual component mass and mounting position.

Components	Quantities	Mass in g	Total mass g	Mounted on layer
Motor and gear unit				
Ball bearing	4	42	168	0
Cylindrical roller bearing	4	47	188	0
Bearing block	8	60	480	0
Belt pully big	4	44	176	0
Belt pully small	4	17	68	0
Electric motors and gears	4	407	1628	0
Motormount	4	162	648	0
Mecanum wheel and axle	4	2585	10340	0
Motorcontroller	4	138	552	0
Housing				
Aluminum plate top	1	1783	1783	3
Wood housing left/right	2	189	378	0–3
Wood housing front/back	2	116	232	0–3
Aluminum profile 20 × 20	10	103	1030	0–3
Aluminum profile 45°	2	220	440	0–3
Connectors and screws	13	17.2	223.6	0–3
Slot stones	13	1.8	23.4	0–3
Screws DIN 7380 M5x80	50	1.9	95	0–3

(continued)

Table 1. (*continued*)

Components	Quantities	Mass in g	Total mass g	Mounted on layer
Base unit				
Aluminum base plate	1	3606	3606	0
Aluminum plate level 2	1	1432	1432	2
Sick laser TIM7x	2	248	496	0
Sick modules	4	147	588	0
LiFePo4 battery	1	4476	4476	0
PC mainboard and power supply	1	1098	1098	2
Nvidia Jetson Nano	1	249	249	2
Switching relais	2	99.7	199.4	2
Network switch	1	248	248	2
Network router	1	233	233	2
DC/DC converter 12V	1	650	650	1
DC/DC converter 5V	1	650	650	1
DC/AC converter	1	630	630	2
Fan	2	41	82	2
Cable	1	4000	4000	0–3
Serial manipulator				
Franka Emika Panda	1	17800	17800	0
Controller	1	3989	3989	1
Endeffector	1	702	702	
Total mass			60069.4	

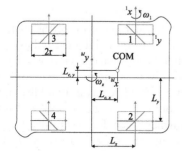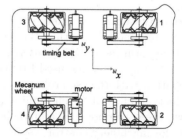

Fig. 4. Simplified geometry of LeoBot including shifted center of mass (left). Base plate including Mecanum wheels, electric motors and timing belt (right).

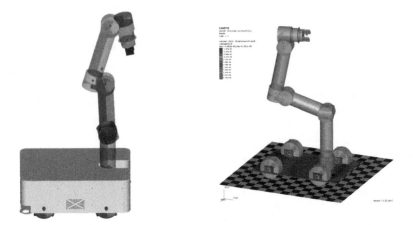

Fig. 5. Different configurations of the manipulator with a COM shift in x and y direction. $[L_{c,x} = 0.059\,\text{m},\ L_{c,y} = 0.01\,\text{m}]$ (left), $[L_{c,x} = 0.003\,\text{m},\ L_{c,y} = -0.004\,\text{m}]$ (right).

2.2   Mechatronic Systems

Mechatronic systems including the installed sensors, safety systems and the drive components are described in this section.

Sensors: The sensors shown in Table 2 can be divided into two groups. One to interact with the environment (e.g. vision, distance, force/torques,...) and one to be used to fulfill basic functionality (e.g. encoder, current sensors,...).

Safety System: The safety system is based on a combination of the included safety stop modules within the Franka Emika Panda serial manipulator and products form the company SICK for the mobile base. This company was chosen due to the fact that the used LIDAR sensors as well as the modbus module are compatible with the used robot operating system (ROS) version. In addition, the products fulfilled all given requirements (e.g. budget, size, accuracy,...). Table 3 shows these components and their safety ratings. The EN ISO 13849 safety norm for machinery control systems is applied and the safety categories (b, 1, 2, 3 and 4) are described there. Figure 7 shows the connections between the SICK components and the rest of the system.

Electric Motor and Timing-Belt. To derive required torque and rotational speed for the electric motor for the platform, assumptions for the max. velocity $v_{max} = 1\,\text{m\,s}^{-1}$, max. acceleration $a_{max} = 1\,\text{m\,s}^{-2}$ and max. mass $m_{max} = 70\,\text{kg}$ are made. Thus, we propose the following method to calculate the needed motor torque for the base platform, select the electric motor and verify the selection using a simplified dynamic model. With $F = m_{max}\cdot a_{max} = 70\,\text{N}$ the inertia force

Table 2. Overview of the used sensors on the mobile manipulator LeoBot.

Sensor Name	Measurement	Information
Intel Realsense D415	Stereo vision and depth	Mounted on end effector
Intel Realsense D435i	Stereo vision, depth and IMU	Mounted on mobile base
SICK laser scanner TIM781S	Distance	Mapping and safety
Franka Emika Panda force/torque	Joint forces and torques	Franka Emika Panda internal
Maxon rotational encoder	Rotations	Mounted on each wheel
Maxon current sensor	Current	Actual current on each wheel
Maxon temperature sensor	Temperature	Motor controller internal

Table 3. Overview of the used SICK safety system components.

Component	Information	Safety Norm, EN ISO 13849
SICK laser scanner	TIM781S	Category B
SICK IO module	FX3-XTIO84002	Category 4
SICK CPU	FX3-CPU	Category 4
SICK Modbus module	FX0 modbus TCP	No safety certification
SICK safety stop	ES11	$B_{10d} = 10^5$ switching cycles

which has to be overcome during acceleration of the mobile manipulator can be calculated assuming only movement on flat ground. Using $T_W = \frac{F \cdot r_W}{4}$ with the wheel radius $r_W = 0.075\,\text{m}$, the desired torque per wheel $T_W = 1.3125\,\text{N m}$ can be determined. With $n_{max} = \dfrac{v_{max} \cdot 60}{r_W \cdot 2\pi}$ the desired maximal wheel rotational speed $n_{max} = 127.32\,\text{min}^{-1}$ can be determined. Using n_{max} and T_W the electric motor with a power of $P_M = 75\,\text{W}$, a nominal torque of $T_N = 0.110\,\text{N m}$ and a rated speed of $n_R = 6870\,\text{min}^{-1}$ is selected. This brushless ec-motor from maxon[3] can be combined with a planetary gear and an incremental encoder. To confirm the selected motor and to ensure the torque safety ratio as well as the needed timing-belt gear transmission ratio i_R has to be calculated. Using the calculations above and the rated rotational speed n_R of the chosen motor as well as the chosen gearbox ratio $i_p = 33$ the rotational speed of the small pulley $n_{N2} = \dfrac{n_R}{i_P} = 208.18\,\text{min}^{-1}$ can be calculated. Together with the maximal desired rotational speed the gear ratio $i_R = \dfrac{n_{N2}}{n_{max}} = 1.64$ of the timing-belt system can be calculated. By the help of the overall efficiency $\eta_{ges} = \eta_P \cdot \eta_B \cdot \eta_L$ including the planet gear efficiency $\eta_P = 0.75$, the timing-belt efficiency $\eta_B = 0.98$ and the bearings efficiency $\eta_L = 0.98$, the resulting wheel torque

$$T_{eff} = T_N \cdot i_P \cdot i_R \cdot c \cdot \eta_{ges} = 3.21\,\text{Nm} \tag{1}$$

---

[3] Maxon a provider of highprecision drive systems, www.maxongroup.com, Motor: EC-i30, 539487, Gear: GP32C, 166938, Encoder: ENC16RIO4096.

can be calculated using the gear transmissions i_R and i_P and the parameter $c = 0.75$ to include additional losses of the mounted Mecanum wheels. Unknown parameters for machine elements and general approach taken from [6]. The safety factor can be calculated with $S = \frac{T_{eff}}{T_W} = 2.4$ which meets a desired safety factor of $S \geq 2$ and can be seen as sufficient for the system. Suitable to the motor-gear-encoder package the Maxon motor controller EPOS4 module was used as discussed in Sect. 2.2. The Mecanum wheels motors and timing belt mounting points can be seen in Fig. 4 (right).

Mecanum Wheels Control: Mecanumwheels are used to enable omnidirectional movability of the platform in the plane without the need of steering, thus allowing for the use of holonomic path planning, see Sect. 3.1. Each Mecanumwheel consists of rollers around the circumference which can rotate freely. The roller axes are arranged here at an angle of $45°$ with the wheel axis [7]. Each wheel is driven by a separate brushless ec motor, which angular velocity is controlled by a Maxon EPOS4 controller utilizing the fieldbus system EtherCAT. The communication with the motor controllers as follower was implemented on main unit (see Fig. 7) in C++ using the Open Source library `Simple Open EtherCAT Master` (SOEM) V1.4.0 [4]. The `leobot_base` node uses four threads to operate. One thread, running with 1 kHz with real-time priority handles the process data objects from the EtherCAT communication (sending control word, target velocity, acceleration/deceleration and receiving status, actual position/velocity/torque) by loading/storing the data in a C-structure and sending/receiving the EtherCAT frames. Another thread reads entries from the follower (EPOS4 controller) *object dictionary* for monitoring (temperature, voltage, and communication statistics), using lower priority and frequency. The ROS communication is split into different threads with medium priority. The subscriber-thread receives the velocity commands $[v_x, \; v_y, \; \omega]^T$ from the ROS system at 1 kHz, calculating the corresponding wheel velocities ω_i using the kinematics equation $\boldsymbol{\omega} = \mathbf{J}\mathbf{v}$ with

$$\mathbf{J} = \frac{1}{r_W} \begin{bmatrix} 1 & -c_c & -(L_x + c_c \cdot L_y) \\ 1 & c_c & (L_x + c_c \cdot L_y) \\ 1 & c_c & -(L_x + c_c \cdot L_y) \\ 1 & -c_c & (L_x + c_c \cdot L_y) \end{bmatrix} \tag{2}$$

and saves ω_i in the structure for the communication. The publisher thread calculates the odometry by integration of the local velocity and publishes the wheel position/velocity/torque as measured by the controller. It is important to use the parameter $c_c = 1$ according to the used Mecanum wheels O-configuration. An analysis of the motion of LeoBot is done by Manzl et al. [5,8], investigations on the movement of an individual Mecanum wheel can be found in [9]. Therein the dynamic simulations where done within the multibody simulation library Exudyn.

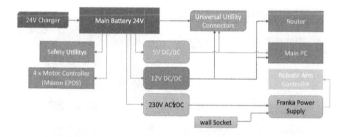

Fig. 6. Overview of different voltage levels.

2.3   Franka Emika Panda

The manipulator Panda of the company Franka Emika has 7 rotational joints as shown in Fig. 1 (right). Therefore, this system is called kinematically redundant. Each joint includes a torque sensor as mentioned in Sect. 2.2 and its total mass is $m_{panda} = 18\,kg$. Main parameters are given in Fig. 2 (left), a detailed parameter description can be found in the work of Gaz et al. [10]. The Franka Emika Panda serial manipulator can be controlled using the Franka Control Interface (FCI). Using the `libfranka` library, joint positions q, joint velocities $\dot{q}$ and joint torques τ are provided within a sample frequency $f_s = 1\,kHz$. `Libfranka` is also used by the standard Franka Emika Panda ROS stack to use the manipulator with moveIt (see Sect. 3.1).

2.4   Electronic System

To achieve a maintenance friendly and save system all connections are installed by the help of a WAGO connection panel[4] with additional fuses for each electronic sub circuit which can be seen in Fig. 3 (left, top view) on the backside plate. Different levels of voltage are necessary for the robot to work properly, therefore two DC/DC switch mode converters are used to generate 5 V and 12 V from the 24 V battery voltage. Furthermore a 230 V inverter is needed to power the Franka Emika serial manipulator. This power supply can also be provided by an external IEC 60320-C13 connector. This allows the use of the robotic arm stationary with its original software. A relay automatically switches the power source for the Franka Emika power supply and shuts down the inverter when a 230 V cable is connected to the robot. A more efficient way to generate the 48 V directly from 24 V battery voltage is, however, not allowed by the Franka Emika controller. The main voltage level is chosen as 24 V which results in an optimal voltage/current distribution with all individual electronic components. DC/DC as well as DC/AC converters are used to achieve different needed voltage levels. Figure 6 shows an overview on the electronic circuit and the different used voltage levels.

---

[4] WAGO, Electronic Interconnections, Interface Electronic and Automation Technology, www.wago.com.

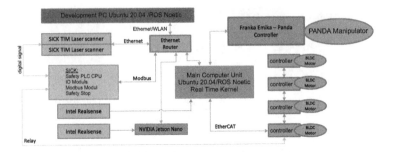

Fig. 7. Communication structure with computation units (blue), sensors (green), safety components (yellow), actuators (orange) and development PC (red). Main unit: Intel I7-6700K CPU with MINI ITX mainboard, second unit: Franka Emika Panda Controller, third unit: NVIDIA Jetson Nano GPU. (Color figure online)

Universal Utility Connector (UUC): The UUC is a multi-pin connector on the back of the robot (see Fig. 3) which provides 24 V, 12 V, 5 V, GND and 10Mbit Ethernet for external hardware which can be mounted on the robot. The 24 V pins are connected to the implemented SICK safety PLC (Programmable Logic Controller, see Sect. 2.2) and are turned off by the emergency circuit.

Batteries and Battery Charger: For power supply a LiPoFe4 battery with 24 V and 20 Ah is used. This lithium-ferrophospat cell battery technology simplifies the usage of the robot while charging the battery. It consists out of eight cells with a nominal voltage of $v_{n,cell} = 3.2 \rightarrow 3.3$ V each. The battery provides nominal current of $i_{max,b} = 25$ A and a peak current of $i_{peak,b} = 37.5$ A for a maximum duration of $t_{peak} = 20$ s. The battery dimensions are 240 mm × 154 mm × 83 mm with a mass of $m_b = 5$ kg. During laboratory tests the average current consumption of 9.8 A was measured and LeoBot can be used for approximately 2 h which is sufficient for the proposed usage in the RoboCup@Work competition as well as during research and education. To ensure a working system also during the charging process a LiPoFe4 charger with direct supply of up to 15 A from the company Victron is used.

3   Leobot Software Architecture and Operating System

All software developments can be found on the GIT repository of the university of Innsbruck RoboCup team *tyrolics*[5], including supplemental material describing the system. All software developments are made under the Berkeley Software Distribution license (BSD). Three computer units as shown in Fig. 7 are used.

---

[5] Team GIT repository https://github.com/leobot-UIBK/LeoBotRoboCup.

3.1 Robot Operating System

LeoBot is a combination of different hardware and software components and since all of these systems need to be interconnected, a common interface is crucial. The communication between different program instances on the individual computation units is established by the used robot operating system (ROS). The robot operating system (version noetic[6]) is used due to its standardization, large and active community and a variety of external libraries and software stacks as well as its large hardware support. To display results of algorithms and sensor data the ROS-tool rviz[7] is used. CoppeliaSim[8] was used to simulate the mobile platform movement to evaluate the path planning functionality [11].

Arm Motion Planning: To establish easy to use arm motion planning the MoveIt [12] framework is chosen. The framework is highly integrated into the ROS ecosystem, directly supports the Franka Emika Panda manipulator and comes with C++ and Python interfaces. A variety of services to perform complex motions, the ability to avoid self collision or collision with pre-defined obstacles are provided. To make the system more robust to unknown or changing environments in addition the point cloud that is generated by the arm depth camera can be fed into the collision map. MoveIt is mainly built to perform pre-planned motions and has only limited real-time support. However, most manipulation tasks in the RoboCup@Work do not need complex feedback during execution, therefore pre-planed motions are sufficient. The Orcos Kinematics and Dynamics Library (KDL)[9] is the standard inverse kinematic solver used in MoveIt. This solver uses a pseudo-inverse Jacobian approach, however, this seems to perform bad with robot arms that have joint limits. The TRAC inverse kinematic solver [13] uses a combination of the method used for KDL, extended by joint limit detection, combined with a Sequential Quadratic Programming (SQP) nonlinear optimization approach. Running both methods and taking the first converging solution has shown to lead to a better solve rate with comparable run-time. As solving type *Distance* is used which minimizes the sum of squared differences to find the solution with shortest motion distance. To have more flexibility and robustness while performing grasps, a custom grasp function has been built on top of MoveIt. Basic motions like approaching a pre-grasp pose are still performed by MoveIt, however it is possible to react adjusted on collisions with the environment between several steps of a grasp motion in different scenarios.

Mobile Platform Navigation: For navigation of the base platform, the ROS navigation stack functionality is used. This includes various algorithms and so-called plugins to set up individual navigation parameters for the mobile manipulator. Detailed tests and parameter setups are described in [11]. Based on the

---

[6] https://www.ros.org/.

[7] rViz, 3D visualizer for ROS.

[8] Robotics simulator CoppeliaSim, www.coppeliarobotics.com/.

[9] https://www.orocos.org/wiki/Kinematic_and_Dynamic_Solvers.html.

sensor data, 2D or 3D occupancy grid maps (costmaps) and predefined maps of the environment are used to evaluate the costs for different path to a specific goal. This occupancy grids are separated into global and local costmaps. Path planning is also separated by a global planner (planes the initial path from start to goal) and a local planner (takes additional information into account). For omnidirectional driven robots it is important to use a suitable planner as for example the teb_local_planner [14]. Which can be configured for omnidirectional movement.

Object Recognition, Computer Vision and Task Planning: To output the correct label and the corresponding bounding box of an object in an image the deep neural network You Only Look Once [15] (YOLO) is implemented on the Nvidia Jetson Nano GPU. Therefore the network is trained by minimizing the error of a predicted box and the corresponding class to a pre-labeled box and class. This also works with multiple objects present in the same picture. To order the given start and goal state of the arena to a (by the LeoBot) executable sequence of tasks the framework ROSPlan [16] is chosen. All possible actions of LeoBot like *move to* or *pick* and their change to the environment can be modeled in a domain file. The start and end state of the arena from the so-called *atwork commander* is parsed automatically into a problem file in pddl format. This problem file can be converted into a plan by considering the domain file. During execution of a plan a dispatcher calls the corresponding implementations of actions specified in the domain file.

4   Conclusion and Outlook

In this paper the technical details of a newly developed mobile manipulator platform including the mechatronic design are discussed. The RoboCup@Work team participated in 2021 successfully in the RoboCup@Work competition, showing the suitability of the presented mobile manipulator to solve the demanded tasks. LeoBots mecanumwheeled mobile base enables omnidirectional movement and by including a Franka Emika Panda serial manipulator, the resulting kinematically redundant system with 10 degrees of freedom can interact safely with its environment. In addition, the utilized manipulator with 7 degrees of freedom enables a reach of 0.855 m, which is significantly lager than the reach of the well-established Kuka YouBot, which has 5 degrees of freedom. Further, the presented system costs approx. 38000 Euros and is thus low-cost compared to equivalent commercial systems. The easy accessible sensor data and the well known parameter are a significant advantage for ongoing and future research. Future work will be done by the researchers to identify the behavior of Mecanum wheels based on laboratory evaluation shown in [5,9]. Furthermore the system will be used to investigate the base positioning problem which comes with mobile manipulators as described in [17].

References

1. Sereinig, M., Werth, W., Faller, L.-M.: A review of the challenges in mobile manipulation: systems design and RoboCup challenges. e & i Elektrotechnik und Informationstechnik **137**(6), 297–308 (2020). https://doi.org/10.1007/s00502-020-00823-8
2. Kraetzschmar, G.K., et al.: RoboCup@Work: competing for the factory of the future. In: Bianchi, R.A.C., Akin, H.L., Ramamoorthy, S., Sugiura, K. (eds.) RoboCup 2014. LNCS (LNAI), vol. 8992, pp. 171–182. Springer, Cham (2015). https://doi.org/10.1007/978-3-319-18615-3_14
3. Zwölfer, A., Gerstmayr, J.: The nodal-based floating frame of reference formulation with modal reduction. Acta Mech. **232**(3), 835–851 (2020). https://doi.org/10.1007/s00707-020-02886-2
4. Manzl, P.: Realtime Movement of a Mecanum Wheeled Robot using the Robot Operating System ROS. Master thesis, University of Innsbruck (2020)
5. Manzl, P., Gerstmayr, J.: An improved dynamic model of the mecanum wheel for multibody simulations. In: IDETC/CIE, vol. 85468, p. V009T09A031. American Society of Mechanical Engineers (2021). https://doi.org/10.1115/DETC2021-70281
6. Mott, R.: Machine Elements in Mechanical Design. Pearson/Prentice Hall, Hoboken (2004)
7. Gfrerrer, A.: Geometry and kinematics of the Mecanum wheel. Comput. Aided Geom. Des. **25**(9), 784–791 (2008). https://doi.org/10.1016/j.cagd.2008.07.008
8. Manzl, P., Sereinig, M., Gerstmayr, J.: Modellierung und experimentelle Validierung von Mecanumrädern. 8. IFToMM D-A-CH Konferenz 2022 (2022). https://doi.org/10.17185/duepublico/75419
9. Bodner, M.: Vermessung und Simulation eines Mecanumrades. Master thesis, University of Innsbruck (2021)
10. Gaz, C., Cognetti, M., Oliva, A., Giordano, P.R., De Luca, A.: Dynamic identification of the franka emika panda robot with retrieval of feasible parameters using penalty-based optimization. IEEE Rob. Autom. Lett. **4**(4), 4147–4154 (2019). https://doi.org/10.1109/LRA.2019.2931248
11. Schöffthaler, F.: Autonomous navigation of a mobile platform in simulation and real-world applications using ROS. Bachelor thesis, University of Innsbruck (2021)
12. Coleman David, T.: Reducing the barrier to entry of complex robotic software: a moveit! case study (2014). https://doi.org/10.48550/ARXIV.1404.3785
13. Beeson, P., Ames, B.: TRAC-IK: an open-source library for improved solving of generic inverse kinematics. In: 2015 IEEE-RAS 15th International Conference on Humanoid Robots (Humanoids), pp. 928–935. IEEE (2015). https://doi.org/10.1109/HUMANOIDS.2015.7363472
14. Rösmann, C., et al.: Trajectory modification considering dynamic constraints of autonomous robots. In: ROBOTIK 2012; 7th German Conference on Robotics, pp. 1–6. VDE (2012)
15. Redmon, J., Divvala, S., Girshick, R., Farhadi, A.: You only look once: unified, real-time object detection. In: IEEE Conference on Computer Vision and Pattern Recognition, pp. 779–788 (2016). https://doi.org/10.1109/CVPR.2016.91
16. Cashmore, M., et al.: Rosplan: planning in the robot operating system. Proc. Int. Conf. Autom. Plan. Sched. **25**(1), 333–341 (2015)
17. Sereinig, M., Manzl, P., Gerstmayr, J.: Komfortzone mobiler Manipulatoren. Sechste IFToMM D-A-CH Konferenz 2020 (2020). https://doi.org/10.17185/duepublico/71180

Cyrus2D Base: Source Code Base for RoboCup 2D Soccer Simulation League

Nader Zare[1]([✉]), Omid Amini[4], Aref Sayareh[5], Mahtab Sarvmaili[1],
Arad Firouzkouhi[6], Saba Ramezani Rad[6], Stan Matwin[1,2], and Amilcar Soares[3]

[1] Institute for Big Data Analytics, Dalhousie University, Halifax, Canada
{nader.zare,mahtab.sarvmaili}@dal.ca, stan@cs.dal.ca
[2] Institute for Computer Science, Polish Academy of Sciences, Warsaw, Poland
[3] Memorial University of Newfoundland, St. John's, Canada
amilcarsj@mun.ca
[4] Qom University of Technology, Qom, Iran
[5] Shiraz University, Shiraz, Iran
[6] Amirkabir University of Technology, Tehran, Iran
{arad.firouzkouhi,saba_ramezani}@aut.ac.ir

Abstract. Soccer Simulation 2D League is one of the major leagues of RoboCup competitions. In a Soccer Simulation 2D (SS2D) game, two teams of 11 players and one coach compete against each other. Several base codes have been released for the RoboCup soccer simulation 2D (RCSS2D) community that have promoted the application of multi-agent and AI algorithms in this field. In this paper, we introduce "Cyrus2D Base", which is derived from the base code of the RCSS2D 2021 champion. We merged Gliders2D base V2.6 with the newest version of the Helios base. We applied several features of Cyrus2021 to improve the performance and capabilities of this base alongside a Data Extractor to facilitate the implementation of machine learning in the field. We have tested this base code in different teams and scenarios, and the obtained results demonstrate significant improvements in the defensive and offensive strategy of the team.

Keywords: 2D Soccer Simulation · RoboCup · Base code

1 Introduction

Soccer is one of the most popular team-based sports in the world. This is a multi-player, real-time, strategic, and partially observable game in which players of each team should cooperate to score more goals. In addition to the cooperative strategy, the players should manage different tactical and technical strategies against their opponent. Designing and implementing this game in a good, realistic graphical simulation environment and encouraging researchers to develop fully autonomous players with human-like skills creates complex challenges for A.I. research. Hence, soccer is considered an exciting environment for developing A.I. and robotic algorithms to solve real-world challenges. The importance

© The Author(s), under exclusive license to Springer Nature Switzerland AG 2023
A. Eguchi et al. (Eds.): RoboCup 2022, LNAI 13561, pp. 140–151, 2023
https://doi.org/10.1007/978-3-031-28469-4_12

of soccer as a game and as a challenging domain for testing the A.I. and machine learning algorithms led to an overreaching vision of a robotic team competing against the best human team by 2050 [1].

The world Cup Robot Soccer Initiative was founded to create a realistic environment similar to real soccer that encourages researchers to employ Robotic and A.I. for solving wide ranges of problems [2]. The first RoboCup was held during the IJCAI-97 [3], and it offered three competition tracks: real robot league, software robots, and expert robot competition. Among them, the Soccer Simulation 2D league (SS2D) [4] provides a wide range of research challenges such as autonomous decision-making, communication and coordination, tactical planning, collective behaviour and teamwork, opponent modelling and behavior predicting [5–12].

In this league, the RoboCup Soccer Simulation Server (RCSSServer) executes and manages a 2D soccer game between two teams of eleven autonomous software programs(agents). It holds the complete knowledge of the game, such as the exact position of every element in the game and their movements. The game further relies on the communication between the server and each agent. Each player receives relative and noisy information about the environment, and based on its logic and algorithms, the agent produces basic commands (like dashing, turning, or kicking) to influence the environment. A visual example of the game is shown in Fig. 1. Another key component of this league is the base code[1] of agents that is responsible for communicating with the server, handling the noisy partial observability of the game, modeling the server world, and making multi-agent decisions throughout the game. Due to the complexity of these tasks, designing an operational base code can astonishingly accelerates the RSS2D teams' progress.

Over the past years, many teams have contributed to the RCSS2D community by releasing their bases, which are mentioned below. One of the first bases was from Carnegie Mellon University, a.k.a "CMUnited" in 2001 [13,14], then a windows-based team was released by "TsinghuAeolus" [15] in 2002. The release of "UvA Trilearn" base [16] in 2003, helped many teams worldwide. "Brainstormers" [17], "WrightEagle" [18] and "Marlik" [19] released their team codes in 2005, 2011 and 2012 respectively. "HELIOS-Base or Agent2D" has been released by the "HELIOS" team from AIST Information Technology Research Institute [20,21]; this is the most important, most relevant, and most frequently used publicly available source code release in soccer simulation 2d. It has been considered as the base code for many prosperous teams such as Cyrus2d [24–26] and Glider2d [27,29]. Later, "Cyrus2D" and "Gliders2D" released their 2014 [22] and 2019 [23] bases respectively that are based on agent2d.

In this paper we are planning to describe and release a more advanced base that is called Cyrus2D in three consecutive versions. We have followed the incremental strategy of evolving base code proposed by [23,30] to exemplify the impact of different approaches and to trace their functionalities. In the first

---

[1] For simplicity, throughout this paper we will use the "base" term instead of base code.

version(v0.0) we combined the newest release of Helios and Gliders bases with some modifications on their parameters. In the second version(v1.0), we have developed this base code to include three A.I.-based components of Cyrus2D that were successfully implemented in this team. Finally, in the third version(v1.1) we took advantage of Pass Prediction Deep Neural Network module for unmarking decision-making. The performance of these versions went on the rigorous evaluations against the Agent2D, Glider2D bases and the obtained outcomes (number of scored and received goals, and the winning rate) proved the prevalence of our base code. The rest of the paper is organized as follows: Sect. 2 we will define the foundation of our base (version zero), in the next section we will explain the deployment of three ideas(Blocking Strategy, Offensive Risk Evaluation, Simple Unmarking Strategy) on the Cyrus2D v0.0 which results in Cyrus2D v1.0. In Sect. 4, we will present the idea of using Pass Prediction in Unmarking Strategy(Cyrus2D v1.1). In the next section, we will compare Cyrus2D base with other Soccer Simulation 2D bases against best three teams in RoboCup 2021. Finally, we talk about our future works.

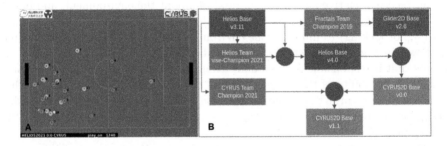

Fig. 1. A: Soccer simulation 2D league. B: The evolution of Helios2D, Glider2D and Cyrus2D base codes

2   Cyrus2D Base Version 0.0

One of the most popular SS2D bases is the Helios Base (agent2d) V3.11 which was released in 2010 [20,21]. This base includes several components such as *librcsc-4.0.0, soccerwindow2-5.0.0 and fedit2-2.0.0*. Gliders and Fractals, who use the Agent2D base, won the championship of RoboCup 2016 and 2019, respectively [27–29]. They also released a simplified version of their teams called Gliders2d base [23,30]. It is an advanced version of Helios base v3.11 with improved formation, passing behavior, and stamina management. It employs a modified version of the Marlik team [19] blocking algorithm, and few unique strategies specifically designed for each team. On the other hand, Helios has started improving its base and components such as *librcsc* based on the new versions of C++ from 2019 [31,32]. In this paper we start by introducing the first version of Cyrus2D base (V0.0). It is established by rewriting the newest version of the Agent2D by merging the latest Gliders2D base (see Figure 1[B]). This base code is fully compatible with the latest version of *librcsc*, but the blocking algorithm and tuning parameters of the Gliders2D base are removed. The Cyrus2D base is released in the Cyrus team repository and will be updated to be compatible with the *rcsserver* and *librcsc*[2]. In order to enhance the functionality of this base, we have implanted three simplified functionalities of Cyrus on this base and we introduce them as the consecutive versions of Cyrus2D base. In the following sections, we will describe these ideas.

3   Cyrus2D Base Version 1.0

3.1   Blocking Strategy [3]

As the environment of SS2D is highly dynamic and unpredictable, an innovative defensive strategy can increase the winning chance of the team. To establish the defensive strategies, we need to understand defensive actions and how players can cooperatively perform to minimize the risk of receiving the goal. Blocking and marking are two main defensive actions that prevent the opposing team from controlling the ball and playing with it. Blocking stops the progress of the opponent's ball holder on the field, and marking prevents the passing of the ball to the opposing team players. Therefore, when one of our agents tries to block the ball holder, the other players should choose to mark the opponent players. In the Cyrus2D base, we implemented multi-agent blocking decision-making. The blocking function or "Blocking Simulator" is called when the opponent owns the ball. It simulates the dribbling behavior of the opponent ball holder called the "dribbling curve" and then finds a position that one of our players can arrive in, before arrival of the opponent's ball holder and (our) players. To simulate the dribbling curve, it predicts the first position of the ball that the opponent's player can kick the ball. In the next step, it predicts the following ball positions of dribbling behavior. The dribbling speed is considered 0.7 m/s. To find the dribbling direction, we evaluate ten positions around the ball position using the reversed formulation of "Field Evaluator" in Helios base. To improve the performance of the Blocking algorithm, we implemented some conditions to prevent players from using extra stamina or going far from their home position.

3.2   Offensive Risk Evaluation [4]

To score more goals, the team's ball holder must move the ball towards the opponent's goal area, and a final striker must shoot the ball towards the goal. Dribbling and passing are examples of possible actions that can lead the ball towards the goal. Henceforth, the ball holder must choose the best action between the possible passes and dribbles. For this purpose, we need to scrutinize our base code and improve the implementation of the offensive strategy.

---

[2] https://github.com/Cyrus2D/Cyrus2DBase.
[3] This Algorithm Is Implemented in Src/bhv_basic_block.cpp.
[4] This Algorithm Is Implemented in Src/chain_action/action_chain_graph.cpp.

The Agent2D base has a decision-making algorithm called *Chain-Action*, which uses a modified version of Breadth-First Search to decide an action for the ball owner in an action graph tree. The Chain-Action has *action generator* modules such as Pass-Generator, Short-Generator, and Dribble-Generator. An *action generator* module receives a state of the game and then generates all possible actions in that state. The Chain-Action also includes a simple *predictor module* that receives a state and an action; then, it generates a new state. It simulates the possible outcome of the game after applying the received action [26].

After predicting a new state, Chain-Action evaluates the state based on the ball position using a module called *Field-Evaluator*. This module receives a state and uses the X coordinate of the ball and its distance to the opponent's goal to measure its value. To expand the tree to the next level, the chain-action chooses a node with the maximum value. An example of this procedure is shown in Fig. 2.

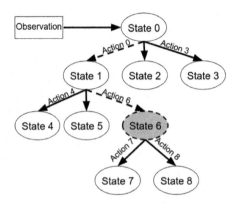

Fig. 2. Example of Chain-Action. A multi-branch tree search is performed. Each edge presents an action and each node corresponds to a state instance.

We improved the Field-Evaluator module by including a term that is subtracted to its calculation algorithms which is called *Offensive Risk Evaluation (ORE)*. This value is calculated based on the minimum number of cycles that the opponent players need to reach the ball in the input state of Field-Evaluator.

The Field-Evaluator first calculates the minimum number of cycles c that the opponent player needs to reach the ball. Then it uses an array with seven elements where the n-Th element would be the ORE term if the opponent reaches the ball in n-Th-cycles. To populate this array, we took advantage of the genetic algorithm[5].

For our task, the genetic representation is a list of seven values. A solution must be an array of seven values between 0 to 50, in descending order, as opponents closer to the ball are more dangerous. The fitness function is the average

---

[5] We reduced the array size to seven because our GA algorithm with several settings found that the eighth and following cells of the best arrays will be 0.

goal difference of Cyrus2D base in 100 games against random opponents from 10 teams of RoboCup competitions in 2021. To initialize the first population, we randomly generated 100 solutions. After evaluating with the fitness function, we generate 80 new children, from 160 parents which are selected randomly with probability based on their fitness score. After cross-over, we update the new children to possible solutions by making sure each value is less than or equal to the value before it. In the next step, some of their genes can be mutated with a low random probability. The mutation is done in a manner that the mutated solution is still considered possible.

If the generated solution is not in descending order, to preserve the validity of the solution, we replace the first occurrence of illegal value with a value lower than the previous element for example if the generated solution looks like *solution* = [10 18 5 4 3 2 1], the validity procedure transforms it to *fixed solution* = [10 9 5 4 3 2 1]. Afterwards, we create the new population of 100 by selecting 20 of the best chromosomes of the previous generation and adding the 80 new children.

We repeat this process until the population converges or until 100 iterations are evaluated.

3.3   Unmarking Strategy [6]

Unmarking is the player's ability to move, avoid being marked, and relocate himself in a space where he could receive a pass from the ball possessor.

In the unmarking algorithm, a player who wants to unmark is called the "unmarker", and the player who will pass the ball to the unmarker is called the "passer". The passer player can be a player who owns the ball or does not have ball possession at the moment, but it is possible to be a ball possessor in the future. An unmarking Strategy identifies the passer, and after identifying the passer, the unmarker should find a position to receive a pass from the passer in future cycles. An effective unmarking should consider the actions of other agents and the cooperation between them.

Cyrus2D base version 1.0 includes a simple unmarking strategy. In this algorithm, all players do unmarking for the ball possessor to receive a pass from him. After identifying the passer, the unmarker simulates ten targets in ten directions around him according to its previous movement. After generating 100 targets, it ignores targets close to teammate or opponent players and targets far from its home position. The home position is the target position of a player that is calculated based on team's formation. The next step simulates eight lead passes from the passer to itself in every target to find which target it can receive a pass. A pass has a score calculated using the "Field Evaluation" formula in the Helios base. The score of each target is calculated based on the scores of possible received passes in the position and the minimum distance of opponent players to the position. Eventually, the target with the maximum score will be selected as the unmarking target.

---

[6] This Algorithm Is Implemented in Src/bhv_unmark.cpp.

4   Cyrus2D Base Version 1.1 [7]

In Cyrus2D base Version 1.0, we improved the offensive strategy, using a novel Blocking behavior, and a simplified Unmarking strategy. In this section, we will explain the improvement on the Unmarking strategy using a module called Pass Prediction. The Pass Prediction module includes a trained DNN, that receives a state of the game, and identifies which player will be the pass receiver in that state. This module enables us to generate a tree that assigns a passer to each player in the future cycles of the game. To generate a data set for training the DNN, we employed Data Extractor module.

4.1   Data Extractor

As in real soccer, passing is one of the possible actions that can lead the ball to the goal. Predicting the pass target player, from the point of view of the ball possessor, has many benefits in defensive and offensive algorithms. In this paper, the ball possessor is the player who can kick the ball in the current cycle or receive the ball in the future cycles.

To predict the behavior of (our) ball possessor, we were required to create the dataset of game states from this player point of view. For this purpose, we embedded a Data Extractor module in each one of the players and then we recorded the features of game states and their corresponding label. The label shows the uniform number of the player who is target of the best pass [12, 25, 26].

To generate a data set, our player (ball holder) feeds the state of the game and its selected pass receiver uniform number to the Data Extractor when the ball is in its kickable area. After that it saves the features and the label in a CSV file. Later this dataset will be used to train the Pass Prediction model.

4.2   Unmarking Strategy with Help of Pass Prediction Module

To improve the Unmarking Strategy and sketching the flow of the game, each player of Cyrus2D base tries to simulate a tree that includes the probable passes and their outcoming states. Each node of the tree contains a state of the game where one of our players is the ball owner in that state. The edges from the current state shows a probable pass in the future. The root node of the tree is the first state of the game where one of our players can kick the ball. A player can not be ball possessor in more than one node of the tree. To create the tree, the unmarker feeds the state of the root node to the Pass Prediction module, and receives the probability of players for receiving a pass. This module includes a trained DNN that can receives features of the game generated by Data Extractor and gives probability of players to receive a pass in the given state. In the next step, the unmarker selects two passes with the maximum probability higher than a limit and inserts them into a list called Pass List.

---

[7] This Algorithm Is Implemented in Src/bhv_unmark.cpp, Src/data_extractor/DEState.cpp and Src/data_extractor/offensive_data_extractor.cpp.

Then after, it pops the pass with highest probability from the Pass List. Next, it simulate the outcome that it send the outcome state to Pass Prediction module and eventually insert best passes from the outcome of Pass Prediction module. This procedure continues until the number of tree nodes is equal to ten or there is not any pass in the Pass List. Figure 3 shows an overview of the Unmarking Decisioning. After termination of this procedure, the umarker agent looks for its corresponding node in the tree, then it chooses the parent player as the "Passer" for the unmarking procedure in order to receive the pass from that player in the future.

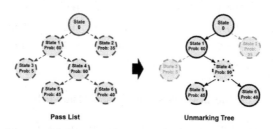

Fig. 3. Overview of the unmark decisioning algorithm. The left tree shows the result states and their points. The Bold circles (full, dotted, and long dashed) in the right tree present the selected best nodes from the candidate list. The dashed circle one indicates the node that the unmarker is the ball owner in its state, and its parent node is the dotted circle.

5   Results

5.1   Training DNN for Cyrus2D V1.1 [8]

For generating a data-set for training the "Pass Prediction DNN", we ran 500 games against Helios Base v3.11 and newest version, Gliders2D base v1.6 and v2.6, Cyurs 2021, Helios 2021, and YuShan 2021. We obtained total 1,429,032 data instances. We split them into two subsets, 85% for training and 15% for testing. The prediction model (DNN) has three layers of 128, 64, 32 and 11 neurons, with RELU activation function and a softmax function at the last layer. The validation accuracy of the trained neural network on the test data was 68.1%. We used Python TensorFlow Keras library [33] for training the model, and we implemented a library called CppDnn [34] to use the trained model in C++. The CppDnn is a C++ library powered by Eigen [37]; this library creates a deep neural network model by reading the weights of a trained DNN model.

To evaluate the impact of the implemented features and algorithms and comparing the Cyrus2D base with the HELIOS base(Ag) and Gliders2d Base(G2D), we ran X-number games between two versions of HELIOS base(3.1.1/newest), two versions of Gliders2d Base (1.6/2.6), six versions of Cyrus2D base (C2D0 = Cyrus2D zero, C2DB = Cyrus2D zero base with Blocking Strategy, C2DR

---

[8] All Scripts for Training Are Available in Scripts/training_unmark.

Table 1. Win rate

Team	H2D 3.11	H2D new	G2D 2.6	Cyrus21	Helios21	YuShan21	Average
H2D 3.11	–	26.2	3.5	0.0	0.0	0.0	4.9
H2D new	73.8	–	9.8	0.2	0.0	0.0	14.0
G2D 1.6	95.5	85.4	30.3	0.4	0.0	0.2	35.3
G2D 2.6	96.5	90.2	–	1.9	0.0	1.0	31.6
C2D 0.0	100.0	98.1	78.5	7.2	0.0	4.3	48.0
C2D B	99.6	97.2	77.0	6.4	0.0	3.6	47.3
C2D R	99.4	97.4	81.0	7.9	0.2	6.3	48.7
C2D U	100.0	99.0	80.8	5.6	0.2	5.8	48.6
C2D 1.0	99.3	98.6	79.9	8.6	0.3	4.6	48.6
C2D 1.1	99.8	99.6	84.1	5.8	0.8	4.9	49.1

Table 2. Goals scored (Goals Conceded)

Team	H2D 3.11	H2D new	G2D 2.6	Cyrus21	Helios21	YuShan21	Average
H2D 3.11	—	1.6(2.7)	0.5(3.0)	0.2(6.2)	0.1(13.0)	0.2(7.5)	0.4(5.4)
H2D new	2.7(1.6)	—	0.7(2.3)	0.3(5.9)	0.1(11.1)	0.3(6.4)	0.7(4.5)
G2D 1.6	3.5(0.7)	2.6(1.0)	0.8(1.4)	0.5(5.3)	0.1(6.5)	0.2(4.8)	1.3(3.3)
G2D 2.6	3.0(0.5)	2.3(0.7)	—	0.5(3.5)	0.1(5.5)	0.2(3.8)	1.0(2.3)
C2D 0.0	4.3(0.2)	2.8(0.3)	1.1(0.4)	0.6(2.8)	0.2(3.6)	0.3(1.9)	1.6(1.5)
C2D B	4.2(0.2)	2.9(0.2)	1.1(0.4)	0.6(2.7)	0.2(3.8)	0.3(2.4)	1.6(1.6)
C2D R	4.1(0.2)	2.9(0.3)	1.1(0.4)	0.6(2.6)	0.2(3.8)	0.3(1.7)	1.5(1.5)
C2D U	4.4(0.2)	3.2(0.3)	1.3(0.5)	0.6(2.9)	0.2(4.0)	0.4(2.1)	1.7(1.7)
C2D 1.0	4.8(0.2)	3.6(0.2)	1.3(0.4)	0.7(2.6)	0.2(3.9)	0.4(2.5)	1.8(1.6)
C2D 1.1	4.4(0.2)	3.2(0.2)	1.2(0.4)	0.6(2.8)	0.2(3.8)	0.3(2.3)	1.7(1.6)

= Cyrus2D zero base with ORE, C2DU = Cyrus2D zero base with UnMarking Strategy, C2DV1.0 = Cyrus2D version one with all of the previous features and C2DV1.1 = V1.0 with pass prediction) against Helios bases [20], Glider2d Base(v2.6) and three of the best teams in RoboCup (Helios2021 [35], YuShan [36], Cyrus2021 [25]).

Table 1 shows the expected winning rate of all version of Cyrus against opponent teams. The winning rate is calculated by $num\_wins/(num\_games - num\_draws)$. Table 2 presents the average number of our scored goals and conceded goals respectively.

The results demonstrate Cyrus2D base v1.1 prevalence over other released bases. For instance the Cyrus2D base wins Helios and Gliders2D bases in more than 99% and 84% of games respectively. The average win-rate of Cyrus2D against best three RoboCup teams is 3.76 (0.2% to 3.8%) percent higher than the winning rate of Helios base against those teams, and 2.86 (2.9% to 3.8%) percent higher than Gliders2D base.

6   Conclusion

In this paper, we aimed to introduce three versions of Cyrus2D base code and their particular features. The first version of Cyrus2D base was created by combining the latest release of Helios Agent2D and Gliders2D bases. For this version, we removed some of the fine tuned parameters. In the next version, Cyrus2D v1.0, we have upgraded the Blocking, and offensive strategy by using the Offensive Risk Evaluation and unmarking behavior. In the Cyrus2D v1.1 we improved the unmarking behavior using the Pass Prediction. To evaluate the performance of Cyrus2D, we ran 500 games against Gliders2D, Helios base, and best three teams in RoboCup 2021. The obtained results shows significant improvement on win-rate, scored goals and conceded goals. For our future work, we are planning to enhance the Cyrus2D base in terms of chain action movement prediction, and marking by using multi-agent decision-making.

Acknowledgements. We acknowledge the support of the Natural Sciences and Engineering Research Council of Canada (NSERC). We thank the HELIOS and Gliders teams for their code bases and extraordinary contributions to the SS2D league.

References

1. Burkhard, H.D., Duhaut, D., Fujita, M., Lima, P., Murphy, R., Rojas, R.: The road to RoboCup 2050. IEEE Robot. Autom. Mag. **9**(2), 31–38 (2002)
2. Noda, I. and Matsubara, H.: Soccer server and researches on multi-agent systems. In Proceedings of the IROS-96 Workshop on RoboCup, pp. 1–7 (1996)
3. Kitano, H., Asada, M., Kuniyoshi, Y., Noda, I., Osawa, E.: Robocup: the robot world cup initiative. In: Proceedings of the 1st International Conference on Autonomous Agents, pp. 340–347 (1997)
4. Kitano, H., Asada, M., Kuniyoshi, Y., Noda, I., Osawa, E., Matsubara, H.: RoboCup: a challenge problem for AI. AI Mag. **18**(1), 73–73 (1997)
5. Noda, I., Stone, P.: The RoboCup soccer server and CMUnited clients: implemented infrastructure for MAS research. Auton. Agents Multi-Agent Syst. **7**(1–2), 101–120 (2003)
6. Riley, P., Stone, P., Veloso, M.: Layered disclosure: revealing agents' internals. In: Castelfranchi, C., Lespérance, Y. (eds.) ATAL 2000. LNCS (LNAI), vol. 1986, pp. 61–72. Springer, Heidelberg (2001). https://doi.org/10.1007/3-540-44631-1_5
7. Stone, P., Riley, P., Veloso, M.: Defining and using ideal teammate and opponent models. In: Proceedings of the 12th Annual Conference on Innovative Applications of Artificial Intelligence (2000)
8. Butler, M., Prokopenko, M., Howard, T.: Flexible synchronisation within robocup environment: a comparative analysis. In: Stone, P., Balch, T., Kraetzschmar, G. (eds.) RoboCup 2000. LNCS (LNAI), vol. 2019, pp. 119–128. Springer, Heidelberg (2001). https://doi.org/10.1007/3-540-45324-5_10
9. Reis, L.P., Lau, N., Oliveira, E.C.: Situation based strategic positioning for coordinating a team of homogeneous agents. In: BRSDMAS 2000. LNCS (LNAI), vol. 2103, pp. 175–197. Springer, Heidelberg (2001). https://doi.org/10.1007/3-540-44568-4_11

10. Prokopenko, M., Wang, P.: Relating the entropy of joint beliefs to multi-agent coordination. In: Kaminka, G.A., Lima, P.U., Rojas, R. (eds.) RoboCup 2002. LNCS (LNAI), vol. 2752, pp. 367–374. Springer, Heidelberg (2003). https://doi.org/10.1007/978-3-540-45135-8_32
11. Prokopenko, M., Wang, P.: Evaluating team performance at the edge of chaos. In: Polani, D., Browning, B., Bonarini, A., Yoshida, K. (eds.) RoboCup 2003. LNCS (LNAI), vol. 3020, pp. 89–101. Springer, Heidelberg (2004). https://doi.org/10.1007/978-3-540-25940-4_8
12. Zare, N., Sarvmaili, M., Sayareh, A., Amini, O., Matwin, S., Soares, A.: Engineering features to improve pass prediction in soccer simulation 2d games. In: Alami, R., Biswas, J., Cakmak, M., Obst, O. (eds.) RoboCup 2021. LNCS (LNAI), vol. 13132, pp. 140–152. Springer, Cham (2022). https://doi.org/10.1007/978-3-030-98682-7_12
13. Stone, P., Asada, M., Balch, T., Fujita, M., Kraetzschmar, G., Lund, H., Scerri, P., Tadokoro, S., Wyeth, G.: Overview of Robocup-2000. In: Stone, P., Balch, T., Kraetzschmar, G. (eds.) RoboCup 2000. LNCS (LNAI), vol. 2019, pp. 1–29. Springer, Heidelberg (2001). https://doi.org/10.1007/3-540-45324-5_1
14. Stone, P., Riley, P., Veloso, M.: The CMUnited-99 champion simulator team. In: Veloso, M., Pagello, E., Kitano, H. (eds.) RoboCup 1999. LNCS (LNAI), vol. 1856, pp. 35–48. Springer, Heidelberg (2000). https://doi.org/10.1007/3-540-45327-X_2
15. Yao, J., Chen, J., Cai, Y., Li, S.: Architecture of TsinghuAeolus. In: Birk, A., Coradeschi, S., Tadokoro, S. (eds.) RoboCup 2001. LNCS (LNAI), vol. 2377, pp. 491–494. Springer, Heidelberg (2002). https://doi.org/10.1007/3-540-45603-1_66
16. Kok, J.R., Vlassis, N., Groen, F.: UvA Trilearn 2003 team description. In: Polani, D., Browning, B., Bonarini, A., Yoshida, K. (eds.) Proceedings CD RoboCup 2003. Springer, Padua (2003)
17. Riedmiller, M., Gabel, T., Knabe, J., Strasdat, H.: Brainstormers 2d - team description 2005. In: Bredenfeld, A., Jacoff, A., Noda, I., Takahashi, Y. (eds.) Proceedings CD RoboCup 2005. Springer (2005)
18. Bai, A., Chen, X., MacAlpine, P., Urieli, D., Barrett, S., Stone, P.: WrightEagle and UT austin villa: RoboCup 2011 simulation league champions. In: Röfer, T., Mayer, N.M., Savage, J., Saranlı, U. (eds.) RoboCup 2011. LNCS (LNAI), vol. 7416, pp. 1–12. Springer, Heidelberg (2012). https://doi.org/10.1007/978-3-642-32060-6_1
19. Tavafi, A., Nozari, N., Vatani, R., Yousefi, M.R., Rahmatinia, S., Pirdir, P.: MarliK 2012 soccer 2D simulation team description paper. In: RoboCup 2012 Symposium and Competitions: Team Description Papers, Mexico City, Mexico (2012)
20. Akiyama, H., Nakashima, T.: HELIOS base: an open source package for the robocup soccer 2d simulation. In: Behnke, S., Veloso, M., Visser, A., Xiong, R. (eds.) RoboCup 2013. LNCS (LNAI), vol. 8371, pp. 528–535. Springer, Heidelberg (2014). https://doi.org/10.1007/978-3-662-44468-9_46
21. Akiyama, H.: Agent2D base code. http://www.rctools.sourceforge.jp (2010)
22. Khayami, R., et al.: CYRUS 2D simulation team description paper 2014. In: RoboCup 2014. Joao Pessoa, Brazil (2014)
23. Prokopenko, M., Wang, P.: Gliders2d: source code base for RoboCup 2D Soccer simulation league. CoRR abs/1812.10202 (2018)
24. Zare, N., et al.: Cyrus Soccer 2D Simulation Team Description Paper: In: RoboCup 2013, p. 2013. Eindhoven, Netherlands (2013)
25. Zare, N., Sayareh, A., Sarvmaili, M., Amini, O., Soares, A., Matwin, S.: CYRUS 2D soccer simulation team description paper 2021. In: RoboCup 2021 Symposium and Competitions, Worldwide (2021)

26. Zare, N., et al.: improving dribbling, passing, and marking actions in soccer simulation 2D games using machine learning. In: Alami, R., Biswas, J., Cakmak, M., Obst, O. (eds.) RoboCup 2021. LNCS (LNAI), vol. 13132, pp. 340–351. Springer, Cham (2022). https://doi.org/10.1007/978-3-030-98682-7_28

27. Prokopenko, M., Wang, P., Obst, O., Jaurgeui, V.: Gliders 2016: integrating multi-agent approaches to tactical diversity. In: RoboCup 2016 Symposium and Competitions: Team Description Papers, Leipzig, Germany (2016)

28. Prokopenko, M., Wang, P.: Disruptive innovations in RoboCup 2D soccer simulation league: from Cyberoos'98 to gliders2016. In: Behnke, S., Sheh, R., Sarıel, S., Lee, D.D. (eds.) RoboCup 2016. LNCS (LNAI), vol. 9776, pp. 529–541. Springer, Cham (2017). https://doi.org/10.1007/978-3-319-68792-6_44

29. Prokopenko, M., Wang, P.: Fractals 2019: Guiding self-organisation of intelligent agents. In: RoboCup 2019 Symposium and Competitions, Sydney, Australia (2019)

30. Prokopenko, M., Wang, P.: Fractals2019: combinatorial optimisation with dynamic constraint annealing. In: Chalup, S., Niemueller, T., Suthakorn, J., Williams, M.-A. (eds.) RoboCup 2019. LNCS (LNAI), vol. 11531, pp. 616–630. Springer, Cham (2019). https://doi.org/10.1007/978-3-030-35699-6_50

31. Akiyama, H.: Agent2D base code. https://github.com/helios-base/helios-base (2010)

32. Akiyama, H.: LibRCSC, component of Agent2D base code. https://github.com/helios-base/librcsc (2010)

33. Martín Abadi, et al. TensorFlow: large-scale machine learning on heterogeneous systems (2015). Software available from tensorflow.org

34. Nader, Z., et al.: CPPDNN: A C++ library to use a trained DNN by Tensor Flow Keras. https://github.com/Cyrus2D/CppDNN

35. Yamaguchi, M., Kuga, R., Omori, H., Fukushima, T., Nakashima, T., Akiyama, H.: Helios 2021: team description paper. In: RoboCup 2021 Symposium and Competitions, Worldwide (2021)

36. Cheng, Z., Zhang F., Guang, B., Wang, L.: YuShan2021 team description paper for RoboCup2021. In: RoboCup 2021 Symposium and Competitions, Worldwide (2021)

37. Guennebaud, G., Jacob, B., et al.: Eigen v3 (2010). http://eigen.tuxfamily.org

Distributed Optimization Tool for RoboCup 3D Soccer Simulation League Using Intel DevCloud

Guilherme N. Oliveira$^{(\boxtimes)}$, Marcos R. O. A. Maximo ,
and Vitor V. Curtis

Autonomous Computational Systems Lab (LAB-SCA), Computer Science Division,
Aeronautics Institute of Technology, São José dos Campos, São Paulo, Brazil
{guilhermegno,mmaximo,curtis}@ita.br

Abstract. Due to the physical limitations of real robots, simulated robotics is an important area of research that opens up a lot of possibilities to study the robots' dynamics and program their behaviors. RoboCup 3D Soccer Simulation league is a tournament to encourage the development of robots that compete in a high-fidelity simulation ambient.

Due to the high complexity of simulated humanoid robots, numerical optimization techniques are often used to determine the best parameters to control their motion sequence. Even the simplest movements, such as walking, stopping and getting up, can have a meaningful impact on a simulated soccer match if well optimized. Such a process is time consuming and has a high computational cost. For this reason, the usage of high performance clusters is a good way to accelerate the optimization, and for this technology to be used, it is necessary to develop a reliable tool that interfaces the cluster, simulation, and optimization.

The main contribution of this work is to provide a distributed optimization tool for the RoboCup 3D Soccer Simulation League based on the Intel DevCloud cluster.

Keywords: Mathematical optimization · Robotics simulator · Parallel computing

1 Introduction

Mobile robotics is a growing area of knowledge. Robots are increasingly assisting humans in various tasks to overcome challenges and increase the speed and efficiency of processes. To encourage university research, several national and international robotics tournaments, such as RoboCup, where robots must perform tasks or play sports in a competitive manner.

The robots that participate in these competitions still have many physical limitations to overcome. For this reason, there are also robot competitions that

---

G. Oliveira was supported by CNPq.

© The Author(s), under exclusive license to Springer Nature Switzerland AG 2023
A. Eguchi et al. (Eds.): RoboCup 2022, LNAI 13561, pp. 152–163, 2023
https://doi.org/10.1007/978-3-031-28469-4_13

deal only with simulations in different levels of realism. The two-dimensional simulations focus on the development of strategies [15], and the three-dimensional simulations have to develop control sequences for the physical parts of the robots besides creating the strategies [8].

Soccer 3D is RoboCup's 3D soccer simulation league, based on the robot simulator SimSpark [22]. In this league, teams of 11 agents each compete against each other in a sophisticated physics simulator. A Soccer 3D match is exemplified in Fig. 1. Each agent is a virtualization of a humanoid robot (NAO) that is controlled at the joint level. Each robot has its own program, i.e. the agent is independent from the other agents and the simulation environment, and the communication between the agents and the server is done through a simple protocol to specify the desired angular velocity of each of the robot's joints.

Fig. 1. A match of RoboCup 3D Soccer Simulation League. The circles and lines are visual representations of robot agent variables used for debugging the code.

To soften the human job when writing a motion sequence for the robot, some techniques are used that convert high-level instructions, such as robot positions in time, into low-level instructions for the robot joints. One technique widely used in simulations is to register various positions of the robot at different times (steps) and, from this, calculate the instructions that should be sent to the joints by interpolation. This technique is known as keyframe, as explained in [9].

Even with such simplifications, designing a sequence of keyframes to directly control the robots' movements is a very complex task. It is possible to use graphic tools to create these keyframes [7], but this task is very time-consuming and the final movement can be unpredictable and sub-optimal. For this reason, many teams use machine learning techniques [14] and optimization algorithms [3] to build and optimize these movements. In the case of the keyframes, the instructions created manually are often used as starting values for the optimization algorithms. For these techniques to be successful, hundreds to thousands of simulations are required to determine parameters that define a good configuration for the keyframe sequence. In robotics, this is usually the slowest step in an optimization due to the high computational cost of the high-fidelity physics simulation.

To illustrate this computational challenge, two motions with less than 10 keyframes that were optimized in the past took an average of 4 days to optimize on a single conventional computer, according to [13]. Another challenge for 3D simulation is the existence of intentional noise created in the server environment, which exists to simulate the non-ideality of the real world. This makes optimization difficult by adding a stochastic factor to the simulations.

Even with these difficulties, simulated environments are good study objects for artificial intelligence, mainly because they do not suffer the same physical limitations of real robots, such as the need for construction and maintenance. It also allows the possibility of accessing and modifying the physics of the simulation. In fact, reinforcement learning enabled the robots to do complex tasks such as running [12] and recovering from pushes [11]. Indeed, deep reinforcement learning has been widely used to improve kick motions [10,20]. Nevertheless, these machine learning techniques can be data inefficient [19], and mathematical optimization algorithms may still have their usefulness in the simulated robotics environment. Additionally, mathematical optimization techniques have been used in practice and have obtained expressive results in increasing the speed and precision of simulated robot movements [17]. In particular, an algorithm that has shown better results than most others for simulated robotics is the CMA-ES [21]. Because mathematical optimizations present good results when solving complex problems, several improvements have been sought in the last decades to make them faster and more effective.

One way to alleviate the problem of optimization time is to develop a tool that performs several simulations in parallel [16]. This tool, when applied to a cluster with many processing units, is able to perform the optimization of systems and tasks in less time [2]. This technique is highly efficient on keyframe optimizations [13].

However, since the processing is divided among several cores in a cluster, it is essential that this tool is able to guarantee the synchronism and the balancing of tasks, foreseeing and fixing eventual errors automatically. In addition, it must ensure that work already done is saved periodically, since task processing is very time-consuming and power or internet failures can halt the process.

By combining this tool with the Intel DevCloud infrastructure, the Soccer 3D community can leverage a large amount of processing power for optimizing motions, high-level behaviors, and many other tasks that can be improved by parameter optimization. Furthermore, this tool may be easily adapted for other leagues.

The rest of this work is organized in the following logical sequence: the Sect. 2 presents the tools needed for optimization. Then, Sect. 3 explains the case study with the simulated humanoid robot. Next, the results of the study are presented in Sect. 4. Finally, Sect. 5 concludes the article and proposes future works in the field.

2 Background

2.1 Evolution Strategy

The evolution strategy chosen as the basis for optimization was the Covariance Matrix Adaptation Evolution Strategy (CMA-ES). Compared to other evolutionary strategies, CMA-ES is more efficient when the object of study is a system with many interdependent variables. Furthermore, it is possible to alter the initial parameters to seek a balance between the speed of convergence and the tendency to get stuck at a local optimum [4].

To implement the evolutionary strategy, it was chosen to use the pycma open source library from the Python package catalog (PyPi), maintained by the original creator of the CMA-ES and available at [5]. This option was picked due to the high number of implemented functionalities and the constant updates that the library receives from several contributors. In addition, the higher the sample efficiency of the algorithm, the less costly the optimizations become, as the total number of simulations that must be performed decreases. Therefore, it is important to give priority to a code that is well developed and has frequent fixes over one that is lean and runs quickly.

The main functions of the library are ask and tell. The function ask returns a multidimensional array with a series of values to be used in the next simulations. With the tell function it is possible to return the fitness of each of these values, i.e. a number that measures how well the robot performed with the implemented parameters. The workflow is represented in Fig. 2.

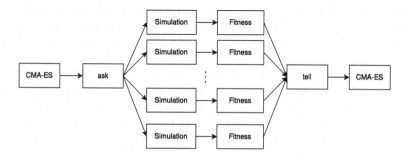

Fig. 2. Example workflow of a CMA-ES evolutionary strategy execution.

2.2   Parallel Computing Control

For the workflow control of the optimization, the software chosen was Dask [18]. It is responsible for ensuring that all the parallel simulations in Fig. 2 occur simultaneously on different processing units.

Dask is an open source library written in Python that is capable of automatically organizing and executing workflows in parallel. Dask also has tools to implement this parallelized computing on servers.

Dask also has a dashboard for viewing past, present, and scheduled tasks, as shown in Fig. 3.

The optimization tool was developed to allow high levels of flexibility. Even the CMA-ES algorithm can be replaced by another iterative optimization method, according to the user's needs.

2.3   Error Prevention

The DevCloud server [6] operates with a waitlist and a maximum time per task and therefore terminates a submitted task if the time limit runs out. Furthermore, both the simulation server and the simulated agents are subject to possible failures. Therefore, measures are needed to minimize errors caused by sudden interruptions of the optimization.

To prevent errors like the ones mentioned from causing the loss of work already done, the open source pickle library, available in Python, was adopted. Pickle is a tool for object serialization, i.e. for converting an object into a sequence of bytes, which in turn can be saved in a file. Serialization is performed on the main CMA-ES object at each fixed number of ask/tell iterations, defaulting to 2 iterations. If the CMA-ES tool is restarted, it will first look for an existing backup file before starting a new optimization.

It is necessary to take into account the possibility of a simulation failing, for various reasons. For this reason a timeout has been implemented in the cost function. A reward of -5 is assigned to the simulation if it fails five times in a row.

The control of the independent processing units is performed by Dask, which is able to redistribute the tasks according to the need of the process and can identify any defective processing unit and rearrange its pending jobs, as seen in Fig. 3.

In Fig. 3, three threads are initially active. Each vertical bar of one color represents a function call, and they are interdependent. At a certain point one thread stops running. Then its tasks are distributed among the other two remaining ones and the program is able to continue running. When another thread is added to the system, the tasks are redistributed again.

2.4   Keyframes

One possible application of the framework is the optimization of a keyframe. The case study will cover the improvement of a kick keyframe.

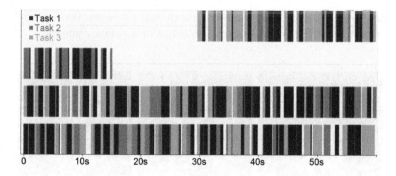

Fig. 3. Dask dashboard on a mock test.

A keyframe [9] is an ordered set of joint angular positions at a given instant, similar to a photography. In a keyframe-based movement, several keyframes are defined and each one is associated to a determined time step. These informations are saved on a JSON file.

To compute the movement, an interpolator uses the data of these keyframes to calculate the joints' position at each time instant. Usually, this interpolator is implemented by an iterator that uses a mix of linear regressions and cubic splines to generate smooth movements, as exemplified in Fig. 4. The joints are then individually controlled by proportional controllers.

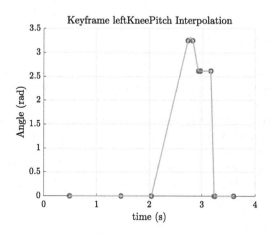

Fig. 4. Reference value for angular pitch of the left knee in a kick motion.

The initial keyframe can be troublesome to the stability of a robot because the initial position of the robot is often very different from the first keyframe. Consequently, the proportional controller's commands can lead to the loss of stability. For this reason, the first keyframe iteration is usually smoothed.

A kick performed by the simulated robot can be controlled by keyframes. The kick motion of Fig. 5, for example, consists of 3 submovements.

Fig. 5. The three keyframe moments of the kick. From left to right: prepFrontKickFast, moveLegBack and moveLegFront.

The three moments of Fig. 5 are divided into 4, 3 and 4 steps respectively, for a total of 11 steps. Each step has 22 different parameters corresponding to angles of the robot joints at a given moment and one extra parameter corresponding to the keyframe's expected duration. Of these 23 parameters, the 4 that were considered the least relevant ones were excluded to increase the optimization convergence speed, as they were related to neck and shoulder angles. The remaining 19 parameters along the 11 steps result in a total of 209 optimizable values.

3   Methodology

The optimization server operates in the Intel DevCloud, whose structure is shown in Fig. 6. DevCloud is a powerful computing environment for edge, AI, high-performance computing (HPC) and rendering workload [6]. It has different Intel CPUs, GPUs, Accelerators and FPGAs available to the public, and it also offers the oneAPI programming environment for developing applications that target multiple hardware architectures.

The number of workers can be adjusted at any time by the user via a file interface. The status of the entire system can be checked via ssh or a web interface on the DevCloud website.

3.1   Kick Keyframe Optimization

Recent changes in the dynamics of robot soccer simulation matches altered the pace of the game. Now, any robot that is at most 0.5 m from the ball can request a Pass command, and it will be granted if there is not an opponent within a 1 m radius circle from the ball. The pass command creates a situation in which the agent has four seconds to do a free kick without the opponents' interference, as any opponent that enters a 1 m radius circle from the ball will be sent back.

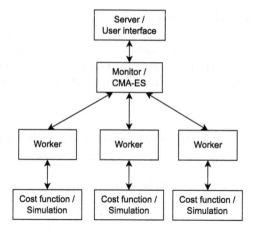

Fig. 6. Structure of the optimization program on DevCloud.

However, the robot must not directly score a goal with this kick. These changes have made the mid-range kick an important element of the team's strategy. Therefore, setting up this kick is the goal of the optimization.

Since there are 209 parameters in total, it would be extremely costly to optimize from scratch. Therefore, a generic kick *keyframe* was used as a seed (initial value). The fitness function of the simulation prioritizes kicks that end up with the ball near the opponent's goal and that have a high maximum altitude. If the kick reaches the objective of leaving the ball in the opponent's penalty area, a large reward is added. A medium reward is given if the agent scores a goal, which is not the kick's main objective. To minimize the influence of random factors on the outcome of the fitness function, each kick was performed six times divided in three different positions, and the total fitness was given as the sum of the fitnesses.

The fitness function is defined according to the equation

$$\text{fitness} = 0.5 \cdot h_{max} + \begin{cases} 50, & \text{if the ball lands on the penalty area,} \\ 25, & \text{if a goal is made,} \\ -d_{goal}, & \text{in other cases,} \end{cases} \tag{1}$$

where h_{max} is the maximum height reached by the ball and d_{goal} is the final distance from the ball to the center of the goal projected on the ground. However, since the default of CMA-ES is to prioritize results with lower values, the fitness function was transformed into a cost function by multiplying it by -1 before sending it to the optimization algorithm.

4   Results and Discussions

The optimization ran for approximately nine hours in the DevCloud with 6 concurrent parallel simulations (matching pycma's default population size), each allocated to a thread on a CPU Intel Xeon 6128 or 8256, and the results are depicted in Fig. 7.

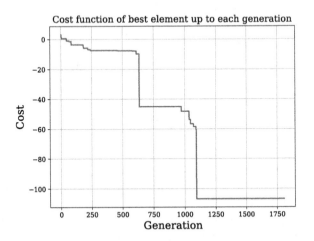

Fig. 7. Evolution of the cost function of the best element of each generation.

The graph in Fig. 7 shows that there is a slow convergence of the cost function. The slow speed of the optimization is expected due to the large number of parameters to be optimized and the random factors involved in the simulations.

By the 1000th generation the robots began to achieve the target of kicking the ball into the penalty area. This result was consolidated, and in the next generations the goal started to be achieved multiple times in the 6 kicks performed with each keyframe. This result demonstrates that the robot is able to kick in the target area consistently, and therefore the goal of the optimization was achieved.

For performance comparison purposes, this same optimization was run for one hour from the same initial conditions in different virtual environments. The first was performed in a sequential manner, i.e. without parallelism, on a personal computer with the Intel Core i7-9750H @ 2.60 GHz. The second was performed in DevCloud by submitting a job requesting three cores, each one assigned to two processes of one thread each, resulting in 6 possible parallel simulations. As the DevCloud's processors have hyper-threading, each simulation was assigned to a single thread. The third was also performed in DevCloud on three cores, but each core had eight processes of two threads each, resulting in 16 parallel simulations per core or 48 in total. The results are arranged in Table 1.

Table 1. Effect of different virtual environments on optimization performance over the period of one hour.

Cores	Total of threads	Total simulations
1	1	281
3	6	912
3	48	1584

The test results show a 225% increase in the total number of simulations performed when parallelism with 3 jobs and 6 threads is used compared to sequential optimization. The number of simulations also increases by 73.7% when the number of threads is increased from 6 to 48. As the number of threads increased, the average value of simulations run per thread decreased. This is mainly due to the fact that the processors are progressively more overloaded in environments with 6 and 48 threads.

To increase the number of simultaneous simulations, it is necessary to increase the population size of the CMA-ES algorithm. By doing so, the speed of each individual simulation decreases and, consequently, the time spent per generation increases, which can slow down the speed of convergence. Therefore, caution is needed when increasing the number of simultaneous simulations, and the optimal number varies depending on the optimization. However, if well managed, this feature can further increase the speed of optimization.

The total computational time is mainly dictated by the evaluation time of each individual in a population. Thus, the major focus of the parallelism is to speed up these evaluations, which are done through physical simulations. To further increase the speed of the evaluations, a new physics simulator specifically optimized for this task has to be developed, which is beyond the scope of this project.

5   Conclusions

The optimization tool created in this project will be of great importance to the ITAndroids team, and it will help in future ITAndoids Soccer 3D projects and researches. Other teams are also encouraged to use the tool, available in the ITAndroids Open repository [1].

There are still many keyframes of different movements that can be optimized for different purposes. However, with the optimization tool ready, this can be done almost routinely by the team. Future work can be developed to study and optimize other movements of the simulated humanoid robot, such as the movement of getting up from the ground after a fall or walking.

The trend is that in the future more and more teams will use AI tools to control their agents, which will make the Soccer 3D scenario more and more competitive and professional.

Acknowledgements. First, G. Oliveira thanks CNPq for making this project possible. The authors also acknowledge the team's sponsors, Altium, Cenic, Intel, ITAEx, MathWorks, Metinjo, Micropress, Polimold, Rapid, SolidWorks, STMicroelectronics, Wildlife Studios, and Virtual.PYXIS.

We also thank the former members of the Soccer 3D team, who dedicated years of their lives to build the foundations that made this project possible.

A special reminder to Bruno and Lucas, members of ITAndroids Soccer 3D in 2019, who were essential for the development of the optimization agent.

References

1. Distribute Optimization Tool on ITAndroids Open repository. https://gitlab.com/itandroids/open-projects/distributed-optimization-tool
2. Alba, E., Tomassini, M.: Parallelism and evolutionary algorithms. IEEE Trans. Evol. Comput. **6**(5), 443–462 (2002). https://doi.org/10.1109/TEVC.2002.800880
3. Depinet, M., MacAlpine, P., Stone, P.: Keyframe sampling, optimization, and behavior integration: towards long-distance kicking in the RoboCup 3D simulation league. In: Bianchi, R.A.C., Akin, H.L., Ramamoorthy, S., Sugiura, K. (eds.) RoboCup 2014. LNCS (LNAI), vol. 8992, pp. 571–582. Springer, Cham (2015). https://doi.org/10.1007/978-3-319-18615-3_47
4. Hansen, N.: The CMA evolution strategy: a comparing review. In: Lozano, J.A., Larrañaga, P., Inza, I., Bengoetxea, E. (eds.) Towards a New Evolutionary Computation. STUDFUZZ, vol. 192, pp. 75–102. Springer, Heidelberg (2007). https://doi.org/10.1007/3-540-32494-1_4
5. Hansen, N., Akimoto, Y., Ueno, Y., Brockhoff, D., Chan, M., ARF1: CMA-ES/pycma on Github. Zenodo (2020). https://doi.org/10.5281/zenodo.3764210
6. Intel: Intel AI DevCloud (2020). https://software.intel.com/pt-br/ai-academy/devcloud
7. Izadi, N.H., Roshanzamir, M., Palhang, M.: AIUT3D motion editor for 3D soccer simulation compatible with SimSpark physics engine. In: 2018 8th Conference of AI Robotics and 10th RoboCup Iranopen International Symposium (IRANOPEN), pp. 32–36 (2018). https://doi.org/10.1109/RIOS.2018.8406628
8. Kögler, M., Obst, O.: Simulation league: the next generation. In: Polani, D., Browning, B., Bonarini, A., Yoshida, K. (eds.) RoboCup 2003. LNCS (LNAI), vol. 3020, pp. 458–469. Springer, Heidelberg (2004). https://doi.org/10.1007/978-3-540-25940-4_40
9. MacAlpine, P., Stone, P.: UT Austin Villa RoboCup 3D simulation base code release. In: Behnke, S., Sheh, R., Sariel, S., Lee, D.D. (eds.) RoboCup 2016. LNCS (LNAI), vol. 9776, pp. 135–143. Springer, Cham (2017). https://doi.org/10.1007/978-3-319-68792-6_11
10. Melo, D., Forster, C., Máximo, M.: Learning when to kick through deep neural networks. In: Anais do VII Simpósio Brasileiro de Robótica e XVI Simpósio Latino Americano de Robótica, pp. 43–48. SBC, Porto Alegre, RS, Brasil (2019). https://sol.sbc.org.br/index.php/sbrlars/article/view/11837
11. Melo, D.C., Máximo, M.R.O.A., da Cunha, A.M.: Push recovery strategies through deep reinforcement learning. In: 2020 Latin American Robotics Symposium (LARS), 2020 Brazilian Symposium on Robotics (SBR) and 2020 Workshop on Robotics in Education (WRE), pp. 1–6 (2020). https://doi.org/10.1109/LARS/SBR/WRE51543.2020.9306967

12. Melo, L.C., Melo, D.C., Maximo, M.R.O.A.: Learning humanoid robot running motions with symmetry incentive through proximal policy optimization. J. Intell. Robot. Syst. **102**(3), 1–15 (2021). https://doi.org/10.1007/s10846-021-01355-9

13. Muniz, F., Maximo, M.R., Ribeiro, C.H.: Keyframe movement optimization for simulated humanoid robot using a parallel optimization framework. In: 2016 XIII Latin American Robotics Symposium and IV Brazilian Robotics Symposium (LARS/SBR), pp. 79–84 (2016). https://doi.org/10.1109/LARS-SBR.2016.20

14. Muzio, A.F.V., Maximo, M.R.A., Yoneyama, T.: Deep reinforcement learning for humanoid robot dribbling. In: 2020 Latin American Robotics Symposium (LARS), 2020 Brazilian Symposium on Robotics (SBR) and 2020 Workshop on Robotics in Education (WRE), pp. 1–6 (2020). https://doi.org/10.1109/LARS/SBR/WRE51543.2020.9307084

15. Noda, I., Matsubara, H., Hiraki, K., Frank, I.: Soccer server: a tool for research on multiagent systems. Appl. Artif. Intell. **12**(2–3), 233–250 (1998). https://doi.org/10.1080/088395198117848

16. Rashid, Z.N., Zebari, S.R.M., Sharif, K.H., Jacksi, K.: Distributed cloud computing and distributed parallel computing: a review. In: 2018 International Conference on Advanced Science and Engineering (ICOASE), pp. 167–172 (2018). https://doi.org/10.1109/ICOASE.2018.8548937

17. Rei, J.L.M., et al.: Optimizing simulated humanoid robot skills (2010)

18. Rocklin, M.: Dask: parallel computation with blocked algorithms and task scheduling, pp. 126–132 (2015). https://doi.org/10.25080/Majora-7b98e3ed-013

19. Samsami, M.R., Alimadad, H.: Distributed deep reinforcement learning: an overview (2020). https://doi.org/10.48550/ARXIV.2011.11012

20. Spitznagel, M., Weiler, D., Dorer, K.: Deep reinforcement multi-directional kick-learning of a simulated robot with toes. In: 2021 IEEE International Conference on Autonomous Robot Systems and Competitions (ICARSC), pp. 104–110 (2021). https://doi.org/10.1109/ICARSC52212.2021.9429811

21. Urieli, D., MacAlpine, P., Kalyanakrishnan, S., Bentor, Y., Stone, P.: On optimizing interdependent skills: a case study in simulated 3D humanoid robot soccer. In: AAMAS, vol. 11, p. 769 (2011)

22. Xu, Y., Vatankhah, H.: SimSpark: an open source robot simulator developed by the RoboCup community. In: Behnke, S., Veloso, M., Visser, A., Xiong, R. (eds.) RoboCup 2013. LNCS (LNAI), vol. 8371, pp. 632–639. Springer, Heidelberg (2014). https://doi.org/10.1007/978-3-662-44468-9_59

Bipedal Walking on Humanoid Robots Through Parameter Optimization

Marc Bestmann$^{(\boxtimes)}$ and Jianwei Zhang

Department of Informatics, University of Hamburg, 22527 Hamburg, Germany
{marc.bestmann,jianwei.zhang}@uni-hamburg.de

Abstract. This paper presents a open-source omnidirectional walk controller that provides bipedal walking for non-parallel robots through parameter optimization. The approach relies on pattern generation with quintic splines in Cartesian space. Additionally, baselines of achieved walk velocities in simulation for all robots of the Humanoid Virtual Season, as well as some commercial robot models, are provided.

Keywords: Bipedal walking · RoboCup · ROS2

1 Introduction

Bipedal walking is one of the biggest challenges in humanoid robotics. Some impressive results have been achieved, e.g. on the Cassie robot [18]. Still, there is, to the best of our knowledge, no simple-to-use open-source controller available that works on a majority of bipedal robots. Commercial robots, e.g. the Darwin-OP [16] or the NAO [15] may come with a walk controller provided by the manufacturer. Some of these widespread platforms may also have open-source solutions available, e.g. the different walk approaches from the RoboCup Standard Platform League (SPL) for the NAO robot. But if a team constructs its own robot, it needs to program its own walk controller or at least modify an existing solution due to differences in the kinematics and dynamics. This is one of the key issues that teams are still facing in the RoboCup Humanoid League [7] and it leads to two problems. First, it creates a high entry barrier for new teams since these need to build working hardware and a complete software stack. For some trivial parts of this software stack, e.g. the connector to the game controller, and even for some more complex parts, e.g. the computer vision, software of other teams can easily be used. Since the robot designs differ, this is not true for the walking, which is arguably one of the most complex parts. Second, even established teams may run into problems with their walking controller if they want to modify their hardware, as this can require changes to the controller or at least new parameter tuning. Therefore, teams may hesitate to introduce hardware changes which is problematic, as this kind of development is one of the goals of the league.

We present an open-source omnidirectional walk controller that is easy to use and works on all non-parallel robots in the Humanoid League Virtual Season (HLVS). Additionally, we provide baselines for the achieved walk velocities

© The Author(s), under exclusive license to Springer Nature Switzerland AG 2023
A. Eguchi et al. (Eds.): RoboCup 2022, LNAI 13561, pp. 164–176, 2023
https://doi.org/10.1007/978-3-031-28469-4_14

that are reached on different robots. These are measured in the standardized simulation environment of HLVS and are therefore easily comparable. Our goal is to allow new teams an easier entry into the league as well as allowing comparison of self-created walk controllers to a baseline for any specific robot. Furthermore, our controller allows to easily test new robot platforms or modifications to existing ones, thus enabling a faster hardware development cycle.

The walking consists of two parts: an open loop pattern generator based on parameterized quintic splines and a closed loop stabilization module. First, the pattern generator describes the desired trajectory of the feet in Cartesian space. Then, PID controllers modify the torso orientation based on IMU data in Cartesian space and optionally step phase modulations can be applied based on either joint torque or foot pressure data. Due to the usage of the standard MoveIt [11] interface, different inverse kinematics (IKs) can be applied, including a generally applicable memetic approach. This facilitates the general usability of the walking as we can abstract from the concrete kinematic structure of the robot. Additionally, the walking does not require a model of the robot's dynamics but a set of parameters for the spline definitions and the PID controllers. We show how these can be optimized automatically by using the Multi-Objective Tree-structured Parzen Estimator (MOTPE). Integrating the walking into an existing code base is convenient, since we provide a ROS 1 and a ROS 2 version as well as direct interfaces in C++ and Python.

For our experiments, it was necessary to create URDF models as well as MoveIt configurations for all evaluated robots. We provide these too, as they may facilitate others in running software on different robots.

Our key contributions are:

– Open source walk controller that is easily usable on any non-parallel humanoid robot
– Baselines of stable walk velocities on different robot platforms
– Collection of ROS 2 URDF description and MoveIt configurations for various humanoid robots
– Comparison of different parameter optimization approaches

2   Related Work

There are many existing approaches to bipedal walking, but we are focusing in this section on approaches that are either applied in the RoboCup domain or similar to our presented approach. Model-free approaches that consist of an open-loop trajectory generator and a stabilizing mechanism are often used in RoboCup [8,23]. These do not require exact dynamic models, which are difficult to obtain for the typical low-cost robots due to sensor noise and delay, joint backlash, and imperfect actuation. The trajectories can be generated in joint space or in Cartesian space, but specifically designed leg representations can also be used [8]. Generating the trajectory in joint space does not require an IK and therefore less computation. On the downside, it can only be applied on robots with a certain joint configuration and is, therefore, less transferable between different robots. While some stabilization methods only work in Cartesian space, there are different methods such as the hip and the ankle strategy that work directly in joint space [3]. As an improvement to Euler angles, fused angles [5] were proposed for improved modeling of balance and applied as the basis for stabilization mechanisms [6].

Typically, a set of parameters needs to be optimized for a walk controller since their performance highly depends on the used parameter set. Therefore, different approaches have been investigated. Shafii et al. used Covariance Matrix Adaptation Evolution Strategy (CMA-ES) [17] to search for optimal walking parameters in simulation [27]. CMA-ES was also used by Seekircher et al. to optimized their model parameters to improve its predictions [26]. Rodriguez et al. used Bayesian optimization for walk parameters with a combination of simulated and real experiments [22]. Silva et al. used reinforcement learning on two of their walking parameters [28]. In the domain of quadrupeds, Saptura et al. used the Nondominated Sorting Genetic Algorithm II (NSGA-II) [13] to optimize the walk parameters [25].

We have used quintic splines with parameter optimization for controlled stand-up motions [29]. While the general idea of our previous work is similar, a different optimization procedure and different objectives are required for bipedal walking. Additionally, our previous work was not evaluated on so many robots and with no standardized simulation parameters, as these have only become available recently.

3   Walk Controller

The presented walk controller expects a goal walk velocity as input, which is typically provided by the path planning or human teleoperation. Based on this, a finite state machine (FSM) decides on the kind of step that needs to be performed and keeps track of the current step phase. Depending on the step type, movements for the foot and torso are defined through Cartesian quintic splines. Different stabilization approaches can be used based on sensor feedback. An overview of the approach can be seen in Fig. 1 and the following sections explain the different parts in more detail.

3.1   Finite State Machine

The walking controller needs to be able to stably start and stop the walking. Therefore, it needs to perform different types of steps, that handle the movement of the torso, and thereby the movement of the CoM, differently. Before lifting a foot for the first step, it starts moving the torso to the side, initializing the lateral swinging motion of the robot. Then the first step is performed with a phase offset between the torso and foot movement. A similar procedure is performed when stopping to walk so that the torso ends up in a centered position. This is modeled with a FSM which leads to a clearer code structure (see Fig. 2). While the robot is walking, it can also perform small dribbling kicks on request by modifying the

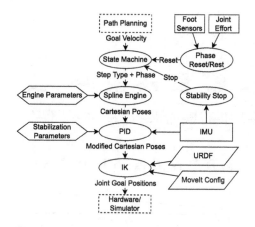

Fig. 1. Overview of the approach with used parameters, sensors and interfacing software parts.

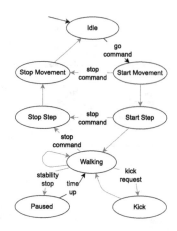

Fig. 2. Visualization of the finite state machine. Transitions are activated if the corresponding condition is met. Orange transitions can only be activated when a step is finished. (Color figure online)

spline to move the foot quickly to the front before setting it down. Generally, the controller may prevent further steps during double support to let oscillations of the robot settle (*Stability Stop*, see Sect. 3.3). A step is normally finished after a fixed time period (defined by one parameter) but the step duration can be modified by the *Phase Reset* and *Phase Rest* (see Sect. 3.3).

3.2 Spline Engine

The core part of the walk controller is the spline engine that generates a fixed cycle gait which is partially based on the IKWalk from team Rhoban [23]. Dependent on the current step type, two sets of six quintic splines are generated to describe the Cartesian pose (x, y, z, roll, pitch, yaw) of the torso and the moving foot in relation to the support foot over the duration of the step. The usage of quintic splines ensures that the trajectories are continuous in the first and second derivatives. This leads to smooth motions which are crucial for a stable walk.

The quintic spline is defined by a list of knots that each specifies the position, velocity and acceleration at a given time. Between these, polynomials are used for interpolation. The values of these knots can be seen as the parameters of the spline engine. Some of these are naturally defined, e.g. the foot is on the ground at the end of the step, and some can be derived from the commanded walking speed, e.g. how far the foot is put forward at the end of the step. The remaining parameters need to be optimized as they depend on kinematic and dynamic properties. A visualization of these parameters is shown in Fig. 3.

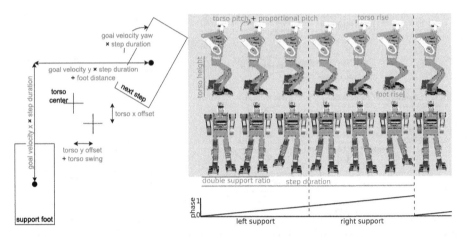

Fig. 3. Illustration on how the parameters (**orange**) and the goal walk velocity (**blue**) influence the motion. On the **left**, the positioning of the next step and the torso is shown. On the **right**, the walk cycle is shown (with exaggerated double support) and the phase variable is illustrated on the **bottom**. Additionally, a not shown parameter defines phase shift of the torso movement towards the foot movement. These are all parameters that are optimized. (Color figure online)

3.3   Stabilization

Different methods for stabilization can be applied to the open-loop pattern that is created by the quintic splines. Two PID controllers modify the torso's orientation in relation to the ground, based on the IMU orientation. For this, we use fused angles [5] which are an orientation representation that is specifically designed for balancing. The PID controllers deal with arising oscillations and external pushes. Another source of forces that lead to instabilities is making ground contact with the moving foot, either by pushing too far downwards into the ground or by starting a new step without having yet made ground contact. To prevent this, we apply phase modulation which either ends a step early or waits for the ground contact based on sensor data. This data can either come from foot pressure sensors, if the robot has those, or from the joint torque feedback, which the widespread Dynamixel servos provide. The latter approach is less reliable due to noise in the sensor data, but is applicable without additional sensors. If the robot gets too unstable, it is also possible to do a stability stop during a double support phase, which completely stops the walk motion, allowing larger oscillations to settle. While this is undesired, as it slows the robot, it is still better than falling. The arms are not used for stabilization, since movements can lead to fouls, i.e. ball holding. On the real hardware, the arms also might get damaged during falls dependent on their position. Therefore, most teams typically keep the arms in a fixed position in the RoboCup Humanoid League.

3.4   Inverse Kinematics

Since the movement is defined in Cartesian space, we need to apply an IK to compute the joint goal positions. To keep our walk controller generally usable, we do not rely on a specific analytic IK. This would need to be adapted to new robot models, and some of the used models, i.e. the Wolfgang-OP [10] and the NUgus [14], do not have an analytic solution because the hip joint axes do not intersect. We use the widespread MoveIt IK interface which provides multiple solvers. For this work, we chose the BioIK [24] solver as it works for all non-parallel robots. It combines genetic algorithms and particle swarm optimization to allow solving of different IK goals. Still, it is fast enough to run in real-time on computationally limited robot platforms.

3.5   Interfacing

We provide multiple interfaces for the walk controller, that allow simple integration into different code bases. The standard interface uses ROS 2 based message passing to provide the motor goal positions and the odometry. Additionally, multiple debug topics are provided that show the internal state of the walk controller and can be visualized with standard ROS tools, i.e. PlotJuggler. This allows simple integration for users of this middleware. We also provide a ROS 1 version. Since the code is written in C++, it is also possible to directly call the corresponding methods without ROS 2 message transfers. Still, either ROS 2 or ROS 1 packages are necessary as dependencies for building the code. Additionally, we provide a direct Python 3 interface to the C++ methods which is also used for the described parameter optimization (see Sect. 4).

4   Optimization

As described above, the definition of the splines does not only rely on the goal walk velocity but also on a set of parameters. Naturally, the performance of the walk control correlates to the quality of these parameters and the optimal values differ for each robot type. It is possible to tune these by hand, but it is time-consuming and might result in a local maximum. Therefore, we apply black-box parameter optimization to automatically tune them in simulation. The different stabilization approaches also require parameters, but these are either easy to find, i.e. the thresholds for the phase modulation, or have an existing method for tuning, i.e. the Ziegler-Nichols method [30] for the PID controllers. Therefore, we do not optimize these parameters automatically and will not cover them in the following section.

4.1   Problem Definition

Generally, such an optimization problem can be expressed as finding the parameter set x^* which maximizes a single objective function $f(x)$ (Eq. 1) or multiple objective functions $f_1(x), f_2(x), ..., f_n(x)$ (Eq. 2). The parameters x need to be part of the set of possible parameters X.

$$x^* = \underset{x \in X}{\mathrm{argmax}} \; f(x) \quad (1) \qquad x^* = \underset{x \in X}{\mathrm{argmax}}(f_1(x), f_2(x), ..., f_n(x)) \qquad (2)$$

For each of the spline parameters, a continuous range is defined to create the set of possible parameters X. This range can be identically for all robots, e.g. the *double_support_ratio*, and is defined either through natural boundaries, e.g. *double_support_ratio* ≥ 0, or our previous experience on which values make sense, e.g. *double_support_ratio* ≤ 0.5. Some ranges are dependent on the robot type since they are heavily influenced by its size, e.g. *foot_distance*. These need to be specified for each robot but can be found easily by trying out maximal reachable poses of the feet. It is important to note, that these ranges do not need to be exact and can be larger than necessary, but it can prolongate the optimization process if they are chosen extremely large. The user might also narrow these ranges to achieve a walking with certain desired properties, e.g. with slow steps or a certain torso height.

A natural objective function for walking is the maximum speed that the controller can achieve without the robot falling. Since our goal is an omnidirectional walk controller, there are three continuous goal velocities $(x, y, theta)$. Thus we have an infinite amount of goals and can not describe this as an objective function. Still, our tests have shown that it is enough to optimize the parameters for the four directions forward, backward, sideward, and turn. If the parameters work for these, combinations, e.g. walking to the front-left while turning right, are also working. For sideward and turn movements, only one direction has to be tested, as the parameters are symmetrical enough. This results in four objective functions $f_f(x)$, $f_b(x)$, $f_s(x)$, and $f_t(x)$.

This multi-objective optimization problem can directly be solved by Multi-objective Tree-structured Parzen Estimator (MOTPE) [20]. But it can also be scalarized a priori and then solved as a single-objective optimization problem, e.g. by using Tree-structured Parzen Estimator (TPE) [9]. When using MOTPE, an a posteriori scalarization is necessary for deciding on a parameter set, since a multi-objective optimization will provide a Pareto front, not a single solution. The following scalarization is used. Its factors are based on the typically achieved maximum walk speeds and ensure that each of the directions is contributing equally to the objective. Without these factors, the optimization might focus on the turn direction as the values are typically higher due to the different unit (rad/s instead of m/s).

$$f(x) = f_f(x) + f_b(x) + 2 * f_s(x) + 0.2 * f_t(x) \qquad (3)$$

4.2   Optimization Process

The optimization process is based on the Optuna library [4] which provides implementations of different optimization algorithms (called *sampler*) and a framework to run the whole optimization process (called *study*). During the execution of a study, several *trials* will be executed. For each, the sampler will propose a parameter set, based on the previously evaluated sets. The library user needs to implement a function that evaluates these parameters and returns their objective values to the study. Multiple trials of a study can also be executed in parallel by using an SQL database to share data between the processes, thus allowing good scalability on a computer cluster.

Basing our code on this library leads to a clear interface that is compatible with different samplers and removes the need to implement any sampler ourselves. The only part that is necessary to implement is the evaluation of a trial. To do this, we use the Webots simulator [19] since the robots models of the other teams are available for this simulator and we can use the exact same artificial grass simulation environment as defined by the HLVS. Still, our code also supports PyBullet [12] and has an interface to implement the usage of other simulators, but this is not further discussed in this paper.

For each of the directions, we do the following procedure to compute the objective value. First, the robot is initialized by performing steps in the air without gravity and then put back onto the ground. This is necessary, to ensure that the robot starts with the correct pose for the parameter set. Then, the robot walks for 10 s in the corresponding direction, including an acceleration and deceleration phase, as well as a complete stop at the end. This procedure is repeated with linearly incrementing speed until the robot either falls or the traveled distance does not increase. When all four directions are evaluated, the maximally reached velocities are returned either as an array, in case of multi-objective optimization, or scalarzied as a single value in case of single-objective optimization (see Eq. 3).

5 Evaluation

To evaluate the presented walk controller, we first compare different samplers that can be used to optimize the parameters. Then we show how well it generalizes to different robot platforms and provide the achieved baseline values. Additionally, we discuss the previous impact of the approach.

5.1 Optimizer

There are multiple different optimization approaches available (see also Sect. 2). Optuna provides samplers for CMA-ES [17], NSGA-II [13], TPE [9], and MOTPE [20]. Additionally, for TPE and MOTPE a multivariate version (MTPE/MMOTPE) is provided. Optuna claims that it outperforms the independent TPE and MOTPE sampler. We compare the achieved objective value of these samplers with a budget of 1,000 trials. Additionally, we provide the baseline of a random search with 1,000 and 10,000 trials to highlight that these values can not be easily found randomly. The experiment was performed on three different robots, the Wolfgang-OP [10], the OP3 [2], and the Bez robot. These robots were chosen, as they have different sizes, foot shapes (with cleats

Table 1. Achieved objective values for different robot-sampler combinations (higher is better)

Name	Wolf.	OP3	Bez
Multi. MOTPE	1.33	1.81	0.24
MOTPE	1.39	1.89	0.72
Multi. TPE	1.25	1.83	0.31
TPE	1.35	1.81	0.55
CMA-ES	0.90	1.53	0.69
NSGA-II	1.39	1.78	0.63
Random 1000	0.64	1.36	0.17
Random 10000	1.09	1.48	0.36
MOTPE Comb.	1.52	1.95	0.74

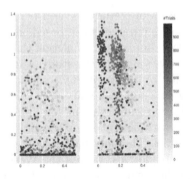

Fig. 4. Comparission between sampled *double_support_ration* parameters (x-axis) of MMOTPE (**left**) and MOTPE (**right**) with achieved objective values (y-axis).

and without), and different degrees of realism in the model. Due to the long time that is needed for the optimization process, each sampler-robot combination was only run a single time. Since the approaches are not deterministic (due to a random initialization), the results may be influenced by randomness.

Of all samplers, the independent MOTPE performs the best (see Table 1). These results are in accordance with our previous work on humanoid stand-up motions [29], where MOTPE also performed better than CMA-ES. While we only compare the samplers in the metric of trials, it is noteworthy that the necessary time for one study is also significantly influenced by the choice of the sampler. Since one parameter set is evaluated by trying to walk with increasing velocity, bad parameter sets that directly lead to a fall are evaluated quicker. For example, the multivariate MOTPE sampler needs only ca. 8 h for 1,000 trials while the independent version needs ca. 24 h on the same machine with an AMD Ryzen 9 5900 × 12-core CPU. We observed that the independent (MO)TPE tries to improve more on the current local maximum that it has found, while the multivariate version explores the parameter space more (see Fig. 4). As it tries out more bad parameter sets due to this, the multivariate version is also faster to compute in our scenario. But, it is not performing well in fine-tuning the best parameter set. Therefore, we propose to use a mixed approach of first optimizing 1,000 trials using the multivariate version and then optimizing further 500 trials with the independent version, as this leads to slightly better results with similar computation time (see *MOTPE Combination* in Table 1).

5.2   Generalization

We evaluate the presented walk controller on all robots of the Humanoid Virtual Season [1], as well as some popular humanoid robots that are included in Webots using the simulation parameters of the Humanoid Virtual Season. To do this,

Table 2. Achieved walking velocities on different robot platforms

Robot Platform	Team/Company	Height [m]	Forward [m/s]	Backward [m/s]	Sideward [m/s]	Turn [rad/s]
Bez	UTRA	0.50	0.21	0.05	0.12	2.45
Chape	ITAndroids	0.53	0.30	0.36	0.10	2.09
Gankenkun	CITBrains	0.65	-	-	-	–
KAREN	MRL-HSL	0.73	0.54	0.48	0.18	3.10
NUgus	NUbots	0.90	0.38	0.42	0.26	1.97
RFC2016	01.RFC Berlin	0.65	0.37	0.40	0.36	1.99
Wolfgang-OP	Hamburg Bit-Bots	0.83	0.48	0.51	0.22	1.89
Darwin/OP2	Robotis	0.45	0.29	0.29	0.12	2.47
OP3	Robotis	0.51	0.45	0.54	0.13	2.01
NAO	Aldebaran	0.57	0.43	0.58	0.25	0.85

for each robot a URDF is needed to solve the IK. These were only available for the commercial robots. For the other robots, the URDF was created using the URDF export function of Webots based on the simulator model. For each robot, we optimized the spline parameters using the above-described approach of doing 1,000 trials using the multivariate MOTPE and then 500 trials using the independent MOTPE. We did not use any of the stabilization methods for this comparison, since they are not necessary in a simulation without external forces.

The reached walk velocities of the best parameter set are shown in Table 2. There is only one robot in the HLVS, the Gankenkun, that does not work since it has parallel-kinematics in the legs. This is generally not supported by URDF and therefore it is not possible to solve the IK. We assume, that the robot would be able to walk if a custom IK is used. All robots without parallel kinematics were able to walk. Naturally, the larger robots, e.g. the Wolfgang-OP, reached higher velocities than the smaller robots, e.g. the Bez. Noteworthy is that the commercial robot models are less realistic, especially the OP3, which has a motor torque of 1,000 Nm in simulation. Therefore, they reach high walk velocities.

5.3   Previous Usages

The Hamburg Bit-Bots used earlier versions of the presented walk algorithm successfully on real hardware at the RoboCup championship in 2018 and 2019. Furthermore, the walk controller won the teen-size push recovery technical challenge in 2019. We made the experience that the optimized parameters are often not directly applicable to our real world robot. Mainly the torso pitch needs to be adjusted. Our assumption is that the model of the robot has an inaccurate weight distribution. Still, adapting the parameters from simulation to the real robot is simpler than optimizing them from scratch. Furthermore, in our experience, it leads to better walking, as humans tend to focus on the first local maximum when optimizing manually.

In simulation, it has been used by the Hamburg Bit-Bots to score third place in the RoboCup 2021, first place in RoboCup Brazil Open 2021, and second place in the HLVS 22. The Hamburg Bit-Bots also used it successfully on a simulated Darwin-OP robot in the running robot competition 2020 and 2021, including a modified controller version for walking on stairs. The team NUbots is using the walk engine since 2019 [14]. The quintic spline engine has been used by Putra et al. as a basis for their walking approach [21].

6   Conclusion

We presented our open-source walk controller that is working on all non-parallel robots in the RoboCup Humanoid Virtual Season, as well as on widespread commercial robots. It reduces the entry barrier for new teams by providing a simple-to-use solution to the bipedal walk problem. The usefulness of this approach has been evaluated in the multiple real and virtual RoboCup tournaments.

In the future, we would like to test the algorithm on further robot models and are, therefore, asking more teams to make their models open-source. The optimization procedure could be improved further by automatically limiting the search space based on the kinematic restrictions of the robot.

The presented software, as well as accompanying videos and URDFs, can be found at: https://bit-bots.github.io/quintic_walk/.

References

1. Robot Models HLVS. https://humanoid.robocup.org/hl-vs2022/teams/
2. Robotis OP3. http://emanual.robotis.com/docs/en/platform/op3/introduction/
3. Aftab, Z., Robert, T., Wieber, P.B.: Ankle, hip and stepping strategies for humanoid balance recovery with a single model predictive control scheme. In: 12th IEEE-RAS International Conference on Humanoid Robots. IEEE (2012)
4. Akiba, T., Sano, S., Yanase, T., Ohta, T., Koyama, M.: Optuna: A next-generation hyperparameter optimization framework. In: 25th ACM SIGKDD International Conference on Knowledge Discovery & Data Mining (2019)
5. Allgeuer, P., Behnke, S.: Fused angles: a representation of body orientation for balance. In: IEEE/RSJ International Conference on Intelligent Robots and Systems (IROS). IEEE (2015)
6. Allgeuer, P., Behnke, S.: Omnidirectional bipedal walking with direct fused angle feedback mechanisms. In: IEEE-RAS 16th International Conference on Humanoid Robots (Humanoids). IEEE (2016)
7. Asada, M., von Stryk, O.: Scientific and technological challenges in robocup. Annu. Rev. Control Rob. Auton. Syst. **3**, 441–471 (2020)
8. Behnke, S.: Online trajectory generation for omnidirectional biped walking. In: IEEE International Conference on Robotics and Automation (ICRA). IEEE (2006)
9. Bergstra, J.S., Bardenet, R., Bengio, Y., Kégl, B.: Algorithms for hyper-parameter optimization. In: Advances in Neural Information Processing Systems (2011)

10. Bestmann, M., Güldenstein, J., Vahl, F., Zhang, J.: Wolfgang-op: a robust humanoid robot platform for research and competitions. In: IEEE-RAS 20th International Conference on Humanoid Robots (Humanoids). IEEE (2021)
11. Chitta, S., Sucan, I., Cousins, S.: Moveit![ros topics]. IEEE Robotics Autom. Mag. **19**(1), 18–19 (2012)
12. Coumans, E., Bai, Y.: Pybullet, a python module for physics simulation for games, robotics and machine learning. http://pybullet.org (2016–2021)
13. Deb, K., Agrawal, S., Pratap, A., Meyarivan, T.: A fast elitist non-dominated sorting genetic algorithm for multi-objective optimization: NSGA-II. In: Schoenauer, M., Schoenauer, M., et al. (eds.) PPSN 2000. LNCS, vol. 1917, pp. 849–858. Springer, Heidelberg (2000). https://doi.org/10.1007/3-540-45356-3_83
14. Dziura, N., et al.: The nubots team extended abstract (2022)
15. Gouaillier, D., et al.: Mechatronic design of NAO humanoid. In: IEEE International Conference on Robotics and Automation. IEEE (2009)
16. Ha, I., Tamura, Y., Asama, H., Han, J., Hong, D.W.: Development of open humanoid platform DARwIn-OP. In: SICE Annual Conference 2011. IEEE (2011)
17. Hansen, N., Ostermeier, A.: Completely derandomized self-adaptation in evolution strategies. Evol. Comput. **9**(2), 159–195 (2001)
18. Li, Z., et al.: Reinforcement learning for robust parameterized locomotion control of bipedal robots. In: IEEE International Conference on Robotics and Automation (ICRA). IEEE (2021)
19. Michel, O.: Cyberbotics ltd. webots^{TM}: professional mobile robot simulation. Int. J. Adv. Robotic Syst. **1**(1), 5 (2004)
20. Ozaki, Y., Tanigaki, Y., Watanabe, S., Onishi, M.: Multiobjective tree-structured parzen estimator for computationally expensive optimization problems. In: Proceedings of the 2020 Genetic and Evolutionary Computation Conference (2020)
21. Putra, B.P., Mahardika, G.S., Faris, M., Cahyadi, A.I.: Humanoid robot pitch axis stabilization using linear quadratic regulator with fuzzy logic and capture point. arXiv preprint arXiv:2012.10867 (2020)
22. Rodriguez, D., Brandenburger, A., Behnke, S.: Combining simulations and real-robot experiments for bayesian optimization of bipedal gait stabilization. arXiv preprint arXiv:1809.05374 (2018)
23. Rouxel, Q., Passault, G., Hofer, L., N'Guyen, S., Ly, O.: Rhoban hardware and software open source contributions for robocup humanoids. In: Proceedings of 10th Workshop on Humanoid Soccer Robots, IEEE-RAS International Conference on Humanoid Robots, Seoul, Korea (2015)
24. Ruppel, P., Hendrich, N., Starke, S., Zhang, J.: Cost functions to specify full-body motion and multi-goal manipulation tasks. In: IEEE International Conference on Robotics and Automation (ICRA). IEEE (2018)
25. Saputra, A.A., Takeda, T., Botzheim, J., Kubota, N.: Multi-objective evolutionary algorithm for neural oscillator based robot locomotion. In: 41st Annual Conference of the IEEE Industrial Electronics Society (IECON). IEEE (2015)
26. Seekircher, A., Visser, U.: A closed-loop gait for humanoid robots combining LIPM with parameter optimization. In: Behnke, S., Sheh, R., Sarıel, S., Lee, D.D. (eds.) RoboCup 2016. LNCS (LNAI), vol. 9776, pp. 71–83. Springer, Cham (2017). https://doi.org/10.1007/978-3-319-68792-6_6
27. Shafii, N., Lau, N., Reis, L.P.: Learning to walk fast: optimized hip height movement for simulated and real humanoid robots. J. Intell. Robotic Syst. **80**(3–4), 555–571 (2015)

28. Silva, I.J., Perico, D.H., Costa, A.H.R., Bianchi, R.A.: Using reinforcement learning to optimize gait generation parameters of a humanoid robot. XIII Simpósio Brasileiro de Automaçao Inteligente (2017)
29. Stelter, S., Bestmann, M., Hendrich, N., Zhang, J.: Fast and reliable stand-up motions for humanoid robots using spline interpolation and parameter optimization. In: 20th International Conference on Advanced Robotics (ICAR). IEEE (2021)
30. Ziegler, J.G., Nichols, N.B., et al.: Optimum settings for automatic controllers. Trans. ASME **64**(11), 759–765 (1942)

A Library and Web Platform for RoboCup Soccer Matches Data Analysis

Felipe N. A. Pereira[✉], Mateus F. B. Soares, Olavo R. Conceição,
Tales T. Alves, Tiago H. R. P. Gonçalves, José R. da Silva, Tsang I. Ren,
Paulo S. G. de Mattos Neto, and Edna N. S. Barros

Centro de Informática - Universidade Federal de Pernambuco, Av. Jonalista Anibal
Fernandes, s/n - Cidade Unviersitária, Recife, PE 50.740-560, Brazil
{fnap,mfbs2,orc,tta,thrpg,jrs8,tir,psgmn,ensb}@cin.ufpe.br

Abstract. An important part of ensuring continuous development and betterment of performance of a robot soccer team is the process of analyzing past matches and generating insights about possible causes for good or bad outcomes the team has in the field. RoboCup robot soccer matches from leagues such as the 2D Simulation League generate log files that can help create insights using data analysis, but since there is currently no active open collaboration between teams for building an ecosystem of analysis tools, this area is underdeveloped - as a community - and could be improved. We propose an open-source data analysis library that contains all of the basic structures needed for implementing any analysis, as well as a collection of ready-to-use, quasi-agnostic analysis that can be used to analyze matches from any soccer league with little or none adaptation. We believe this can be a common ground for developers from any team to work together in the advancement of technology and lower the barrier of entry into the data analysis realm for teams that are not yet involved in the area. We also demonstrate how this library can be leveraged as a software component for other projects, by building a custom web platform that utilizes it.

Keywords: Performance metrics · Data analysis · Web development · Open source · Data visualization

1 Introduction

RoboCup's 2D Soccer Simulation League (SIM2D) and Small Size League (SSL) are two robot soccer leagues in which a log file is generated and made available at the end of a match containing rich information about the game such as position of all the game entities at all times, game states, players states and scores.

This log file is a valuable record of a team's execution, but presents a challenge to extract relevant insights by directly looking at it in its raw form, since it contains huge amounts of data that would be very hard to look into in an

© The Author(s), under exclusive license to Springer Nature Switzerland AG 2023
A. Eguchi et al. (Eds.): RoboCup 2022, LNAI 13561, pp. 177–189, 2023
https://doi.org/10.1007/978-3-031-28469-4_15

efficient analytic way, as a human. Hence, to transform data into intelligence, it is necessary to create software capable of processing all that information to present it in a more human-readable way. However, since there are no general standards on how to operate these analyses or an active ecosystem of tools made for this specific purpose in the RoboCup community, each team develops their own isolated solutions, which prompts to a lot of re-work and missed collaboration opportunity, limiting the rate of the possible technological advances in the area.

The amount of data generated by a single match of RoboCup 2D Soccer Simulation League (SIM2D) is about 25MB. In the Small Size League (SSL), this log file can get up 500 MB. Despite being different in size, these files are quite similar in what they store, but since there is no common ground or platform that analyzes and manages this data, most of it is unused or left aside. A common tool for collecting and organizing this information can bring tremendous value to a team's understanding of their robots or simulated agents.

Such a tool should be diverse, adequate and compatible across environments and its development should be as context-agnostic as possible, meaning that all of its internal components and systems should be designed with general purpose and modular components, so that little to no adaptation is needed to correctly function in different contexts.

A Python package can be used as a component for building other software. This format is beneficial for sharing and coupling with other tools, for instance, a modern web application that serves as a Graphical User Interface (GUI). This type of usage improves the user experience by providing a more practical way of interacting with them [2,4], and in the context of this paper: a soccer library. We demonstrate how a thoughtful design of such a platform can boost the library's usability and enable a productive, efficient, and friendly environment for extracting valuable insights from soccer matches.

We believe that a unified platform for data analysis development encourages the community to collaboratively grow the field and encompass other related areas such as Machine or Reinforcement Learning. Thus, taking into consideration the utility of a unified codebase, the community growth possibilities and the value brought by having data analyzed, our contributions in this paper are the following.

1. The release of an open source data analysis Python library called "Soccer Analyzer" for SIM2D and SSL. It proposes to be an open standard so that any RoboCup community member can use its built-in analysis, collaborate to improve the existing ones, or create new ways of exploring data from robot soccer matches.
2. Built-in analysis algorithms compatible with SIM2D and SSL.
3. An evaluation of a model for expected goals in soccer matches using SIM2D log files.
4. An evaluation of the released library capabilities as a module for building other software by implementing a web platform that provides the user a web application as a friendlier way of interacting with the Soccer Analyzer library.

2   Related Work

Initiatives around gathering and analyzing data in sports have been developed in real life [5,6] and simulated environments [14]. Within the SIM2D community, we have found work from teams that developed heuristics to use the data generated from the game events (logs), although all of the projects found were category-specific and also built with other purposes besides transforming data into information. Albeit useful, this approach lacks scalability, reuse, and modularity, which hampers the development in the area.

One of the first implementations of a log analyzer application found is from the ITAndroids team in 2013 [7]. The software aimed to detect players' positions in order to detect formation patterns, its purpose was to be an auxiliary tool and was created for this single usage. No entry point of development was provided for future extensions of the mechanism.

Another example is the Namira Log Analyzer [1], a robust application developed by the Namira team in 2020 that receives the generated log files from SIM2D matches and returns team and player analysis such as the amount of complete passes and interceptions along with its accuracy, a shoot counter that differentiates between three types of shots (goal shots, width or on length shots), total stamina usage, stamina used per distance, etc. This approach is more generic than ITAndroid's, providing multiple algorithms that generate analyses and the results are given as a JSON file which is a very useful format to transfer data between applications. Other purposes were given to this tool, Namira developed a tournament simulator named Namira TPAS, which is an interface capable of running a batch of custom matches with ease, and utilizing the log analyzer to process the games statistics. Although very useful, the tool was not developed in a way that provides a simple programming interface for those who want to contribute to the functionality already provided and also it is not very customizable.

A similar tool was MT2018's data miner [14]. They developed a way to extract data from the SIM2D log files using regular expressions in Python, the resulting data was output in a CSV file and stored in a MySQL database for future consulting. Their goal was to mine the data and use them as input for machine learning algorithms. But to build software over regular expressions can considerably hinder the scalability and reuse of the code, due to its dense and hard to read syntax which is not ideal for a large codebase that needs clarity and readability.

In SSL analysis, the RoboFEI-SSL team built a log analyzer [10] with a graphical interface capable of displaying the field's current status, the players' positions, and the ball direction, as well as referee messages in the log. The application has a lot of features and filters, providing the user a custom experience for visualizing only the intended information and can be used in real time matches. This approach provides the end user a more sophisticated interface, but due to the use of QT5 for graphical interface, it restricts the application to a non-friendly design if compared with more advanced front-end frameworks like React, that allow richer interactions with the user.

We have also created in the past a Log Analyzer for SIM2D [8], but it suffered from the same limitations as some of the cited projects above: it wasn't scalable, since the GUI code was intertwined with the data analysis code, meaning that creating new analysis with components already created was not possible; every analysis was independent and did not communicated or shared information with other parts of the code. Also it was focused on only SIM2D, so it couldn't be used to analyze matches from other leagues.

3 Soccer Analyzer

To assist a constant development cycle, a modular, reusable, and easily maintainable codebase is a must for fast scaling and implementation. Speed and clarity are of the essence to having data-oriented development when on a search to evaluate Key Performance Indicators (KPIs). Soccer Analyzer[1] (SA) is a common structure designed exactly for that. It aims to be a foundational codebase that provides building blocks for creating analyses using Python and Pandas whilst presenting each of them as a resource for more advanced implementations.

This structure was created with the purpose of being a common and shared codebase upon which the RoboCup community can construct data analysis on top of any football-related robotic category. This initiative follows the tendency of building common architecture designs [12] for Robocup competitions. Having a common library that anyone can use, expands collaboration possibilities and lessens the barriers for teams to start investing in the field of data analysis. Besides that, it takes away the focus of the infrastructure required to create such investigations and directs the attention where it is needed: in analyzing the data.

The common structures created are elements such as the field, the ball, players/robots, a match, and characteristics of these elements such as width, height, speed, score, and category among many others. These components can be manipulated at will to adapt to different contexts enabling their usage in various robot soccer categories.

Specific analyses are also common structures, ball possession can, for instance, always be calculated in the same manner - considering the closest player to the ball - and to achieve this information only position and time data are needed. This analysis can be agnostic to the specific category that is being evaluated, other analysis may not work in the same manner; SIM2D stamina analysis makes no sense in SSL considering that robots don't have stamina. The library does not restrict its usage though, if a team using Soccer Analyzer decides to use stamina as a feature to examine a robot battery, for instance, they will still be able to do so. The package provides the tools, the method in which one wants to use the tools provided is completely up to their development choices.

Currently, Soccer Analyzer supported contexts are Soccer Simulation League (SIM2D) and Small Size League (SSL), and analyzing the log files from these

---

[1] https://github.com/robocin/SoccerAnalyzer/.

leagues works out of the box. Support for other contexts, such as other robot soccer leagues, or even human soccer can be implemented systematically mapping the provided modules or creating new ones.

3.1   Internal Architecture

The Soccer Analyzer architecture was carefully thought to be simple and able of attaching new modules. The modules are operations, tools, collections, or algorithms that bring results or assist other modules in any way required as can be seen in Fig. 1. All components of the SA are called modules, with the most important ones being the Match, MatchAnalyzer and every Analysis module.

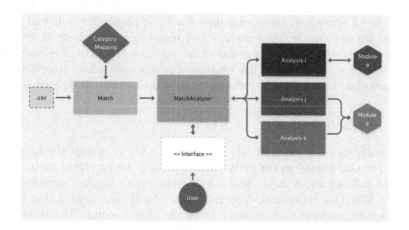

Fig. 1. Soccer analyzer architecture

A Match object contains the most information possible of a given game match: geometric dimensions such as width, height, interest spots, players, field, events, final score and category. In essence, a match of any category can be represented in this structure. This object can be expressed as a block of information that can be populated and consulted by any module. A match construction is dependent on its category, a SIM2D match has different properties than a SSL match, but a category match is not dependent on the match object because the attributes are interchangeable and adaptive. This approach directs and prioritizes the information that is being represented and the various forms it can be portrayed instead of structure necessary to describe it.

To create a match the necessary components are: a .csv file of a real match, which contains information extracted from .log files in SSL and .rcg or .rcl files in SIM2D; and a object with the mappings of each category attributes with the contents in these files. The structure chosen for this purpose was an enumerator. Besides mapping, the enumerator type determines which category is being analyzed and having this information changes the behaviour of control modules like the MatchAnalyzer. The MatchAnalyzer is the central piece of SA workflow, the processing of almost every other module is centered around the usage of MatchAnalyzer. It controls and triggers other regions of the framework and its responsibilities are to manage the Match objects and create the available Analysis objects of the given Match.

The Analysis modules are responsible to ingest the logs and process their information running algorithms to generate the results, which are the analyses themselves. This module can make use of other auxiliary components or even other analyses. The architecture provides visibility of external scopes to read information but not to alter, this lessens the points of incorrectness that could propagate to the entire system given that the resources and information are shared across the software. All Analyses inherit from an abstract analysis class that defines an interface for computation and data sharing. This means that each one, across all types and categories, have the same access methods and private methods, which helps implementation and usage from anywhere in the code. The general structure of SA contains other modules than the ones mentioned above. The current release of SA contains packages providing algorithms and data structures that enable the software to develop and visualize those analyses.

3.2   Inputs and Outputs

SA internally uses a Python library called Pandas [3] to manage the log's information. For this library to correctly parse the data, the logs must be in comma separated values (csv) format, which is not the native logging arrangement of SSL and SIM2D competition. The raw data of SSL comes in a .log file and SIM2D server produces one rcl and one rcg file after a match. To convert these files into csv format, we used two internal softwares previously developed by the RobôCIn team, one for each category.

When SA executes, it provides a runtime interface with all information it was able to process from a given csv input. The interface provides output methods that generate pieces of information in different forms, depending only on the needs of the developer. Currently, the output formats are: a JSON file, which is a standard for data consumption in web applications; a populated Python dictionary and a simple description text with the metrics calculated.

3.3   Analysis

The analysis modules are an important part of SA contributions. Their elaborations were guided with category agnosticism in mind, meaning that one analysis algorithm should bring correct results to data from any category. This approach brings an inherent complexity to implementation, but delivers a much improved user experience for anyone using the library as a data analysis engine. More details on the analyses can be viewed in Table 1.

Table 1. Soccer analyzer implemented analysis

Analysis list	
Analysis	Description
Ball History	Collects the ball position columns in the data, returning a tuple containing all ball positions in the x and y axis during the entire match
Foul Charge	The algorithm detects the positions where fouls were committed during the game, along with their quantities and proportion between both teams
Penalty	Detects the game times in which penalties, or shootouts, occurred by monitoring the playmode and returns two lists containing the occurrences for each team
Playmodes	Summarizes quantitatively which playmodes occurred in a given match and returns this information as a list of distinct playmodes
Stamina	A simple extraction of stamina attribute in SIM2D players during the game cycles.
Ball Possession	Processes spatial data of all entities in the game relative to the ball and determines the most likely entity to have the ball possession in each cycle.
Time after events	Measures the amount of time passed since the occurrence of a desired event
Expected goals	A model for evaluating goal probability using angle to goalposts, distance from shot to goal and number of players nearby the shot

3.4 Expected Goals (xG)

Results in soccer, more so than in any other sport, can be greatly influenced by random factors. In order to score, one must first attempt a shot at goal. Assessing a shoot performance could simply entail taking a look at the total shots and shots on target. While these are useful metrics for dealing with chance creation, they do not tell the whole story as not all shots bring the same value when it comes to scoring. This is where expected goals (xG) comes into play. xG is a metric that approximates the probability that a shot will result in a goal based on a number of factors such as the distance from where the shot was taken, angle with respect to the goal line, the game state (what is the score), if the shot came during a counter attack and more. xG can therefore serve as a gauge of how good a team is at creating chances and limiting the opponents' chances. It can also be used to analyze a player's ability to create shooting opportunities in dangerous areas and how well it takes its chances.

To create the xG model for 2D soccer, a dataset was constructed to store variables and needed attributes. To create the dataset the team used the Robocup Archive database [9]. Then, we processed over 400 log files from matches between the years 2019 and 2021 interested in variables such as shot location (x and y coordinates), distance from shot location to goal, number of players near the shooter, angle to goal posts and more. With the obtained dataset the team generated more attributes, such as distance squared, distance from center of the pitch, etc to use in the model testing phase.

All models used are based on the logistic function and were built using the Generalized Linear Model Regression from statsmodels module [13]. After testing different model builds with different parameter combinations we concluded that the best performance was achieved using **angle**, **distance** and **players nearby** since they all achieved a P value of 0 in the model, which means that they are all statistically significant at 1% level, and in further testing, using the log likelihood and the ROC curve, the model built with cited parameters was the best performer.

4 Web Platform

4.1 Motivation

A GUI for Soccer Analyzer: In order to be highly reusable and purpose-focused, the Soccer Analyzer library is designed to have purely textual output; this means that, for rendering plots, another software is needed. This second software would use Soccer Analyzer as its core for data analysis, and render plots based on the library outputs, while also providing a more user-friendly and organic way of interacting with the library.

Hence, there is the possibility for existing a second application, responsible for being the front-facing part of the software, and a facilitator that ensures a better user experience. In this section we will describe the development and outcomes of such a platform, as a way of demonstrating the Soccer Analyzer library potential to serve as a component for other software.

4.2 Front-End and Back-End

Technologies and Implementation: Our front-end application was made using the React framework. In the back-end, we run a server with Python and Flask. By providing the output in JSON format natively, the Soccer Analyzer library integrates well with such a framework, since JSON is a very commonly used standard in the web applications space.

User Experience: One of our biggest priorities was ensuring that the user experience was at the center of our design process, which was crucial in order to align expectations with all parties involved. Hence, we worked closely with RobôCIn SIM2D team, having discussions, watching games together to brainstorm what could be useful points of interest to be analyzed, understanding their needs and making sure we had their expected features and whether the solutions we've made actually helped solve the problems raised by them.

Data Visualization: Since providing a user-friendly way of visualizing data about a match is one of the platform's main objectives, the plotting aspect of the application is highly important. The way the graphics are presented to the users has the power to influence their interpretation of the information displayed, thus it has a critical role in the process of creating new strategies, testing them, and evaluating their results interactively, as well as in correctly understanding the players' and team's behaviour. The process of developing the visualizations on the front-end web platform is facilitated by the Soccer Analyzer library ability of functioning in standalone form. It helps with quick prototyping by utilizing packages such as Matplotlib for sketching visualizations, while also needing very little adaptation in order to interface with the other software being developed for showing the final graphics with web plotting libraries and frameworks.

General Dashboard: The General Dashboard is a screen responsible for showing general analysis and statistics about how the teams' performances compare. Its function is to provide a macroscopic view of what is happening in recent played matches, the statistical patterns of notable players across multiple teams and leagues and which teams are doing a good campaign in which championships, for example.

Team Profile: The team profile is made of two sections: the general team information and the team analysis screen. The general team information section shows some of the most important information about the team in a compact form: number of played matches, wins, losses, draws, etc., are laid out in plots that help achieve a sense for the team's average situation in the competitions, in any of the available Leagues.

The team analysis screen has been through two major iterations, the first one featuring a list of the available matches and a list of the available analyses, and an area in which the chosen analysis applied to the chosen match was displayed; and the second one (still in development) is an infinite white canvas in which it is possible to instantiate and move around any kind of media, such as images, videos, PDF's, and most importantly, the output from the analysis, both in textual and in graphical form.

Integration of Front-End, Backend and Soccer Analyzer: Since Soccer Analyzer is built from the ground up with the aim of being used as a software component, it is really simple to integrate it in any application. Our current architecture involves having a Python Flask server that imports the Soccer Analyzer package and, upon receiving HTTP requests from the front-end app, chooses which imported data analysis function to run and give back to the client its response.

5 Experiments and Real World Use

The reported library has been used by the RobôCIn team during past SIM2D competitions, and assisted in the decision-making process when determining how the team's code should be modified in order to achieve better results in those contests. Its aid helped the team win first place in the Latin American Robotics Competition (LARC) of 2021.

One of the first data analysis experiments conducted while in a competition was simple but insightful: by plotting a line graph of all teams' mean stamina along the time axis together with visual indicators on when a goal occurred, we realized that significant drops in stamina happened consistently after a goal.

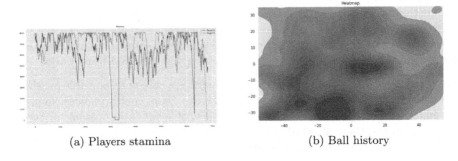

(a) Players stamina   (b) Ball history

Fig. 2. Plots in jupyter notebook using SoccerAnalyzer

One of the insights in this situation was that, if we could avoid this drop in stamina for at least some of the players, the mean stamina of our team would be higher than the other team after each goal, which could be an advantage in certain situations.

In 2021, we realized that our goalie was grabbing the ball with its hands and thus making fouls whenever a teammate would back off the ball. In order to stop this behaviour, a code change was made but its efficacy couldn't be easily verified by manually watching various matches. We used the Soccer Analyzer library to build an analysis that was able to read multiple log files and detect when the goalie as doing this erratic behaviour. With this information in hand, we were able to quantitatively compare the occurrences before and after the change and evaluate it more precisely.

In 2022, a Jupyter Adapter module, shown in Fig. 2, was implemented in SA, integrating plotting functions targeted to be used inside a Jupyter Notebook. This feature was designed to incorporate the mindset of data driven development inside RobôCIn's SIM2D team planning, development and testing cycles. The feature was used as a playground for quick analyses of stored games and is gradually being inserted into the team.

Some analysis have a inherit error factor due to its algorithm calculation approach. For instance, ball possession minimal distance to ball might incorrectly assign possession in a cycle where the ball is travelling from a pass or kick, or when two robots are disputing the ball, since there is no evaluation or processing being done to handle these corner cases. This error can be mitigated using another heuristic but cannot be completely correct, since there is no ground truth data about this specific analysis. In contrast, foul charge, play-modes, stamina, penalty, and time after events have a ground truth value which can be manually counted by watching the game or reading the game logs; within the SA architecture these last mentioned analysis had the exactly same results as the ground truth values.

5.1   The xG Model

The xG value is calculated inside the *shooting analysis* module every time a shot is identified, the module utilizes the model described in Sect. 3.4 passing the parameters angle (from shot location to goal posts), distance from shot location to goal and number of players near the shooter, and receiving the goal probability value between 0 and 1. The results can then be used to plot goal probability per region or to highlight a team's chance quality in games (see Fig. 3). The xG value can also be obtained through the course of a game or a sequence of games to help assess the chance quality a team or a player is achieving and how well it is taking those chances comparing actual goals scored and the expected value.

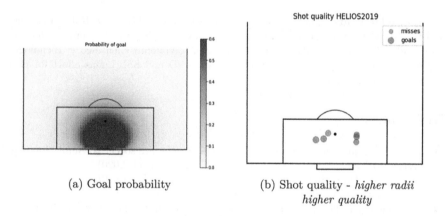

(a) Goal probability

(b) Shot quality - *higher radii higher quality*

Fig. 3. Goal probability and shot quality

6   Conclusions and Future Work

SIM2D team algorithms tend to be very complex to develop and hard to make improvements to, and game analysis helps significantly with this process because it allows organizations to detect the strong and weak points of the team. The main use of Soccer Analyzer is to provide support to SIM2D and SSL games, possibly extending to other categories in the future, for teams to better understand and detect where is important to make changes in order to improve performance.

Soccer Analyzer is directly useful as a library for simple programs and also functions well as a software component for developing other complex software. By releasing it as an open-source project, we hope for it to become the go-to data analysis toolkit in the RoboCup community, help teams introduce themselves to the data analysis realm and create a platform in which different teams can collaborate and share code towards the advancement of the field.

We are already working on adapting Soccer Analyzer for other RoboCup Leagues, and in the long-term we plan on adapting it for analyzing soccer played by humans. Another long-term goal is generalizing the library even further, so it can be an analysis tool for any kind of sport that shares fundamental similarities with soccer. We have started to design a new serverless architecture for the Web Platform that will ensure its scalability and ease of maintenance, and we aim to ship this new infrastructure together with the software's next version. Further development, test and polishing of the Web Platform will also be conducted in order to prepare it for public release, as we have plans of giving access for the RoboCup community to our Web Platform.

In the next release, we plan on providing the Web Platform with a code editor that can be used to create custom analysis directly inside the browser, enabling users to have more control over their data analysis session. We also plan on integrating the Web Soccer Monitor project [11] as a component inside the Web Platform to enable visualization of past matches through log files.

Acknowledgements. The authors would like to acknowledge the RoboCIn's team and Centro de Informática - UFPE for all the research support. The first author was also funded by the Conselho Nacional de Desenvolvimento Científico e Tecnológico (CNPq). They also appreciate all the Simulation 2D and SSL teams effort for their open-source contributions.

References

1. Asali, E., Negahbani, F., Bamaei, S., Abbasi, Z.: Namira soccer 2d simulation team description paper 2020. arXiv preprint arXiv:2006.13534 (2020)
2. Benyon, D.: Designing User Experience. Pearson, Harlow (2019)
3. pandas dev: pandas, June 2022. https://pandas.pydata.org/
4. Hartson, R., Pyla, P.S.: The UX Book: Process and Guidelines for Ensuring a Quality User Experience. Elsevier (2012)
5. Hendriks, K., et al.: Football data analysis. Technische Universiteit Eindhoven (2016)
6. Memmert, D., Raabe, D.: Data Analytics in Football: Positional Data Collection, Modelling and Analysis. Routledge (2018)
7. Muzio, A.F.V., Lira, C.P., Ramos, L.F.M., Ferreira, R.R.: ITAndroids 2D CBR 2013 TDP
8. Pereira, F.N., Soares, M.F., Barros, E.N.: A data analysis graphical user interface for robocup 2d soccer simulation league. In: 2020 Latin American Robotics Symposium (LARS), 2020 Brazilian Symposium on Robotics (SBR) and 2020 Workshop on Robotics in Education (WRE), pp. 1–6. IEEE (2020)
9. RoboCup: Robocup official repository (2022). https://archive.robocup.info/
10. RoboFEI: Loganalyser robofei SSL (2021). https://gitlab.com/robofei/ssl/LogAna lyserRoboFei-SSL

11. RobôCIn: Web soccer monitor, September 2021. https://github.com/robocin/Web SoccerMonitor
12. RobôCIn: Soccer common, May 2022. https://github.com/robocin/soccercommon
13. statsmodels: statsmodels module, June 2022. https://www.statsmodels.org/sta ble/
14. Yang, Z., et al.: Mt2018: team description paper. In: RoboCup 2018 Symposium and Competitions, Montreal, Canada (2018)

Web Soccer Monitor: An Open-Source 2D Soccer Simulation Monitor for the Web and the Foundation for a New Ecosystem

Mateus F. B. Soares[(✉)], Tsang I. Ren, Paulo S. G. de Mattos Neto, and Edna N. S. Barros

Centro de Informática - Universidade Federal de Pernambuco, Av. Jonalista Anibal Fernandes, s/n - Cidade Unviersitária, 50.740-560 Recife, PE, Brazil
{mfbs2,tir,psgmn,ensb}@cin.ufpe.br

Abstract. The 2D Simulation League (SIM2D) is one of the most accessible RoboCup leagues, since there are no hardware costs included, and codebases from which you can build a new team are readily available for free. The league, however, still has a notable barrier of entry: its setup. The setup process for the SIM2D environment can be daunting for newcomers, especially for people new to programming or that don't have access to a linux based distro. In this sense, if there is interest in lowering the barrier of entry of the SIM2D league, and with it, of robotics in general, it would be helpful to have tools with minimal setup, or that don't require any setup at all. This article reports on an Open-Source monitor for SIM2D games, "Web Soccer Monitor", that runs entirely on the browser and doesn't require a setup to function. It is useful in itself as it simplifies the experience of utilizing a monitor from the user's perspective, while also providing developers with a more modern and agile framework in which to implement new features, but it also serves as the foundation in which the RoboCup community can start building an entire ecosystem of SIM2D web tools, which would lower even more the barrier of entry and would, among other things, facilitate the creation of new categories, such as a fully-fledged SIM2D Junior League.

Keywords: Web development · Open source · Data visualization · 2D soccer simulation league

1 Introduction

In 2021, a conversation about the development of a web-based simulator for SIM2D began spreading between the league members and organizers. A possible use-case for it would be to host SIM2D games for a new Junior league, since a web simulator would simplify the setup aspects of the SIM2D environment and thus facilitate the participation of Junior competitors in the category.

In that context, we decided to tackle part of this challenge by designing and implementing from scratch a web-based monitor inspired by the already existent

© The Author(s), under exclusive license to Springer Nature Switzerland AG 2023
A. Eguchi et al. (Eds.): RoboCup 2022, LNAI 13561, pp. 190–199, 2023
https://doi.org/10.1007/978-3-031-28469-4_16

and widely used monitors rcssmonitor [12] and soccerwindow2 [10], both desktop programs that need a setup in order to work.

While in the design process, we realised that by tackling this problem, we are not only helping to possibly create new RoboCup leagues, but also lowering the barrier of entry for newcomers in the current SIM2D scene, and improving the experience of current members. By developing this software, we are creating an application that has the potential to be the foundation for a suite of new SIM2D tools, all designed to run within an internet browser.

In this paper we report the development and community involvement of an open source web-based SIM2D league monitor, and we also talk about the potential it has for being a stepping stone for the flourishment of a new ecosystem in the SIM2D RoboCup community, which might be influential to other leagues as well.

2   Related Works

A number of SIM2D monitors exist. The software rcssmonitor [12] and soccer-window2 [10], for example, are two widely used monitors in the scene, however, they're not web-based (require setup). There is also a project by the RoboFEI team called "LogAnalyserRoboFei-SSL" [13] that provides, among other things, a monitor for visualizing matches; this project, although, is for SSL only, and runs locally as a desktop application.

The RoboCup Archive [7] uses JaSMIn [9] as a web monitor for replaying past log matches, and it is a great project, but might be heavy to run on lower-end computers due to its use of 3D graphics powered by opengl. Also, because of its 3D nature, it is harder to develop new features for JaSMIn, when compared with a 2D counterpart, namely, the 2D canvas API provided by the browser. Another hindrance for its continued development and community engagement is the use of vanilla JavaScript instead of a framework such as React [6], which is a widely know library that helps developers program faster and manage their projects more efficiently. Hence, it is justified to build another tool, focused on 2D graphics and implemented with a robust web framework, therefore enabling us to offer the community a greater variety of choices (2D or 3D monitoring) when a browser based monitor is needed. JasMIn and this new 2D monitor could then be used alongside as different views inside other browser based projects.

3   Web Soccer Monitor

3.1   Technology and Architecture

Web Soccer Monitor uses the React framework [6] for building its core and accessory UI (User Interface) components. It uses the React Konva library [11] for rendering the 2D graphics of the monitor itself in a HTML canvas, which is lightweight and should be easy to run on most computers.

It is capable of processing a compacted log file uploaded by the user by sending it to a back-end, responsible for extracting and converting the *.rcg.gz file into a JSON format that is then sent back to the front-end so the monitor can start playing it.

The back-end is a simple python Flask server that uses gzip [8] to uncompress the file and the rcss-log-extractor [14] script to translate it to an intermediary .csv format.

Although having a back-end works, it is not ideal for a reusable component to have the necessity of connecting to a specific external service for basic functioning. Thus, we are working on making Web Soccer Monitor a front-end only application: the idea is that it should be able to function as a front-end software by itself, meaning that no connection with a back-end should be mandatory (Fig. 1).

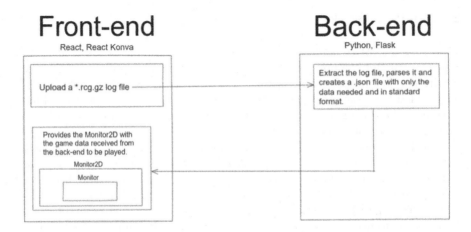

Fig. 1. Web soccer monitor architecture diagram

3.2   Component-Based Modularity

Since it is crucial for the Web Soccer Monitor component to be reusable, it is important that it respects a design pattern that takes into consideration that a component must be as independent as possible, and, hence, every piece of data that can be contained into the component itself must be handled locally, but all of the data that needs external access must be received as a parameter.

In the current implementation, there is a general "Monitor" component (shown in Fig. 2), which is responsible for rendering the graphics, and a SIM2D specific "2DMonitor" component (shown in Fig. 3), which provides a custom score bar at the top, plus a connection with SIM2D specific controls. This is done in such a way as to modularize everything to the maximum and enable the creation of new functionalities with less work, by reusing what already existing components can offer.

This design choice is what could enable us to create a "SSLMonitor" component (for Small Size League), for example, without having to re-write all the functionality related to rendering the graphics, since the Monitor component already exists and could be used in conjunct with SSLMonitor or any other new component. In this specific case, one example of a value passed as a parameter down to the Monitor component is the function that draws the background: since different leagues have slightly different pitches, the background drawing function must be defined in the more specific component, such as 2DMonitor or SSLMonitor, and passed down to Monitor, which will just execute it on every frame before rendering the players and the ball, for example.

A contrary example to this is the scale value for the camera view (the zoom the camera has, relative to the game pitch): it is completely controlled within the Monitor component, so its value can be handled inside the component and no input is needed for this control to happen. The current time of a playing match, otherwise, is an information that both the Monitor and its parent component (2DMonitor or SSLMonitor) need to know, so it must be defined in the parent component and passed down to the child (Monitor).

All of this ensures that the components are as modular, reusable, and independent as possible. It also simplifies development, by creating a simple framework in which the relationships between components follow clear rules, and creating new components is mostly a matter of understanding its position in the component hierarchy, and how it handles different types of data.

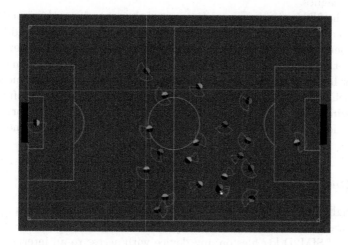

Fig. 2. Web soccer monitor general monitor component

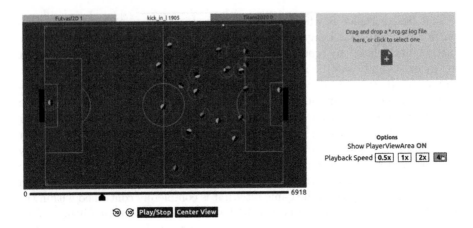

Fig. 3. Web soccer monitor SIM2D monitor component

3.3 Features

Currently, the main feature is the ability to replay SIM2D matches, by uploading a log file in the *rcg.gz format. Support for real time viewing of matches is being worked on, as well as support for replaying the log files of other categories, such as SSL, which would eventually make Web Soccer Monitor a tool for multiple RoboCup leagues.

Inside the 2DMonitor component, some of the controls are: horizontal bar for jumping to any specific time in the game, along with fast forward and backward time jumpers of 10 cycles, a play/stop button and a "Center View" button that resets the camera position and scale, and the playback speed. An option for showing or hiding the view area of the players is also available, with more similar features being developed, some directly inspired by other existing software such as soccerwindow2 [10], and some entirely new, such as ball trajectory and real time statistics, i.e., ball possession percentage, which could improve the watching experience.

3.4 A Ready-to-Use Website

Even though Web Soccer Monitor is thought of being a software component that other software can be built with, it is a useful tool just by itself, since it enables the replay of SIM2D log files on any device with access to an internet browser, without the need for installing anything, which can facilitate the viewing of match replays whenever a computer with SIM2D environment is not available, such as while using smartphones.

Hence, it is important to have a ready-to-use website from where Web Soccer Monitor is made accessible, which has been done and made available as a link in the description of Web Soccer Monitor official repository [15].

3.5   A Simpler Setup for Offline Use

Although by having a hosted website running Web Soccer Monitor (Fig. 4) frees users from any kind of setup, if teams want to setup their own instance, the only needed steps are to clone the project's repository, use the Node Package Manager and Python Pip utilities to automatically install the project's dependencies and then run the front end with Node Package Manager and back end with Python, which is simpler than having to manually install all of the needed packages, such as needed from currently available monitors.

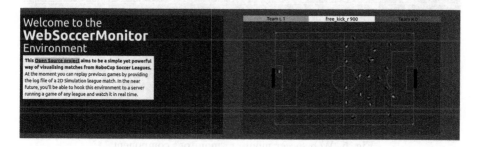

Fig. 4. Web soccer monitor website greeting banner

4   Open Source Movement

4.1   Community Involvement

Since the beginning, this has been a project centered towards open source, which is reflected in both the design philosophies talked about in Sect. 3.2 and also the implementation code itself. With an effort to making sure that the code is easy to understand and well documented, as shown in the Fig. 5, we try to make collaborating to the project as interesting and engaging as possible, which is a crucial point in ensuring the health and aliveness of the project in the long term.

Since its release, in late 2021, some interactions have already happened between the community. For instance, an extensive days-long discussion about how to implement real-time visualization for SIM2D within Web Soccer Monitor engaged 4 people and occurred over the span of 3 different GitHub repositories [1–3], including the repository for the rcsserver [5], and also extending itself to the SIM2D Discord server [4].

As work continues, we believe and hope that this kind of interaction and engagement continues to grow, as more members of the SIM2D and general RoboCup community find value in the project at hand, and decide to also collaborate.

```
export default function Monitor(props: {
  /**
   * This is the component that shows the field and all the entities on a canvas. It uses React Konva library
   *
   * It receives config values and React states with their setters, in order to enable reusability and
   * two-way communication between the Monitor component and the outside context, respectively
   */
  config: {
    // category (if it is 2D, SSSL, VSS)
    category: string;
    // parent div to set Monitor's height and width to the same as theirs (the div parent)
    parentDivId: any;
    // height of the topbar
    topbarHeight: number;
    // if true, draws show player view area
    showPlayerViewArea;
    // size of the wedge that represents the player view area
    playerViewAreaSize: number;
    // color of the background everywhere
    backgroundColor: string;
    // total number of players (sum of both teams)
    totalNumberOfPlayers: number;
    // maximum number of frames in a match, in 2D this is equivalent to "showtime"
    maxNumberOfFrames: number;
    // all players refs (DOM references for updating states)
    allPlayersRefs: any;
```

Fig. 5. Web soccer monitor "monitor" component

4.2 Flourishment of a New Ecosystem

With the further development of Web Soccer Monitor and other open-source web tools in the scene, we strongly believe that there is an opportunity for starting the creation of a new ecosystem inside the SIM2D community that could also influence and reach other RoboCup categories. Having a powerful web monitor is the first step towards implementing a web simulator capable of running fully fledged simulations on the browser, requiring no infrastructure from the user's point of view and facilitating the occurrence of matches between teams across the world just by providing each team's binary files, which in itself is extremely useful, but could also give rise to another plethora of useful tools, platforms and frameworks.

5 Conclusions

Web Soccer Monitor is an open-source web-based monitor that aims to be a convenient and powerful alternative for visualising matches from RoboCup soccer leagues. It currently supports replays of SIM2D log files and already has some traction within the community.

We believe it can help bootstrap an ecosystem of web tools that would ultimately benefit the community and possibly aid the creation of new RoboCup leagues, while lowering the barrier of entry of newcomers to the robotics space and, thus, bringing more people to the scene.

6   Future Works

Web Soccer Monitor (WSM) is still in its infancy. In order for it to reach a level of quality necessary for being a widely used software, a number of features must be included, and a good benchmark for that would be implementing all of the features available in an already complete and very capable software such as soccerwindow2 [10].

Real-time visualization is already in the works, and should be shipped in the next major version of WSM. It will be a challenging task, but will bring WSM closer to a tool that real teams would use on a daily basis.

Also expected to be shipped with the next major version is support for other categories such as Small Size league and a binary file that can be run completely locally, outside the browser and as a standalone software, which would remove the need for utilizing the Node Package Manager and Python Pip utilities to install dependencies and use WSM offline.

Another pair of important works that needs to be done is user validation (gather feedback from users about their interactions with WSM for registering purposes and to help iterate on the project's development) and a practical research on the performance of the software, which could help further discuss the implications of choosing a 2D game visualization and reassure users that are apprehensive of utilizing web based tools due to performance concerns.

In the long-term, having a web-based monitor capable of viewing a game happening in real time can promote the creation of a platform for running matches on the browser, a different project, that would enable users to run and watch matches without the need of any setup at all, just by uploading their team's binaries. This way, the existing 2D server simulation code (rcssserver) [5] could be run on a back end on the cloud, enabling, on higher level, a full web based simulation, removing the necessity of running not only the monior, but also the server locally, which would lower even more the barrier of entry for newcomers. A robust implementation of such a platform could also have an in-browser code editor, that could be used to program a team's logic, generate a binary file and run matches all inside the browser.

This third project could use both WSM and JaSMIn [9] for the visualization part of the game, which would provide the user with the ability of 3D visualization for leagues that heavily rely on a 3-dimensional representation (such as SIM3D or the Humanoid League) and 2D visualisation of leagues that are, by nature, 2-dimensional (SIM2D) or that don't rely as much on 3D aspects (Very Small Size League, which although doesn't exist on RoboCup, is very common on, for example, Latin American robot soccer competitions). In some leagues that are naturally 3D, having a 2D representation could also prove to be useful, by enabling teams to look at a game with a different perspective and set of tools. Developers of WSM and JaSMIn could communicate in order to understand better what are the limitations and advantages of 2D and 3D monitors, and create together a general standard for a web monitor, which could boost the usefulness and re-usability of both projects.

By bringing all aspects of the simulation experience to the web, it is possible to innovate and create things that wouldn't be practical in the past. Some interesting ideas that could be built upon this more robust "development-simulation-visualization" ecosystem in the web would be

- **Automatic daily championships:** teams upload the latest versions of their binaries to their account in a platform where hundreds of matches are played between dozens of teams around the world in parallel, generating hundreds of log files each day that could prove to be very useful for generating datasets for statistics, data analysis, and machine learning purposes, while also helping teams that don't have a lot of available computing resources to run high amounts of matches in batch, so they can get insights in their team's performance.
- **Automatic team version evaluation:** similar to the idea above, but focused on running newer versions of teams against older versions of their own code, in order to generate enough statistics about the interactions between all versions as a way to understand if new changes in the code are making the team more or less performant than before.
- **Robot soccer interactive learning platform:** a collection of interactive articles that could teach the basics of robot soccer and give the users small tasks that would require making changes in the in-browser code editor in order to achieve their completion by evaluating if the changes made in the code resulted in the desired behaviour of teams and individual players.

Acknowledgements. The authors would like to acknowledge the RoboCIn's team and Centro de Informática - UFPE for all the research support. The authors appreciate all the Simulation 2D teams for their open-source contributions.

References

1. I follow the steps, and when i click on "start", nothing happens 1. https://github.com/kawhicurry/WebSoccerMonitor-plain/issues/1
2. Implement a udp <-> webrtc bridge (or other way of direct connection between rcssserver and a browser application) 90. https://github.com/rcsoccersim/rcssserver/issues/90
3. Implement connection to rcssserver and enable real-time game visualization 17. https://github.com/robocin/WebSoccerMonitor/issues/17
4. rcss2d communication channels. https://ssim.robocup.org/soccer-simulation-2d/joining-2d/
5. rcssserver. https://github.com/rcsoccersim/rcssserver
6. React. https://reactjs.org
7. Robocuparchive. https://archive.robocup.info/
8. loup Gailly, J.: Gzip. https://www.gnu.org/software/gzip/
9. Glaser, S.: Jasmin. https://gitlab.com/robocup-sim/JaSMIn
10. Helios: soccerwindow2. https://github.com/helios-base/soccerwindow2

11. Lavrenov, A.: Reactkonva. https://github.com/konvajs/react-konva
12. rcsoccersim: rcssmonitor. https://github.com/rcsoccersim/rcssmonitor
13. RoboFei: Loganalyserrobofei-ssl. https://gitlab.com/robofei/ssl/LogAnalyserRobo
    Fei-SSL
14. RobôCIn: rcss-log-extractor. https://github.com/robocin/rcss-log-extractor
15. RobôCIn: websoccermonitor. https://github.com/robocin/WebSoccerMonitor

Champion Papers Track

RoboBreizh, RoboCup@Home SSPL Champion 2022

Cédric Buche[1,4(✉)], Maëlic Neau[1,2,3,4], Thomas Ung[1,4], Louis Li[1],
Tianjiao Jiang[1], Mukesh Barange[5], and Maël Bouabdelli[5]

[1] CROSSING, CNRS IRL 2010, Adelaide, Australia
buche@enib.fr
[2] Lab-STICC, CNRS UMR 6285, Brest, France
[3] College of Science and Engineering, Flinders University, Adelaide, Australia
[4] ENIB, Plouzané, France
[5] LITIS, INSA Rouen Normandie, Rouen, France

Abstract. This paper presents the approach employed by the team
RoboBreizh to win the championship in the 2022 RoboCup@Home Social
Standard Platform League (SSPL). RoboBreizh decided to limit itself to
an entirely embedded system with no connection to the internet and
external devices. This article describes the design of embedded solutions
including manager, navigation, dialog and perception. We present results
from the competition showing up the value of our proposal.

1 Introduction

The progress of Artificial Intelligence (AI) and specifically deep learning algorithms has been outstanding in recent years. However, some advanced algorithms require significant hardware resources to perform local processing. In
the RoboCup@Home SSPL, the platform is the PEPPER robot, which today
is far from cutting-edge in terms of computing power. Therefore integrating
state-of-the-art algorithms with the robot in the competition becomes a real
challenge. A common alternative solution is to use remote algorithms via cloud
services, such as dialogue algorithms proposed by Google [4]. However, transmit-
ting information via the internet in the RoboCup is very challenging due to some
environmental factors such as internet stability and its limited bandwidth being
shared with other participants. The team RoboBreizh decided to limit itself to
an entirely embedded system which deprived Pepper of internet access and the
many AI solutions offered online, notably by the famous tech companies. The
limited conditions in the arena were not the sole reason for deploying the fully
embedded approach. Indeed, it was also inspired by the practical applications for
such robots, which are often designed for elderly or dependent persons who do
not necessarily have good internet access. The constant transmission of data for
online analysis also involves substantial energy and environmental cost. Finally,
a fully embedded solution has the major advantage of data confidentiality, especially for images and sounds, which are processed locally in the robot rather than

© The Author(s), under exclusive license to Springer Nature Switzerland AG 2023
A. Eguchi et al. (Eds.): RoboCup 2022, LNAI 13561, pp. 203–214, 2023
https://doi.org/10.1007/978-3-031-28469-4_17

being sent to the cloud. It also protects household users against unauthorised access to the robot, guaranteeing the respect of privacy.

This paper presents the embedded solutions employed by the team Robo-Breizh winning the 2022 RoboCup@Home Social Standard Platform League. Figure 1 illustrates a general overview of the proposal. It details the proposed architecture to solve the competition tasks, software used and interconnection between modules. The article is organized as follows. Firstly, Sect. 2 explains the details of using the embedded software architecture as an alternative to the classical NaoQi API for Pepper. Then, we outline the embedded manager module in Sect. 3, that tackles interconnection between modules. Next Sect. 4 describes the embedded perception module, including the detection of objects, persons, colors, distances, poses as well as age prediction. Following the embedded navigation module is presented in Sect. 5. The embedded dialog module is described in Sect. 6 providing details in speech detection, speech recognition and Natural Language Processing (NLP) solutions. Section 7 presents results in the RoboCup@Home challenge 2022. Finally, future works and improvements are discussed in Sect. 8.

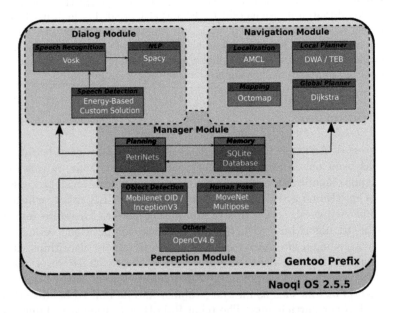

Fig. 1. RoboBreizh's architecture, including used functionalities/technologies.

2   Embedded Software Architecture

The first challenge of working with the Pepper Robot is the integrated NaoQi OS version 2.5.5, based on a 32 bits version of Linux Gentoo that restricted the number of libraries that could be installed. Also, Pepper's native API was written in Python2 and it does not include commonly used libraries in modern libraries, such as PyTorch and TensorFlow or Robot Operating System (ROS). The second limitation is the unavailability of root access to the robot's OS. To overcome those two limitations, we created a continuous integration pipeline based on the work from [15]. Even though Gentoo is an old system, it offers some interesting solutions such as the package manager *Portage* and the *Gentoo Prefix Project* [2]. A Gentoo prefix is an offset version of Gentoo that could be installed alongside another OS without root permission. In practice, we use an integration pipeline based on a Docker Image of a 32 bits Gentoo prefix[1]. At runtime, the required components of the prefix are extracted from the Docker Image and pushed directly to the robot, alongside the NaoQi OS. This solution resolved the root access problem but the limitations of Gentoo (32 bits and the *Portage* package manager instead of *Aptitude*) remained. As a workaround, we decided to cross-compile libraries, including ROS, in Docker using a dump of the Pepper OS[2] and our Gentoo prefix. We initially used the ros-overlay project[3] to cross-compile ROS Noetic. Since Python3 in the native Python API from Pepper is not available, we built on top of the LibQi API a new version that could run onboard and allows to still have access to the resources of the robot[4]. As we also concern about computational speed and hardware optimisation, we decided to use dedicated inference engines such as Tensorflow Lite [1] and ONNX[5] cross-compiled for the Pepper CPU Intel ATOM as well as other machine learning tools such as OpenCV4.6[6] and Kaldi [16] running with Python3 on Pepper hardware.

3   Embedded Manager Module

The manager is the core module of our architecture. It is regarded as a decision maker to choose which actions to perform and trigger the related components for achieving a desired goal. For instance, the actions taken by the manager could be navigating to a specified location, communicating with humans, detecting and manipulating objects in the surrounding environment. The manager is implemented using a Petri Net solution. It is a directed graph of places and transitions that allows parallel processing and versatile finite-state machines. Petri Nets are composed of markers that triggered a transition to move from place to place. Multiple markers could be requested for a transition acting like a semaphore. The possibility of having multiple markers in the graph makes it non-deterministic meaning it is suitable for concurrent processing. For each task, a plan was defined (Fig. 2) to describe the different steps of actions that could be encountered as well as sub-plans that would be triggered upon certain conditions. Having sub-plans allows to have a readable and modular approach. To execute our Petri Net with ROS, we use a package named Petri Net Plans

---

[1] https://github.com/awesomebytes/gentoo_prefix_ci_32b.

[2] https://hub.docker.com/r/awesomebytes/pepper_2.5.5.5.

[3] https://github.com/ros/ros-overlay.

[4] https://github.com/Maelic/libqi-python.

[5] https://github.com/onnx/onnx.

[6] https://github.com/itseez/opencv.

[24], which allows to link a set of places and transitions to the reference of C++ functions. Each function could then act as an independent process firing one or multiple ROS nodes and returning a state to the Petri net to change state. Having a returning state is primordial to adapt the next action and the rest of the task depending on its result. Moreover, the manager is capable of re-planning its behaviour in the situation of failure of certain actions or in blocking situations where the robot could not proceed further. One issue we encountered with using different processes was managing shared data between ROS nodes. To resolve this issue, we use SQLite[7], an embedded persistent storage solution that does not require any server to run.

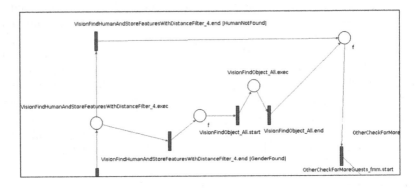

Fig. 2. Example of a Petri Nets: part of the plan for the "find my mates" task.

4   Embedded Perception Module

4.1   Objects

Object Detection. This module provides a set of ROS services with parameters, including a distance filter to prevent detecting objects outside of the arena. The classes of objects to find is accepted as parameters in the service /object_detection. All objects are detected by default (Fig. 3). A home-made combination of pre-trained Single Shot Detector (SSD) [12] with InceptionV3 [20] (600 classes) and MobileNet [8] (80 classes) are used. Based on such detection, specific post-processing are made. For instance, to tackle the task "stickler for the rules", a "shoe on/off and drink on hand" service is available as /shoes_and_drink_detection. To detect the presence of drinks, we interpolate the 3D pose of detected objects with the 3D pose of wrist body joints detected by the pose estimation detector. In addition, the service /seat_detection provides information regarding whether a specific seat is available or taken associated with the surrounding complex situations, such as the presence of multiple chairs and sofas.

---

[7] https://www.sqlite.org.

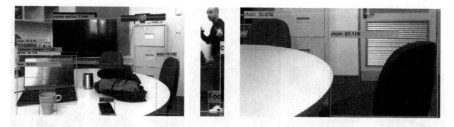

Fig. 3. Detection of object (left), shoes (center in pink) and seats (right in green) (Color figure online)

Color Estimation. Based on detection with the RGB camera, color estimation is computed. Similar to [18], a K-means clustering algorithm is applied to extract some dominant colors from a detected image. RGB pixel values are separated into five clusters for each bounding box of a detected object. Then, the raw pixel values are replaced by the RGB values of their corresponding cluster centroids. Finally, the closest X11 color is used to deliver a color name (Fig. 5).

4.2 Person

Person Detection. This module is associated with the services /person_ detection and /person_detection_posture. A tailored home-made combination of SSD with Inception, SSD with MobileNet and MoveNet Mulitpose[8] is used to detect persons (Fig. 4). SSD with MobileNet is used to get genders of persons. Depending on the distance to a target, Movenet is used for short range detection and SSD for long range detection.

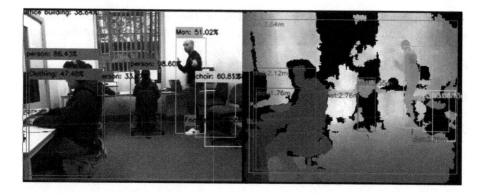

Fig. 4. Combining models to detect person/gender (left) and distances (right).

---

[8] https://www.tensorflow.org/hub/tutorials/movenet.

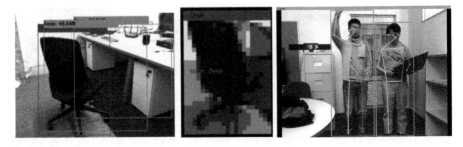

Fig. 5. Color detection (center) and the corresponding raw input image (left). Real time onboard multi pose estimation (right).

Age Estimation. Human faces images are cropped using SSD MobileNet detector. Then age estimation is computed based on the cropped image. Convolutional neural network model (CNN) based caffemodel [10] and TensorFlow Lite model (initially for Raspberry Pi)[9] are used for estimating human age.

Pose Detection. A CNN-based model (MoveNet Multipose) predicts human joint locations of multiple persons from an RGB image (Fig. 5). The model is able to detect 17 body joints of up to 20 persons in the same frame. It is the best trade-off between speed and accuracy for edge devices (originally designed to run on smartphones) [9]. The job is done using /person_detection_posture service. Similarly, /wave_hand_detection service applies the pose detector to detect a hand waving information from a customer in the restaurant task.

4.3   Localization and Distances

The positions of object and person relative to the front camera are estimated using the pinhole camera model [19] with the consideration of the intrinsic camera matrix for raw (distorted) images: $\begin{bmatrix} f_x & 0 & c_x \\ 0 & f_y & c_y \\ 0 & 0 & 1 \end{bmatrix}$.

Considering X_{obj}, Y_{obj} as center of an 2D bounding box from the RGB front camera and considering Z_{obj} as center of an 3D point cloud in connection with X_{obj}, Y_{obj} from the depth camera, the pinhole model of a camera is computed as

$$point_z = Z_{obj}; \quad point_x = \frac{(X_{obj} - c_x) * Z_{obj}}{f_x}; \quad point_y = \frac{(Y_{obj} - c_y) * Z_{obj}}{f_y}$$

Then the distance, in respect to the robot, is computed and could be used to provide filters for object or person detection.

---

[9] https://github.com/radualexandrub/Age-Gender-Classification-on-RaspberryPi4-with-TFLite-PyQt5.

4.4   Onboard Implementation

The perception module is designed to provide fast information in an embedded implementation. The combination of hardware optimisation and the choice of models designed for edge devices such as MobileNet allowed us to detect a person and object efficiently. We accessed images from the Pepper cameras directly through the NaoQi API thus we have the flexibility of choosing a resolution of 640×480 px (slower but more accurate detection) or 320×240 px (faster but less accurate) depending on our needs. Metrics are available on Table 1.

Table 1. Performance of the embedded perception using 640×480 px RGB images.

Model	Onboard computing
Object detection (SSD Inception + MobileNet)	2800 ms ± 300 ms
Person detection (SSD MobileNet)	800 ms ± 200 ms
Person detection (MoveNet Multipose)	1600 ms ± 100 ms
Pose estimation (MoveNet Multipose)	1900 ms ± 500 ms
Color estimation (K-mean, per object)	500 ms ± 450 ms
Age estimation (Caffemodel)	270 ms ± 30 ms

5   Embedded Navigation Module

The navigation module uses the ROS navigation stack [23] with Adaptive Monte-Carlo Localization (ACML)[10] for localization, the Dijkstra algorithm as global planner over the global costmap and the Dynamic Window Approach (DWA) as local planner[11] over the local costmap. We use a corrected version of the naoqi_driver ROS package[12] to access lidars and depth camera data. The information provided from Pepper's lidar is insufficient for computing costmaps, therefore PointClouds obtained from the depth camera are added as inputs, allowing Pepper to detect objects that could be out of range from its lidars sensing capability, such as detecting tables or chairs. In addition, RoboBreizh uses Spatio-Temporal Voxel Layer [13] as a 2D local costmap plugin continuously adding a new layer of weight of temporary obstacles at every runtime of detecting moving obstacles in a dynamic environment. With the contribution from the voxel layer and DWA, our Pepper robot is able to safely navigate in a known environment as well as avoiding dynamic obstacles. Timed Elastic Bands local planner TEB [17] was also tested as local planner to dynamically avoid obstacles, however, such methodology is resource intensive. For Simultaneous Localization And Mapping (SLAM) we planned to use Octomap [7], but nevertheless, drifts happened in the competition which made mapping unreliable using this technique. Thus, we corrected the final map manually.

---

[10] http://wiki.ros.org/amcl.

[11] http://wiki.ros.org/dwa_local_planner?distro=noetic.

[12] https://github.com/Maelic/pepper_naoqi_ros.

6   Embedded Dialog Module

The embedded dialog module allows Pepper to detect human speech and analyze its semantic meaning. Robot's native API provides a tool to record the sound and a grammar parsing library. However, these tools are limited in terms of maintainability and flexibility. Therefore RoboBreizh customises an embedded dialog system. The first part is sound processing, which consists of listening to the sound, detecting and recording voices from human. Next, the recorded sound is transformed into text using speech recognition. Finally, a Natural Language Processing algorithm is implemented to catch the intentions of the user.

Sound Processing. In noisy situation, processing raw audio signal to separate noise from actual speech is resources consuming. This is why we implemented a simple speech detection algorithm that will trigger the Automatic Speech Recognition (ASR) only when necessary. This algorithm can also be tweak to process sound coming from a specific location when the operator position is known for instance. The only reported approach for sound processing in RoboCup@Home is HARK[13] [6,14] that is used for sound source detection and localization. Unlike HARK, our solution is low-resource and optimize to process the Pepper raw audio signals, without any conversion. The pseudo code is presented in Algorithm 1, taking inspiration from[14]. We use Pepper's 4 microphones to compute a weighted value of the energy level of current audio signal. It defines changes in sound intensity to estimate whether a human is speaking. We set up different weights for different microphones to adjust the detection when the speaker location is known. An average of this energy value is compared to a threshold (parameter reevaluated in real-time in order to adapt to the ambient noise). The process run as a loop that continuously evaluates the average energy values and moves from state to state: silence → possible speech → speech → possible silence → silence. Once the state of possible silence exited, the current sound buffer is written to be processed by the speech recognition.

Speech Recognition. Speech recognition should not be significantly resource-intensive and should work offline. Given these conditions, the considered speech recognition options were PocketSphinx, Kaldi, Vosk[15] and Deepspeech. In [3], the speech recognition implementation from Kaldi, Pocketsphinx, Picovoice, and Google were tested using the measurement metric "word error rate" (WER). We performed the same test, and Vosk got better accuracy across the offline solutions when measuring the WER with different sentences, 101 times. The lightest model available is used ("vosk-model-small-en-us-0.15").

Natural Language Processing (NLP). A straight forward but tedious approach to parse commands is to write the grammar of a define language.

---

[13] https://hark.jp/.

[14] http://wiki.ros.org/speech_recog_uc.

[15] https://alphacephei.com/vosk/.

Algorithm 1: Detect User Speech

---

input :

counterSpeech, thOffset, counterSilence
FrontMicImportance, LeftMicImportance
RightMicImportance, RearMicImportance
hh

/* Initialisation */

1 STATE = SILENCE
2 firstTime = True
3 rstCounterSpeech = counterSpeech
4 rstCounterSilence = counterSilence

/* Main loop */

5 **while** *True* **do**

/* update energy */

6 $\quad$ energy = (getFrontMicEnergy()*FrontMicImportance) +
(getLeftMicEnergy()*LeftMicImportance) +
(getRightMicEnergy()*RightMicImportance) +
(getRearMicEnergy()*RearMicImportance) / 4

7 $\quad$ **if** *firstTime* **then**
8 $\quad\quad$ ymin_prev, ymax_prev, ymed_prev = energy
9 $\quad\quad$ firstTime = False
10 $\quad$ **end**

/* update ymed */

11 $\quad$ **if** *energy > ymax_prev* **then**
12 $\quad\quad$ ymax = energy
13 $\quad$ **else**
14 $\quad\quad$ ymax = hh * ymax_prev + (1 -hh) * ymed_prev
15 $\quad$ **end**
16 $\quad$ **if** *energy < ymin_prev* **then**
17 $\quad\quad$ ymin = energy
18 $\quad$ **else**
19 $\quad\quad$ ymin = (1 -hh) * ymin_prev + hh * ymed_prev
20 $\quad$ **end**
21 $\quad$ ymed = (ymin + ymax) / 2
22 $\quad$ **if** *STATE == SILENCE* **then**
23 $\quad\quad$ **if** *energy > ymed_prev + thOffset* **then**

/* update threshold */

24 $\quad\quad\quad$ STATE = POSSIBLE_SPEECH
25 $\quad\quad\quad$ threshold = ymed_prev + thOffset
26 $\quad\quad\quad$ counterSpeech = rstCounterSpeech - 1
27 $\quad\quad$ **end**
28 $\quad$ **end**
29 $\quad$ **if** *STATE == POSSIBLE_SPEECH* **then**
30 $\quad\quad$ counterSpeech -= 1
31 $\quad\quad$ **if** *energy > threshold and energy > ymed* **then**
32 $\quad\quad\quad$ **if** *counterSpeech <= 0* **then**
33 $\quad\quad\quad\quad$ counterSpeech = rstCounterSpeech
34 $\quad\quad\quad\quad$ STATE = SPEECH
35 $\quad\quad\quad\quad$ startRecording()
36 $\quad\quad\quad$ **else**
37 $\quad\quad\quad\quad$ STATE = POSSIBLE_SPEECH
38 $\quad\quad\quad$ **end**
39 $\quad\quad$ **else**
40 $\quad\quad\quad$ STATE = SILENCE
41 $\quad\quad$ **end**
42 $\quad$ **end**
43 $\quad$ **if** *STATE == POSSIBLE_SILENCE* **then**
44 $\quad\quad$ counterSilence -= 1
45 $\quad\quad$ **if** *energy > threshold* **then**
46 $\quad\quad\quad$ STATE = SPEECH
47 $\quad\quad$ **else if** *counterSilence == 0* **then**
48 $\quad\quad\quad$ STATE = SILENCE
49 $\quad\quad\quad$ stopRecording()
50 $\quad\quad$ **else**
51 $\quad\quad\quad$ STATE = POSSIBLE_SILENCE
52 $\quad\quad$ **end**
53 $\quad$ **end**
54 $\quad$ **if** *STATE == SPEECH* **then**
55 $\quad\quad$ **if** *energy < ymed and energy < threshold* **then**
56 $\quad\quad\quad$ STATE= POSSIBLE_SILENCE
57 $\quad\quad\quad$ threshold = ymed
58 $\quad\quad\quad$ counterSilence = rstCounterSilence - 1
59 $\quad\quad$ **else**
60 $\quad\quad\quad$ STATE = SPEECH
61 $\quad\quad$ **end**
62 $\quad$ **end**
63 **end**

Spacy model [5] is a solution to find relations between words. It is used for Part-of-Speech Tagging to retrieve global and local dependency trees. Once these relations are set, we developed a parsing algorithm that explored each dependency tree in the sentence and parse the command into intention and arguments. The generated intent is a dictionary that delivers simple instructions (e.g. "navigate to the kitchen" triggered the output {"intent": "go"; "destination": "kitchen"}).

7 Performance

RoboBreizh ranked 1^{rst} in the competition. In the first stage, RoboBreizh received a score of 913 points. In stage 2, RoboBreizh received 1363 points ranked at 1^{rst} and at the end won the final with 0.775 points. (Table 2).

Table 2. Final competition board (SSPL).

Place	Team	Stage 1	Stage 2	Finals
1	RoboBreizh	913	1363	0.775
2	Sinfonia Uniandes	334	634	0.458
3	LiU@HomeWreckers	87.33	87.33	

Find My Mates (450 Pts). The robot got to the operator, waited for a signal and then moved towards the centre of the living room. Then Pepper ran multiple detections. If a person was detected, information about him and adjacent objects were stored in the database. Then the robot looked for another person until he found 3 individuals. Once all visual information was retrieved, Pepper came back to the operator and described every person and their surroundings.

Receptionist (425 Pts). During this task Pepper needed to go to the entrance, listened to guests' names and drink, and then lead them to the living room. Once navigated to the living room, the robot had to present the guest and the host to each other. Then the robot detected available seats and offered one to the guest. Finally, it came back to the entrance and repeated the process.

GPSR (0 Pts). Pepper had to parse a complex order from the host and defined a plan accordingly. The robot was able to understand one command properly and delivered a relevant plan as output. However, AMCL relocalized Pepper at the wrong place during navigation, which muddled up the end of the task.

Stickler for the Rules (450 Pts). RoboBreizh decided to focus on the forbidden room rule. Pepper rotated on itself until it detected someone, then verified its position. A person was detected in the forbidden room and Pepper kindly asked the person to leave. Then the robot waited and verified the presence of the detected person in the room. If the detected person was still in the room, Pepper would ask the person to leave again.

Finals. RoboBreizh created a scenario to showcase the robot's abilities, interacting with the team leader. During this presentation, we showed that all our modules were running onboard the robot. PEPPER uses k-means to determine objects' color and provided information regarding posture and age of persons. RoboBreizh showed how command parsing system works in detail The team explained why having everything onboard was an interesting approach and how difficult it was to make it practically works.

8   Conclusion and Future Work

In this paper, we offered an overview of the architecture developed by the team RoboBreizh to perform at the SSPL RoboCup@Home contest. After introducing developed modules, we presented the application of such proposals during the competition in 2022. For its 24th edition, the city of Bordeaux in France has been chosen to organize the RoboCup competition next year. The objective of RoboBreizh is to do its best to make further improvements and win another championship next year in our native country.

Future works will examine a proposal for implicitly detecting and understanding users' intentions and needs. Traditional approaches to this problem in robotics use explicit signals from the user such as voice [21] or gesture [22]. However, the deployment of service robots in assisting activities of daily life is leading the way to more implicit interactions with autonomous agents [11].

Acknowledgment. This work benefits from the support of Brest City (BM), CERV-VAL, Brittany and Normandy regions.

References

1. Abadi, M., et al.: TensorFlow: a system for large-scale machine learning. In: 12th USENIX Symposium on Operating Systems Design and Implementation (OSDI 2016), pp. 265–283 (2016)
2. Amadio, G., Xu, B.: Portage: bringing hackers' wisdom to science. arXiv preprint arXiv:1610.02742 (2016)
3. Brinckhaus, E., Barnech, G.T., Etcheverry, M., Andrade, F.: Robocup@home: evaluation of voice recognition systems for domestic service robots and introducing Latino dataset. In: Latin American Robotics Symposium, Brazilian Symposium on Robotics and Workshop on Robotics in Education, pp. 25–29 (2021)
4. Cohen, A.D., et al.: LaMDA: language models for dialog applications (2022)
5. Honnibal, M., Montani, I.: spaCy 2: natural language understanding with bloom embeddings, convolutional neural networks and incremental parsing. To Appear **7**(1), 411–420 (2017)
6. Hori, S., et al.: Hibikino-Musashi@Home 2017 team description paper. arXiv preprint arXiv:1711.05457 (2017)
7. Hornung, A., Wurm, K.M., Bennewitz, M., Stachniss, C., Burgard, W.: OctoMap: an efficient probabilistic 3D mapping framework based on octrees. Auton. Robots **34**, 189–206 (2013)

8. Howard, A.G., et al.: MobileNets: efficient convolutional neural networks for mobile vision applications. arXiv preprint arXiv:1704.04861 (2017)
9. Jo, B., Kim, S.: Comparative analysis of OpenPose, PoseNet, and MoveNet models for pose estimation in mobile devices. Traitement du Signal **39**(1), 119–124 (2022)
10. Levi, G., Hassner, T.: Age and gender classification using convolutional neural networks. In: IEEE CVPR Workshops (2015)
11. Li, S., Zhang, X.: Implicit intention communication in human-robot interaction through visual behavior studies. IEEE Trans. Hum.-Mach. Syst. **47**(4), 437–448 (2017)
12. Liu, W., et al.: SSD: single shot multibox detector. In: Leibe, B., Matas, J., Sebe, N., Welling, M. (eds.) ECCV 2016. LNCS, vol. 9905, pp. 21–37. Springer, Cham (2016). https://doi.org/10.1007/978-3-319-46448-0_2
13. Macenski, S., Tsai, D., Feinberg, M.: Spatio-temporal voxel layer: a view on robot perception for the dynamic world. Int. J. Adv. Robot. Syst. **17**(2) (2020)
14. Oishi, S., et al.: Aisl-tut@ home league 2017 team description paper. In: RoboCup@ Home (2017)
15. Pfeiffer, S., et al.: UTS unleashed! RoboCup@Home SSPL champions 2019. In: Chalup, S., Niemueller, T., Suthakorn, J., Williams, M.-A. (eds.) RoboCup 2019. LNCS (LNAI), vol. 11531, pp. 603–615. Springer, Cham (2019). https://doi.org/10.1007/978-3-030-35699-6_49
16. Povey, D., et al.: The Kaldi speech recognition toolkit. In: IEEE Workshop on Automatic Speech Recognition and Understanding. IEEE Signal Processing Society (2011)
17. Rösmann, C., Feiten, W., Wösch, T., Hoffmann, F., Bertram, T.: Efficient trajectory optimization using a sparse model. In: 2013 European Conference on Mobile Robots, pp. 138–143. IEEE (2013)
18. Saraydaryan, J., Leber, R., Jumel, F.: People management framework using a 2D camera for human-robot social interactions. In: Chalup, S., Niemueller, T., Suthakorn, J., Williams, M.-A. (eds.) RoboCup 2019. LNCS (LNAI), vol. 11531, pp. 268–280. Springer, Cham (2019). https://doi.org/10.1007/978-3-030-35699-6_21
19. Sturm, P.: Pinhole camera model. In: Ikeuchi, K. (ed.) Computer Vision, pp. 610–613. Springer, Boston (2014). https://doi.org/10.1007/978-0-387-31439-6_472
20. Szegedy, C., Vanhoucke, V., Ioffe, S., Shlens, J., Wojna, Z.: Rethinking the inception architecture for computer vision. In: Proceedings of the IEEE Conference on Computer Vision and Pattern Recognition, pp. 2818–2826 (2016)
21. Tada, Y., Hagiwara, Y., Tanaka, H., Taniguchi, T.: Robust understanding of robot-directed speech commands using sequence to sequence with noise injection. Front. Robot. AI **6**, 144 (2020)
22. Waldherr, S., Romero, R., Thrun, S.: A gesture based interface for human-robot interaction. Auton. Robot. **9**(2), 151–173 (2000)
23. Zheng, K.: ROS navigation tuning guide. In: Koubaa, A. (ed.) Robot Operating System (ROS). SCI, vol. 962, pp. 197–226. Springer, Cham (2021). https://doi.org/10.1007/978-3-030-75472-3_6
24. Ziparo, V.A., Iocchi, L., Nardi, D.: Petri net plans. In: Fourth International Workshop on Modelling of Objects, Components, and Agents, pp. 267–290 (2006)

RoboCup2022 KidSize League Winner CIT Brains: Open Platform Hardware SUSTAINA-OP and Software

Yasuo Hayashibara[1]([⊠]), Masato Kubotera[1], Hayato Kambe[1], Gaku Kuwano[1], Dan Sato[1], Hiroki Noguchi[1], Riku Yokoo[1], Satoshi Inoue[1], Yuta Mibuchi[1], and Kiyoshi Irie[2]

[1] Chiba Institute of Technology, 2-17-1 Tsudanuma, Narashino, Chiba, Japan
yasuo.hayashibara@it-chiba.ac.jp
[2] Future Robotics Technology Center, Chiba Institute of Technology, 2-17-1 Tsudanuma, Narashino, Chiba, Japan

Abstract. We describe the technologies of our autonomous soccer humanoid robot system that won the RoboCup2022 Humanoid KidSize League. For RoboCup2022, we developed both hardware and software. We developed a new hardware SUSTAINA-OP. We aimed to make it easier to build, harder to break, and easier to maintain than our previous robot. SUSTAINA-OP is an open hardware platform. As the control circuit, we selected a computer with higher processing power for deep learning. We also developed its software. In terms of image processing, the new system uses deep learning for all object detection. In addition, for the development of action decision-making, we built a system to visualize the robot's states and solved many problems. Furthermore, kicking forward at an angle action is added as a new tactical action. In RoboCup2022, even when the robots were facing each other with the ball between them, by this action the robot succeeded in getting the ball out in the direction of the opponent's goal.

Keywords: Open hardware platform · Deep learning · Visualize

1 Introduction

We describe the technologies of our autonomous soccer humanoid robot system which is the winner of the RoboCup2022 Humanoid KidSize League. In RoboCup2022 Bangkok, we won the first prize in the soccer 4on4 and the drop-in challenge. The results are indicated in Table 1. For RoboCup2022, we developed both hardware and software. We developed a new hardware SUSTAINA-OP. We aimed to make it easier to build, harder to break, and easier to maintain than our previous robots GankenKun [1]. SUSTAINA-OP is an open hardware platform, and its design data is available to the public [2]. Several research groups have previously proposed open platform robots for KidSize League. For example, DARwIn-OP [3] is the pioneer regarding open platforms and has been used by many RoboCup participants and researchers. Recently, Wolfgang-OP [4] was developed to address the increasing difficulty of Humanoid League regulations.

© The Author(s), under exclusive license to Springer Nature Switzerland AG 2023
A. Eguchi et al. (Eds.): RoboCup 2022, LNAI 13561, pp. 215–227, 2023
https://doi.org/10.1007/978-3-031-28469-4_18

Compared to those robots, the main features of SUSTAINA-OP are as follows: parallel link leg, high computational power for machine learning, and a structure that does not require bending. As the control circuit, we selected a computer NVIDIA Jetson Xavier NX 16 GB [5] with higher processing power, so that high computational processes such as deep learning can be processed in real-time in the robot. We also developed its software. In terms of image processing, the previous system used a color lookup table, but the new system uses deep learning for all object detection. In addition, for the development of action decision-making, we built a system to visualize the robot's states and solved many problems. For example, we were able to identify and solve a problem seen at RoboCup2019 of not being able to approach the ball in a straight line. Furthermore, kicking at an angle action is added as a new tactical action. In RoboCup2022, even when the robots were facing each other with the ball between them, by this action, the robot succeeded in getting the ball out in the direction of the opponent's goal.

CIT Brains is a team of Chiba Institute of Technology (CIT) from 2007. The aim of the development is not only research, but also education. Most of our team members are undergraduate students. Through developing the autonomous humanoid robot system, they study many hands-on robot technologies. Students continue to meet daily throughout the year to discuss and develop their robots. None of the students who participated in RoboCup this year had an experience of the previous competitions. We believe that the reason why our team, which was almost newly established, was able to win the competition is that they have been meeting remotely every day to progress their development.

Table 1. Result of CIT brains in RoboCup2022 humanoid kidsize league

Category	Results
Soccer 4 on 4	1st place (11 teams participated) 5 wins – 0 loss Total goals: 55 goals – 1 loss
Drop-in challenge	1st place (33 points)

2   Mechanics

We have developed a new robot, SUSTAINA-OP, our sixth-generation robot hardware for RoboCup2022. It was designed by Masato Kubotera. For RoboCup2022, we have built 6 robots with the same configuration. SUSTAINA-OP is available on GitHub as an open hardware platform with a CC BY-NC-SA 4.0 license [2]. We provide the robot's 3D model and printed circuit board data. The 3D robot models can be viewed in a browser using Autodesk Viewer [6]. We provide the data as an open hardware platform to promote developing humanoid robots for RoboCup, with the hope that our platform can foster future engineers. The contents of the open hardware platform will be further enhanced in the future. Figure 1 shows the history of typical KidSize robots developed by CIT Brains. CIT Brains has been participating in the Humanoid League KidSize since RoboCup2007, with six major generations of robots and numerous minor versions.

Figure 2 shows the robot configuration of SUSTAINA-OP, and Table 2 shows the specifications. The mechanism follows that of our conventional robots, but is easier to assemble, harder to break, and easier to maintain. For example, the structure is such that parts can be assembled by fitting them into holes without bending, which requires a high level of skill. This mechanism is intended to be able to maintain and operate the robot with a minimum of engineering knowledge; the reason behind this is that our student members graduate and leave the team typically within a few years before they are fully experienced.

An example of ease of assembly is described below. The previous mechanisms use thrust bearings in the hip yaw joint. Although they are relatively inexpensive and easily available, the gap adjustment is difficult. If the gap is too narrow, the frictional force increases and the joint cannot move smoothly. If the gap is too wide, the rigidity of the hip joint decreases and gait becomes unstable. In the SUSTAINA-OP, we eliminated the need for gap adjustment by using cross-roller bearings. An example of being harder to break is that the ball bearings have been replaced with oilless plain bearings. In addition, TPU bumpers made by a 3D printer were attached to the front and back of the waist, chest, and shoulders to soften the impact of falls. As a result, the robots played 9 matches in RoboCup2022 and had about 50 falls, but no robot was removed from the field due to hardware failure. CIT Brains has been participating in RoboCup since 2007, and in each match there has been at least one robot that has been unable to move, including wiring disconnections and CPU resets, and has had to leave the competition. In RoboCup2022, however, none of these problems occurred. This result indicates that SUSTAINA-OP has achieved its goal of hardware that is less prone to failure.

3   Control Circuit

SUSTAINA-OP is equipped with NVIDIA's Jetson Xavier NX 16 GB as the control circuit. The previous robot, GankenKun, was equipped with an NVIDIA Jetson TX2. Comparing their maximum processing power, the TX2 has 1.33TFLOPS (FP16) while the Xavier NX has 6TFLOPS (FP16). In our system developed for RoboCup2022, all image processing was performed by deep learning, as described in the following sections, which could increase the computation cost. Hence, a new computer was adopted as preparation for adding more computationally demanding processing in the future. Figure 3 shows the configuration of the control circuit.

The camera interface of SUSTAINA-OP has been changed from the USB to MIPI CSI-2. In our previous robot GankenKun, the USB camera was sometimes disconnected when the robot was given an impact force such as when it fell down. The software automatically reconnected the camera so that recognition could be continued as much as possible, but the recovery system sometimes did not work. After switching to a camera with MIPI CSI-2 connection, such a problem no longer occurs. As mentioned above, in RoboCup2022, none of our robots were picked up due to hardware problems.

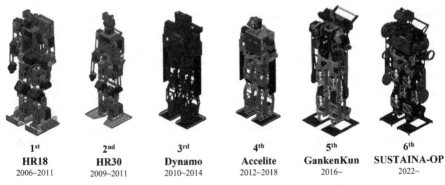

1st	2nd	3rd	4th	5th	6th
HR18	**HR30**	**Dynamo**	**Accelite**	**GankenKun**	**SUSTAINA-OP**
2006~2011	2009~2011	2010~2014	2012~2018	2016~	2022~

Fig. 1. The history of the KidSize robots in CIT Brains

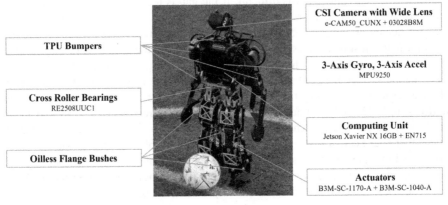

Fig. 2. Structure of the humanoid robot SUSTAINA-OP

Table 2. Specifications of the humanoid robot SUSTAINA-OP

Category	Specification
Robot Name	SUSTAINA-OP
Height	646.61 mm
Weight	5.18 kg
Walking speed	Max. 0.33 m/s
Degrees of freedom	19
Actuators	Kondo B3M-SC-1170-A x 10 pcs Kondo B3M-SC-1040-A x 9 pcs

(continued)

Table 2. (*continued*)

Category	Specification
Sensors	TDK InvenSense MPU-9250 (The included Magnetometer is not used) e-con Systems e-CAM50_CUNX
Computing unit	NVIDIA Jetson Xavier NX 16 GB AVerMedia EN715
Battery	Hyperion HP-G830C2800S3 LiPo 3S1P 2800 mAh

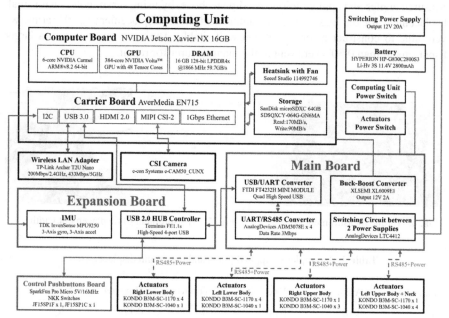

Fig. 3. Overview of the control circuits [2]. Red parts indicate our developed parts [7–9] (Color figure online).

4 Quality Control of Servo Motors

We believe that one of the factors that kept the robot walking stably in RoboCup2022 was the quality control of the servo motors. In RoboCup2021, which was conducted using a simulator, we created a device to measure the characteristics of the servo motors shown in Fig. 4 in order to create our own model. Using this device, we measured the torque of each servo motor. As a result, we could not find any motor with significantly reduced torque, including used servo motors. Therefore, the results of these torque studies were not used in the development of RoboCup2022.

Next, the error in the output of each servo motor was measured using originally developed software [10]. The error represents the difference in output when a weak force is applied. Figure 5 shows the measurement. The output errors were measured at servo motor angles of -90, -45, 0, 45, and 90deg. The Kondo B3M-SC-1040-A and B3M-SC-1170-A used in SUSTAINA-OP are multifunctional servo motors that can measure the angle of the output. For example, we set 0 deg as the target angle and apply weak external clockwise and counterclockwise forces. By measuring the degree of angular displacement with an angle sensor in the servo motor, the width of the output axis displacement can be measured. This range indicates the range in which the joint angle changes with a small force, and when used in a leg joint, it is related to the displacement of the toe position. Even small displacements in angle are magnified at the toe position. Therefore, it is important to make this as small as possible, in other words, to prevent the toes from being moved by external forces as much as possible. Table 3 shows the measured angles and applied joints. As shown above, by using servo motors with small displacement for the legs and servo motors with large displacement for the arms, we were able to construct a robot that can walk stably. The servo motors with a larger displacement than 0.50° were sent to the manufacturer for repair.

Fig. 4. Torque measurement device   **Fig. 5.** Measurement of displacement width

Table 3. Examples of applied joints and maximum displacement

Ankle pitch	Ankle roll	Knee pitch	Hip roll	Hip yaw	Arm/Neck
0.17°	0.18°	0.36°	0.17°	0.17°	0.36°

5 Walking Control and Motion

For walking control, we use the same gait pattern generator [11] as the previous robot GankenKun. However, by changing the hardware to SUSTAINA-OP, the robot appeared to almost fall over when turning while moving forward. This was due to the fact that the lateral width of the sole had narrowed from 92.5 mm to 80.0 mm from the previous robot GankenKun. Since this gait pattern generator simply adds a crotch yaw axis rotation to the forward gait pattern as the toe trajectory, the center position of the foot sole and the target ZMP (Zero Moment Point) [12] are not always at the same position. When the foot is large, as in the case of the previous robot GankenKun, the displacement between the center position of the foot and the target ZMP does not affect the gait so much. The narrow width of the sole of SUSTAINA-OP in the early stage of development have brought this problem to the surface. In RoboCup2022, we solved this problem by increasing the width of the sole. In the future, the gait pattern generator should be changed so that the target ZMP is centered on the sole.

To solve such problems related to gait, we developed a system that automatically collects robot behavior in response to arbitrary commands. The gait control module of this system generates a gait pattern by receiving from other modules the values of the number of steps, angle, forward stride length, side step length, and period. We constructed a system that can automatically replay those commands written in yaml. This enables the reproduction of the robot's behavior when given a series of walking commands, including the timing of command transmission. Figure 6 shows an example of a set of commands written in yaml. This system shortens the time required to find and solve problems, such as the problem described above. The system was also used in the refactoring of the gait control module. By verifying that the motion is the same before and after various scenarios, we were able to prevent bugs from occurring.

Figure 7 shows the kick motion being played back. Generating a stable kicking motion without falling over is important in soccer. In RoboCup2022, the soccer player kicked 360 times during the 9 matches, and fell only 3 times except for collisions with robots. One of the reasons might be the small error between the leg joint angles and the target angles as described above. Another factor is the standardization of operations. The system replays the motion by linearly complementing the angles of each joint specified at each keyframe. Since the joint angles for each keyframe are set manually, differences in the success rate of kicks occurred depending on the operator. In order to prevent falls during kicking, a manual for motion creation was prepared and shared among the members of the team. As a result, we found that all six robots can kick with the same motion data, so only the initial posture is adjusted.

```
command:
  - command_start_cnt: 500
    command_end_cnt: 1000
    walk_command: M
    num_step: 0
    period: 0
    stride_x: 0
    stride_y: 0
    stride_theta: 5
  - command_start_cnt: 1400
    command_end_cnt: 2500
```

Fig. 6. Example of yaml describing a series of commands

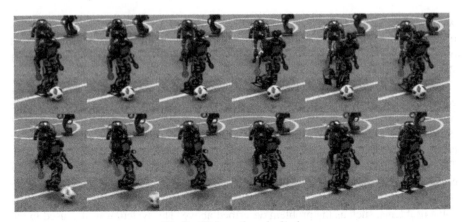

Fig. 7. Kick motion playback

6 Perception

6.1 Computer Vision

Our image processing system is twofold. One is the object detector using YOLO [13], and the other is pixel-level semantic segmentation, both based on deep learning. In the previous competition, we employed deep learning for detecting the ball, goal posts, and white lines but for other objects including opponent robots and the green area of the soccer field, we employed a simple method that mainly uses a color lookup table created in advance. However, the color-based method often causes misrecognition. In particular, the robot sometimes misrecognized its own shoulders and arms as those of other robots, resulting in unnecessary obstacle avoidance behavior. This year, we extended our system to detect all types of objects using deep-learning-based methods.

In RoboCup2022, we attempted to recognize robots using YOLO. The challenge was that, unlike balls and goals, there are many types of robots, making it difficult to collect enough images for training the detector. We employed several enhancements to easily collect training images: first, instead of using two robot classes (teammates and opponent robots), we employed a unified robot class for all robots and added a post-processing classifier to distinguish the robot teams by their color marker. Second, we added a feature to save captured camera images while the robot is playing soccer.

By continuing to collect images during the competition, we succeeded in building a large dataset. Figure 8 shows an example of the collected images. We selected 9410 images collected at the venue of RoboCup2022 and combined them with the dataset collected at the Chiba Institute of Technology. The dataset included images of the balls provided at the venue in addition to several types of balls prepared in advance. The total number of images in the dataset was 18426, of which the annotated numbers were 14881, 24035 and 20730 for the ball, post, and robot, respectively. The images collected during the competition were uploaded to our institute server so that our 11 offshore support members could immediately make annotations and expand the dataset. Several types of balls were used in RoboCup2022, and the computer vision was able to stably detect them. Figure 9 shows the object detection including the other team robot. Figure 10 shows a scene from a RoboCup2022 match, in which the robot was avoiding obstacles as it weaved between enemy robots. This kind of obstacle avoidance behavior was also observed in many other matches.

We employed deep-learning-based semantic segmentation to detect green areas and white lines on a soccer field. In the previous competitions, the semantic segmentation was used only for white lines, and the green areas were detected using the color lookup table; however, we have found the method unreliable under natural light, which was introduced in RoboCup2019, and varying sunlight conditions during RoboCup2021/2022 Virtual Season. Therefore, we expanded the semantic segmentation to detect not only white but also green. In RoboCup2022, the robot was able to stably detect the white line and green in all three fields regardless of the time of day, and there was no significant error in self-localization. Figure 11 shows an example of white line and green detection result.

Fig. 8. Example of dataset

Fig. 9. Results of object detection

6.2 Self-localization

We employ Monte Carlo Localization for self-localization. Landmark measurements from the abovementioned computer vision system are fused with motion predictions from kinematic odometry using a particle filter.

The landmarks employed are goal posts and white lines on the field. We employed YOLO for detecting goal posts and semantic segmentation for detecting white lines. Since white line detection can be affected by lighting conditions, we improve the robustness by interpreting only the edges between white and green segments as white lines.

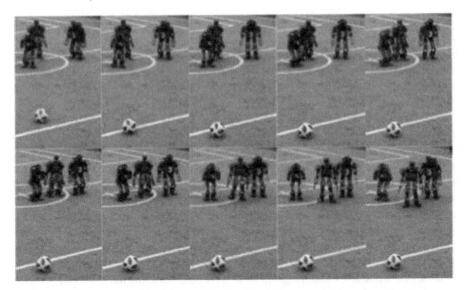

Fig. 10. Avoiding a robot during a game

Fig. 11. Example of white line and green detection by deep learning (Color figure online)

The 3D position of the detected objects using the kinematic camera poses, assuming that all objects are on the ground.

7 Visualization

Visualization is important in developing robot intelligence. We developed a system that collects and visualizes robots and game status during development. Figure 12 shows an example of the information obtained by our visualization system. We installed a wide-angle camera on the ceiling of our soccer field to capture images of the entire field. In addition, we developed software that displays the internal information of the robots and visualizes the data communicated between robots for information sharing. The following information can be viewed with this software.

1) Location and orientation of the robot
2) Selected role and action

3) Moving target position

4) Detected ball position

5) Period of processing loop

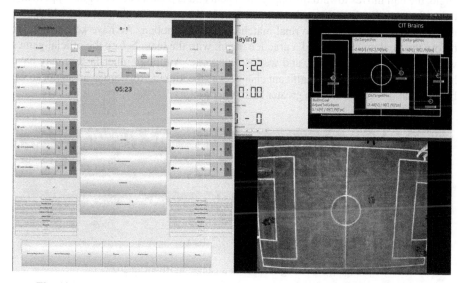

Fig. 12. System to display images of the field and internal information of the robots

This information made it clear at a glance what kind of data the robot chose to act on. This enabled the development of decision-making for actions to be made more easily. For example, in RoboCup2019, there was a problem that the robot could not approach the ball in a straight line. By using this system, we were able to discover and solve the problem of occasional processing delays that were causing unstable walks towards the ball.

8 Action Decision Making

In this system, the HTN (Hierarchical Task Network) planner [14] is used for action decision-making instead of GOAP (Goal Oriented Action Planning), which we have used in the past. Both planners have a mechanism in which, by describing the elements of actions, the planner automatically selects and executes the elements of actions that will achieve the goal, such as "putting the ball in the goal". Each robot is assigned a different role in the game, and different objectives are set for each role, such as forward, defender, and keeper. These roles are changed depending on the position of the ball and other factors. These mechanisms allow for flexible role changes. When these roles were written in GOAP, it was relatively difficult to find and correct bugs and to create the desired action pattern because of the complexity of the cost setting during planning. However, after using the HTN planner, it became relatively easy.

In RoboCup2022, we have added a new feature: the ability to kick a ball forward at an angle. This function was added because SUSTAINA-OP's hip motion range is larger than that of our previous robot, and deep learning enables stable detection of the robot. Figure 13 shows the kick. In the past, when robots faced each other like this, they often kicked the ball in front of each other, resulting in a stalemate. By kicking forward at an angle, the ball moves toward the opponent's goal and is positioned out of the opponent's field of view, allowing us to gain an advantage in the game. We think the new kick was the key factor for winning the final match with team Rhoban. We have kept it a secret before the final match; we had a practice match with the team before the official games with the secret kick disabled, and we lost the game. Although other factors may have had an influence, the most significant change was the kicking forward at an angle, so we believe that the influence was greater than other factors.

Fig. 13. Kicking forward at an angle in RoboCup2022

9 Conclusion

In this paper, we present our autonomous soccer robot system that won the RoboCup2022 KidSize Humanoid League. For RoboCup2022, we put a lot of development efforts in both hardware and software. As a result, we were able to build a stable system that was able to compete in RoboCup2022 without serious problems in both hardware and software. The hardware used in RoboCup2022 is available to the public, and part of the software is also available to the public. We hope to continue to provide information on the development of autonomous humanoid robots in the future.

References

1. OpenPlatform. https://github.com/citbrains/OpenPlatform
2. SUSTAINA-OP. https://github.com/citbrains/SUSTAINA-OP
3. Ha, I., Tamura, Y., Asama, H., Han, J., Hong, D.W.: Development of open humanoid platform DARwIn-OP. In: SICE Annual Conference 2011 (2011)
4. Bestmann, M., Güldenstein, J., Vahl, F., Zhang, J.: Wolfgang-op: a robust humanoid robot platform for research and competitions. In: IEEE-RAS 20th International Conference on Humanoid Robots (2021)

5. NVIDIA. https://www.nvidia.com/
6. SUSTAINA-OP Autodesk Viewer redirect page. https://citbrains.github.io/SUSTAINA-OP/autodesk_viewer.html
7. Main Board Ver. 2.2. https://github.com/MasatoKubotera/MainBoard_ver2_2
8. EN715 Expansion Board Ver. 1.1. https://github.com/MasatoKubotera/EN715_Expansion Board_ver1_1
9. Start Stop Switch Ver. 3.0. https://github.com/MasatoKubotera/StartStopSwitch_ver3_0
10. B3MServoChecker. https://github.com/citbrains/B3MServoChecker
11. GankenKun_webots. https://github.com/citbrains/GankenKun_webots
12. Kajita, S., Hirukawa, H., Harada, K., Yokoi, K.: Introduction to Humanoid Robotics, vol. 101. Springer, Berlin (2014)
13. Redmon, J., Divvala, S., Girshick, R., Farhadi, A.: You only look once: unified, real-time object detection. In: Proceedings of the IEEE Conference on Computer Vision and Pattern Recognition (2016)
14. Georgievski, I., Aiello, M.: An overview of hierarchical task network planning. arXiv preprint arXiv:1403.7426 (2014)

Champion Paper Team AutonOHM

Marco Masannek and Sally Zeitler[✉]

University of Applied Sciences Nuremberg Georg-Simon-Ohm,
Kesslerplatz 12, 90489 Nuremberg, Germany
marco.masannek@th-nuernberg.de, sally.zeitler@gmail.com
https://www.th-nuernberg.de/en/faculties/efi/research/laboratories-
actively-involved-in-research/mobile-robotics/robocupwork/

Abstract. This paper presents the team AutonOHM and their solutions
to the challenges of the RoboCup@Work league. The hardware section
covers the robot setup of Ohmn3, which was developed using knowledge
from previous robots used by the team. Custom solution approaches
for the @Work navigation, perception, and manipulation tasks are discussed in the software section, as well as a control architecture for the
autonomous task completion.

1 Introduction

The RoboCup@Work league, established in 2012, focuses on the use of mobile
manipulators and their integration with automation equipment for performing
industrial-relevant tasks [1].

The competition is divided into several tests of increasing difficulty. Points are
awarded for reaching workstations, picking up and placing objects. Additional
points are given for arbitrary surfaces, containers and special workstations such
as shelves. Arbitraries are unknown prior to the competition and can be anything
from grass to aluminium foil. Manipulating incorrect objects, losing objects, and
colliding with visual obstacles such as barrier tape lead to point deductions. One
restart per test is allowed, which is triggered instantly upon collision with arena
elements. After restart, all points are reduced to zero. If the restart is triggered
due to a collision, all following points for this test are multiplied by 75%.

The first test is the Basic Manipulation Test. Five objects need to be detected
and transported from one table to another. The next three tests are Basic Transportation Test 1–3, where the robot has to grasp several objects from multiple
workstations. With each test, more objects and workstations are used. Decoy
objects, arbitrary surfaces, different table heights as well as physical and visual
obstacles are introduced. Additionally, the robot needs to grasp and place on a
shelf and place objects into containers.

The following two tests are speciality tests. During the Precise Placement
Test, the robot has to place objects into object-specific cavities. For the Rotating
Turntable Test, the robot needs to pick the correct objects from a rotating
turntable. The finale combines all previous challenges, all table heights and table
configurations are used as well as decoy objects, arbitrary surfaces, containers,
visual and physical obstacles.

© The Author(s), under exclusive license to Springer Nature Switzerland AG 2023
A. Eguchi et al. (Eds.): RoboCup 2022, LNAI 13561, pp. 228–239, 2023
https://doi.org/10.1007/978-3-031-28469-4_19

2   AutonOHM

The AutonOHM-@Work team at the University of Applied Sciences Nuremberg Georg-Simon-Ohm was founded in September 2014. In 2017, the team was able to win both the German (Magdeburg) and the World Championship (Nagoya) title. With the knowledge and experience gained in the former tournaments, the team was also able to defend both of these titles in 2018.

In late 2018 most of the members finished their studies, which is why the team had to be rebuilt in 2019. Since then, the team consists of a small group of "core" members and changing short-term members. In 2021 the AutonOHM-@Work team won the SciRoc Challenge 2021 - Episode 5: Shopping Pick & Pack, as well as the World Championship title in the RoboCup Worldwide competition. Since late 2021 the team has welcomed new members.

Furthermore, the team defended the World Championship title in the RoboCup Worldcup 2022.

3   Hardware Description

We are using a customized Evocortex [3] R&D platform with the smallest form factor available. The platform is equipped with an omnidirectional mecanum drive, an aluminum chassis capable of carrying loads up to 100 kg and a Li-Ion Battery with a nominal voltage of 24 V and roughly 12.5 Ah capacity. In our configuration, the platform does include any sensors, power management or computation units, which means it only serves as our base. Every further component needed was mounted in or on the chassis.

3.1   Sensors

Lidars. Mapping, navigation and the detection of physical obstacles is performed by three SICK TiM571 2D Lidars. One each is mounted at the front and the back of the robot scanning 180°. As this creates dead zones at the robot's sides, a third sensor was mounted centred at the bottom of the robot, resulting in a full 360° scan of the robot's surroundings.

Cameras. We use an Intel RealSense D435 3D-camera for the object perception. It is attached to the manipulator so that it can be positioned above the workstations to detect the surface and the position of the objects.

For barriertape detection, multiple ELP USB fisheye cameras can be mounted around the robot, which enables a 360° view. During the competition, we usually rely on a single fisheye camera because we have observed that all the barriertape is still detected.

Fig. 1. 360° fisheye camera setup

Fig. 3. Robot bottom

Fig. 2. Image of our Ohmnibot

Fig. 4. Laser scan area

3.2   PC

The newly introduced neural networks require a GPU for computation onboard of the robot. As embedded GPU chips such as the Nvidia Jetson TX2 do not provide enough processing power for the task optimization and navigation algorithms, we designed a custom PC solution consisting of an AMD Ryzen 3700x processor, a mini-ATX mainboard and a low power Nvidia GTX1650 graphics card, which is connected to the mainboard with a riser cable. This enabled us to build a flat case with both the mainboard and the graphics card safely mounted inside. The form factor of the case makes it possible to slide it into the robot's back, similar to a server rack.

3.3   PSU

We developed a custom PSU circuit board containing emergency switches for the actuators, a main power switch and high efficiency voltage controllers for 5 V and 12 V. It is equipped with a custom designed plug system with selectable voltage, so every peripheral device can be connected using the same plug type. In addition to that, we use an adjustable DC-DC controller for the power supply of the manipulator, as its power consumption exceeds the limits of the onboard controllers. For the custom PC system, we use a standard 250 W automotive ATX power supply.

3.4 Manipulator

Arm. As our budget did not allow the purchase of an applicable robot arm, we had to develop a custom solution. Industrial pick and place applications are often solved with SCARA robot arms. However, the SCARA concept was not exactly suitable for our purpose, which is why we combined the idea of a cylindrical robot arm with joint arms.

The concept utilizes linear gears to control the z- and x-axis of the arm. In combination with the first rotational z joint, the TCP can be moved to every point (x, y, z) given within the operation area. For more flexibility, two additional rotational joints (y and z) were added between the TCP and the linear x-axis to compensate for the object and arm orientation. The actuators we used are simple Dynamixel MX-106 and AX-64 motors, which were available in our laboratory. They have enough power to control each axis, with the linear z axis being able to lift up to 5 kg.

Most of the parts used were 3D printed using PETG material, including some main mounting parts and all gears. The main bearing, the linear rail and the full extension tray rails have to be purchased. Including the actuators, our current configuration sums up to about 2,500 EUR. We are planning to release the plans once the arm is fully developed, so that any student or research facility can rebuild the arm for educational purposes.

Gripper. The gripper concept also utilizes 3D printed linear gears to convert the rotational force of a motor into linear movement of the fingers. It is based on a single Dynamixel AX-12 motor connected to the driving gear. The power transmission enables the motor to grasp objects with its full torque, rather than it being reduced by a lever with its length conditioned by the gripper fingers. The fin-ray fingers are custom printed out of rubber filament, making them soft and enabling them to close around grasped objects. They are also more wide than standard FESTO fin-ray fingers. This gripper concept is currently being revised to allow the use of force feedback.

Fig. 5. Gripper

4 Software Description

We use Linux Ubuntu 18.04 and ROS Melodic [4] as our operating systems. A custom software architecture was created to simplify the overall structure and to regain system flexibility. Our new design is displayed in Fig. 6.

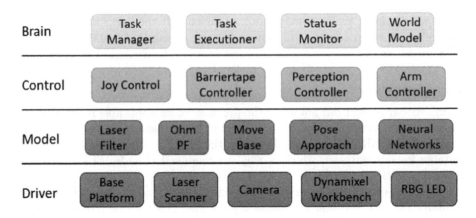

Fig. 6. Software architecture - BCMD

The idea derives from the Model-View-Controller software design pattern, which is adjusted to the usage of the ROS framework. Regarding the frequent use of hardware, an additional driver layer is added below the model layer. Models that need data from hardware, e.g. sensor data, can get them from the individual driver programs. The view layer is realized with each program using interfaces to RVIZ or simple console logging, which makes custom implementations obsolete. Components that require additional control features, such as the robot arm, have dedicated controllers providing simple interfaces for the brain layer, which is responsible for the actual task interpretation and execution. The individual layer components will be explained in the following sections.

4.1 Driver

The driver layer only contains actual hardware control programs, such as the sensor interfaces. The idea here is that the whole layer can be replaced with simulation tools such as Gazebo.

Base Platform. The base platform driver converts incoming cmd_vel messages into wheel rpm and calculates the odometry from obtained rpm. It stops the robot automatically if the incoming commands time out to prevent uncontrolled movements. An additional twist_mux node throttles incoming commands from the joy_controller, move_base and the pose_approach.

Laser Scanner. Three sick_tim nodes provide the interface to the scanners with given IP address and scan area configuration. However, as the Lidar is prone to measurement errors such as shadows or reflections, custom laser filters are applied to the raw data for later computation.

Camera. We use the Intel Realsense SDK with the provided ROS wrapper. The fisheye cameras are accessed via the ROS usb_cam package [20].

Dynamixel Workbench. The tower arm is controlled with a controller instance of the dynamixel_workbench package. It provides a trajectory interface to control multiple motors at once, which we use for trajectory execution. As our gripper also uses a dynamixel motor, but needs extended access to motor variables (e.g. torque), a dedicated controller instance is used for the gripper controls and feedback.

4.2 Model

Our models contain all algorithms used to challenge the problems of the tasks in the @Work league. This includes localization, navigation and perception. The task planner is not included as a model but in the brain layer because it is more convenient to attach it directly to the task manager, as discussed in Sect. 4.4.

Laser Filter. As mentioned in Sect. 4.1, we filter the raw laser data before computing. The first filters are simple area filters to delete the robot's wheels from the scan. The second filter is a custom jumping point filter implementation. We faced problems with reflections of the alu profile rails used for the walls of the arena, which caused the robot to mark free space as occupied. The filter calculates the x- and y-position for each scan point and checks if there are enough neighbors in close range to mark a point as valid. All points with less than n neighbors in the given range will be handled as measurement errors and therefore deleted.

Ohm PF. For localization in the arena, we use our own particle filter algorithm. Its functionality is close to amcl localization, as described in [5,13], with optional support for other sensor types such as cameras. The documentation can be found in German under [22]. The algorithm is capable of using multiple laser scanners and an omnidirectional movement model. Due to the Monte Carlo filtering approach, the localization is robust and accurate enough to provide useful positioning data to the navigation system. Positioning error with the particle filter is about 6 cm, depending on the complexity and speed of the actual movement.

Move Base. We use the ROS navigation stack [10] for global path planning and the local path control loops. Path cost calculations are performed by using the costmap 2D plugins. The base layer is a 2D laser map created with gmapping [11,12]. On top of that, we use a barriertape map layer which contains all detected barriertape points. For local obstacle avoidance, we added an obstacle layer which includes laser data from all three laser scanners. All layers are combined in the final inflation layer. Global path planning is computed with the mcr_global_planner [17] while the path is executed using the TEB local planner [6–9]. As the local planner is not able to precisely navigate to a given goal pose, we set the goal tolerance relatively high. Once we reached our goal with move_base, we continue exact positioning with our custom controller, the pose_approach.

Pose Approach. The pose_approach package utilizes a simple PID controller to move the robot to a given pose. It utilizes the robot's localization pose as input and the target pose as reference. As the controller does not consider costmap obstacles, the maximum distance to the target is 20 cm to prevent collisions. A laser monitor algorithm checks for obstacles in the current scan and stops the robot if necessary.

Fisheye Rectification. The ra fisheye images need to be rectified to be used as input for the detection network. A specific image_pipeline fork [21] is used, which contains this functionality.

NN - Barriertape For the barriertape detection, we use a U-Net with manually labelled datasets. The ROS node receives raw input images and returns a masked binary image. We have ported the network node from Python to C++ to increase the detection rate from 5 Hz up 20 Hz.

NN - Objects. The detection and classification of objects is done with a Tiny-YOLO-v3 network. The node receives a raw input image and returns a vector with the ID, bounding box and confidence of all objects that were found. As our dataset would require more than 10,000 labelled images, which would require a high amount of time to create, we have implemented an automated dataset creation method using Blender and Python. It basically changes environments, illumination, camera and object pose as well as object appearance in pre-defined bounds. The script creates rendered images as well as bounding box, segmentation and 6DoF labels. With this data generation method, data which is quite similar to the original scene can be created, as well as rather abstract data (Fig. 7). We are currently also working on data generation for deformable objects, such as the objects used in the SciRoc Challenge 2021 - Episode 5: Shopping Pick & Pack [19].

Using an original to artificial image ratio of 1:10, we achieved a detection reliability of over 90% for most scenes. Our data generation scripts are public and free to use [15]. The trained network is converted to TRT-Engine using code from the TRT-YOLO-App from the Deepstream Reference Apps [16]. This increases performance as the CUDA cores will be used more efficient, and makes a detection rate of up 60 Hz possible. In the future, other network types such as segmentation networks and 6DoF networks will be explored.

4.3   Controller

Model nodes that require additional control features are connected to control nodes, which then provide interfaces for the brain layer. They use our robot-_custom_msgs interfaces to share information about the subtask, workstation, or objects. Nodes may have specific subtask types implemented into their behaviour to react optimized.

Joy Control. We use a PS5 joystick to move our robot manually (e.g. for mapping). For this, we have implemented a custom teleop_joy_node with similar functionality. We also plan to implement the usage of the PS5 feedback functions such as rumble.

Barriertape Control. The barriertape controller is a custom mapping implementation for visual obstacles. It throttles the input images to the barriertape network and computes the masked images. Looping through multiple cameras enables us to perform 360° barriertape detection.

Received masked images are converted into a point cloud with a predefined density. This pointcloud is then transformed from the individual camera frame into the global map frame. Afterwards, all new points are compared to the existing map points. New barriertape points that are already occupied are ignored to save computation. As we faced problems with image blur and therefore resulting non-precise barriertape detection, we also compute pixels that mark free space (no barriertape detected). They are compared to existing points, which get deleted if they overlap.

The whole map is converted into an occupancy grid and then published periodically, so it can be included in the costmap of the move_base node. The node is controlled via service calls, which enable or disable the detection loop. The map is always published once the node finishes the init process.

Arm Control. As the kinematic model of the tower arm has only one solution for a given TCP position, we developed a custom arm controller node instead

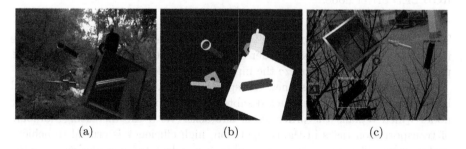

| (a) | (b) | (c) |

Fig. 7. Abstract image (a) corresponding mask label (b) Abstract image with bounding box label (c)

of using moveIt. It is possible to adjust the amount and type of joints and links via ROS parameters, only the inverse kinematics solution has to be adjusted for new arms. Using predefined danger zones, the arm executes a self calculated trajectory to the target pose considering the individual motor parameters. The arm is controlled via ROS services or a development GUI for debugging. When using the services, the arm executes a full task using the given information, which means, in case of a pick task, it moves the TCP to the object position, closes the gripper, and stores the object. After the subtask finishes, feedback of the exit status is returned to the caller.

Perception Control. The perception control node is responsible for the workstation analysis and object detection from a given scene (3D Pointcloud and RGB image). First, the surface equation of the workstation is calculated using the RANSAC [14] algorithm. If a valid result is obtained, raw images are sent to the object perception network (Sect. 4.2). All found objects are then localized using the pinhole camera model, the workstation plane and the bounding box pixels. Finally, the position is transformed into the workstation frame and saved. For moving objects, multiple positions are recorded and then used to calculate the movement equation with RANSAC.

4.4 Brain

The brain layer provides nodes which contain the intelligence of the robot, which means the tracking of itself, its environment and the received tasks.

Worldmodel. All data obtained about the robot's environment is stored in the worldmodel database. This includes the map, all workstation positions and all detected objects on the workstations. The data can be accessed using service calls.

Status Monitor. The status monitor keeps track of the robot itself. It saves the current pose, inventory and state. The associated color code is sent to the RGB LED driver node.

Task Manager. The robot can receive tasks from multiple sources, such as the RefBox or voice commands. In order to process different input formats, different parsers are used to standardize the input for the task manager.

When the robot receives a new transportation task, it is analysed and planned before the execution. All extracted subtasks are managed by the task_manager node, which replans the order of all subtasks. With the increasing numbers of transportation tasks in the competition, high efficiency is crucial to achieve perfect runs. The score of a single subtask is calculated considering expected duration, points, and the risk of failure. These factors may change if certain conditions are met, for example, the navigation time is set to zero if the robot already is at the given position.

Before even starting the planning of subtasks, the received task is analysed for impossible tasks. This would be the case if the target workstation is unknown or unreachable, or an object is lost. All subtasks that cannot be executed are moved to a deletion vector.

A self developed planning algorithm then calculates the raw score of the remaining subtask vector, followed by a simple nearest neighbour search (NN). This result is then fed to a recursive tree calculation method, which searches for the optimal solution. A branch is only fully calculated if the score sum does not exceed the best solution found with the NN. This way, we have achieved an overall planning time for the BTT3 challenge (14 subtasks) of around 10 s. For subtask numbers below 12 the planning only takes 2 s. If the task load exceeds 14 tasks, we skip the recursive strategy, as planning time grows exponentially and therefore cannot produce results in the given time frame of a run.

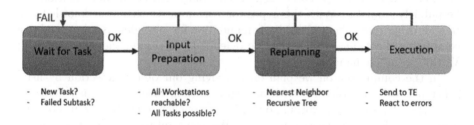

Fig. 8. Task manager states

After planning, every subtask is sent to the task executioner (Sect. 4.4). If the execution was not successful, the task is moved to a failed subtask vector and deleted from the current working STV. The short planning times enable us to replan every time a subtask fails, or new data is available. This is necessary because even simple changes can cause serious errors in the intentional plan. If certain paths are blocked, the navigation time for transportation tasks can increase dramatically, causing a huge loss of efficiency. A final garbage collection checks all deleted and failed subtasks for plausibility again and adds retries for possible subtasks.

Task Executioner. Subtasks that are sent to the Task Executioner get run through an interpreter to extract the actions that are necessary for the task execution. All actions are performed in custom states, which can be adjusted via parameters at creation. The interpreter uses information from the status monitor, the worldmodel and the given subtask to create substates accordingly. The resulting state vector is iterated until finished or failed. While executing, the node reads and modifies the data in the status monitor and worldmodel package. This way, every change is immediately available for all other nodes too.

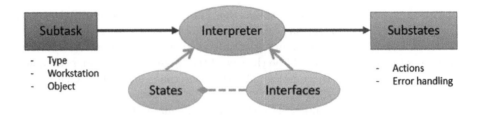

Fig. 9. Task executioner - subtask interpretation

5   Conclusion and Future Work

During the last season, we optimized our robot concept and further extended it.
The robot arm concept has been reworked and improved, a new display has been
implemented as well as voice-feedback functionality. We've started to bundle
our knowledge in a repository [22], where many aspects of our solution will
be publicly available. With this repository, we want to share our knowledge to
provide other researchers with a basic foundation for autonomous robots and task
completion. Last but not least, we've defended our title in the 2022 RoboCup
WorldCup in Thailand.

In the coming season we plan on improving our object detection by using
other network architectures such as segmentation, 6DoF and grasp detection
networks. We are also reworking our gripper concept to enable the use of force
feedback, which will improve our system. We also want to introduce performance
monitoring, allowing us to identify bottlenecks and plan future improvements.
Finally, we plan to extend the above-mentioned repository by adding more documentation as well as some theses from our students.

We are very much looking forward to the upcoming season, where we aim to
defend our title once again at RoboCup 2023 in Bordeaux.

References

1. RoboCup Atwork Website. https://atwork.robocup.org/. Accessed 20 Jan 2021
2. Norouzi, A., Schnieders, B., Zug, S., Martin, J., Nair, D., Steup, C., Kraetzschmar,
   G.: RoboCup@Work 2019 - Rulebook. https://atwork.robocup.org/rules/ (2019)
3. EvoCortex Homepage. https://evocortex.org/. Accessed 20 Jan 2021
4. Stanford Artificial Intelligence Laboratory. Robotic Operating System (2018).
   https://www.ros.org
5. Dellaert, F., Fox, D., Burgard, W., Thrun, S.: Monte carlo localization for mobile
   robots. In: Proceedings 1999 IEEE International Conference on Robotics and
   Automation (Cat. No. 99CH36288C), Detroit, MI, USA, vol. 2, pp. 1322–1328
   (1999). https://doi.org/10.1109/ROBOT.1999.772544
6. Rösmann, C., Feiten, W., Wösch, T., Hoffmann, F., Bertram, T.: Trajectory modification considering dynamic constraints of autonomous robots. In: Proceedings of
   the 7th German Conference on Robotics, Germany, Munich, pp 74–79 (2012)

7. Rösmann, C., Feiten, W., Wösch, T., Hoffmann, F., Bertram, T.: Efficient trajectory optimization using a sparse model. In: Proceedings of the IEEE European Conference on Mobile Robots, Spain, Barcelona, pp. 138–143 (2013)
8. Rösmann, C., Hoffmann, F., Bertram, T.: Integrated online trajectory planning and optimization in distinctive topologies. Robot. Auton. Syst. **88**, 142–153 (2017)
9. Rösmann, C., Hoffmann, F., Bertram, T.: Planning of multiple robot trajectories in distinctive topologies. In: Proceedings of the IEEE European Conference on Mobile Robots, UK, Lincoln (2015)
10. ROS navigation. http://wiki.ros.org/navigation. Accessed 20 Jan 2021
11. Grisetti, G., Stachniss, C., Burgard, W.: Improved techniques for grid mapping with rao-blackwellized particle filters. IEEE Trans. Rob. **23**, 34–46 (2007)
12. Grisetti, G., Stachniss, C., Burgard, W.: Improving grid-based SLAM with rao-blackwellized particle filters by adaptive proposals and selective resampling. In: Proceedings of the IEEE International Conference on Robotics and Automation (ICRA) (2005)
13. Thrun, S., Burgard, W., Fox, D.: Probabilistic Robotics, Massachusetts Institute of Technology (2006)
14. Fischler, M.A., Bolles, R.C.: Random sample consensus: a paradigm for model fitting with applications to image analysis and automated cartography. Commun. ACM **24**(6), 381–395 (1981). https://doi.org/10.1145/358669.358692
15. Github: DataGeneration. https://github.com/ItsMeTheBee/DataGeneration. Accessed 22 Jan 2021
16. GitHub: NVIDIA deepstream reference apps. https://github.com/NVIDIA-AI-IOT/deepstream_reference_apps. Accessed 20 Dec 2019
17. GitHub: bitbots mas_navigation. https://github.com/b-it-bots/mas_navigation. Accessed 20 Jan 2021
18. Alphacei Vosk. https://alphacephei.com/vosk/. Accessed 20 Jan 2021
19. Dadswell, D.: "E05: Pick & Pack," SciRoc, 17-May-2021. https://sciroc.org/e05-pick-pack/. Accessed 13 Jan 2022
20. Github: usb_cam. https://github.com/ros-drivers/usb_cam. Accessed 13 Jan 2022
21. Github: image_pipeline. https://github.com/DavidTorresOcana/image_pipeline. Accessed 13 Jan 2022
22. Github:docs_atwork. https://github.com/autonohm/docs_atwork. Accessed 20 Jan 2022

RoboCup 2022 AdultSize Winner NimbRo: Upgraded Perception, Capture Steps Gait and Phase-Based In-Walk Kicks

Dmytro Pavlichenko[✉], Grzegorz Ficht, Arash Amini, Mojtaba Hosseini,
Raphael Memmesheimer, Angel Villar-Corrales, Stefan M. Schulz,
Marcell Missura, Maren Bennewitz, and Sven Behnke

Autonomous Intelligent Systems, Computer Science, University of Bonn,
Bonn, Germany
pavlichenko@ais.uni-bonn.de
http://ais.uni-bonn.de

Abstract. Beating the human world champions by 2050 is an ambitious goal of the Humanoid League that provides a strong incentive for RoboCup teams to further improve and develop their systems. In this paper, we present upgrades of our system which enabled our team NimbRo to win the Soccer Tournament, the Drop-in Games, and the Technical Challenges in the Humanoid AdultSize League of RoboCup 2022. Strong performance in these competitions resulted in the Best Humanoid award in the Humanoid League. The mentioned upgrades include: hardware upgrade of the vision module, balanced walking with Capture Steps, and the introduction of phase-based in-walk kicks.

1 Introduction

The Humanoid AdultSize League works towards the vision of RoboCup: A team of robots winning against the human soccer world champion by 2050.

Fig. 1. Left: NimbRo AdultSize robots: NimbRo-OP2 and NimbRo-OP2X. Right: The NimbRo team at RoboCup 2022 in Bangkok, Thailand.

© The Author(s), under exclusive license to Springer Nature Switzerland AG 2023
A. Eguchi et al. (Eds.): RoboCup 2022, LNAI 13561, pp. 240–252, 2023
https://doi.org/10.1007/978-3-031-28469-4_20

Hence, the participating teams are challenged to improve their systems in order to reach this ambitious goal. Physical RoboCup championships have not been held in the previous two years, due to the COVID-19 pandemic. For RoboCup 2022, we upgraded our visual perception module with better cameras and more powerful GPUs as well as an improved deep neural network to enhance game perception and localization capabilities. We improved our gait based on the Capture Steps Framework to further enhance the stability of walking. Finally, introduction of phase-based in-walk kicks allowed for more ball control in close-range duels with opponents. We used three NimbRo-OP2X robots alongside with one NimbRo-OP2. Our robots came in first in all competitions of the Humanoid AdultSize League: the main Soccer Competition, Drop-in Games, and the Technical Challenges. Consequently, our robots won the Best Humanoid award in the Humanoid League. Our team is shown in Fig. 1. The highlights of our performance during the RoboCup 2022 competition are available online[1].

2   NimbRo-OP2(X) Humanoid Robot Hardware

For the competition in Bangkok, we have prepared four capable robots: one NimbRo-OP2 [2] and three NimbRo-OP2X robots. Both platforms are fully open-source in hardware[2] and software[3], with several publications [6,7] containing beneficial information on reproducing the platform. The robots, shown in Fig. 1 share similarities in naming, design and features, but are not identical. Both platforms feature a similar joint layout with 18 Degrees of Freedom (DoF), with 5 DoF per leg, 3 DoF per arm, and 2 DoF actuating the head. In contrast to other humanoid robots competing in the Humanoid League, the legs have a parallel kinematic structure that locks the pitch of the foot w.r.t. to the trunk [4]. The actuators receive commands from the ROS-based control framework through a CM740 microcontroller board with a built-in six-axis IMU (3-axis accelerometer & 3-axis gyro). As the hardware is based on off-the-shelf consumer-grade technology, the robots can be acquired at a fraction of the cost of similarly sized research-platforms [4] such as: ASIMO [15], HRP-2 [9], and HUBO [12].

Apart from updating the OP2 actuator controller firmware to allow current-based torque measurements, its hardware was not changed from the RoboCup 2019 competition [13]. The OP2X robots were upgraded to accommodate for the weak points from the 2019 competition. After the field size increase, perceiving objects from larger distances was challenging, as they would often be represented by single pixels. To address this issue, the 2 MP Logitech C905 camera was replaced by a 5 MP C930e camera. With an increase of the image capture resolution, the object perception improved greatly. The increase in resolution necessitated more computing power, for which we upgraded the GPU of each OP2X robot to the Nvidia RTX A2000. Applying these modifications was straightforward due to the modular design of the robot.

---

[1] RoboCup 2022 NimbRo highlights video: https://youtu.be/DfzkMawtSFA.

[2] NimbRo-OP2X hardware: https://github.com/NimbRo/nimbro-op2.

[3] NimbRo-OP2X software: https://github.com/AIS-Bonn/humanoid_op_ros.

3   Visual Perception of the Game Situation

Our visual perception module benefits from both hardware and software improvements. First, upgrading the camera, equipped with a wide-angle lens, and GPU; second, supporting a dynamic resolution and enhancing our previous deep convolutional neural network to recognize soccer-related objects, including the soccer ball, goalposts, robots, field boundaries, and line segments.

The upgraded GPU enabled us to use higher resolution images for the network input without decreasing the frame rate. Moreover, the new camera provided images with a high-quality and a wider horizontal field-of-view, compared to our previous Logitech C905 camera. We compare the two cameras by images captured from the same scene in Fig. 2.

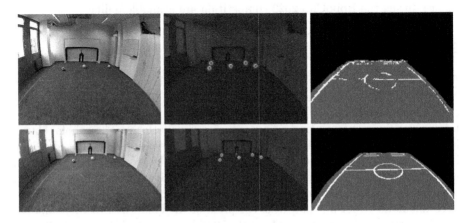

Fig. 2. An image example taken by Logitech C905 camera, heatmap, and segmentation predictions of NimbRoNet2 (first row). The same scene from the Logitech C930e camera, heatmap, and segmentation predictions of the new model (second row).

Inspired by our previous perception model (NimbRoNet2) [13], we used a deep convolutional neural network which is adapted from [1]. Our model has an asymmetric U-Net architecture [14] which consists of a pre-trained ResNet18 model [8] as an encoder and a decoder with a feature pyramid structure, where we only use the 1/4 resolution, and three skip connections. This lighter model reduces inference time to 13 ms by using bilinear upsampling in comparison to the 19 ms runtime of NimbRoNet2. In addition, the new model yields better performance, as illustrated in Fig. 2. The visual perception network has two prediction heads: object detection and segmentation. For object detection, we represent the object locations by heatmaps. We then apply a post-processing step to retrieve the locations from the estimated heatmaps.

We employ the mean squared error loss for training the heatmaps that represent the location of the balls, goalposts, and robots. Additionally, we use the cross entropy loss for training to predict the segmentation of background, field, and lines. The total loss is hence the linear combination of the two mentioned losses. As announced in RoboCup 2022, instead of using a single ball, 17 balls[4] were selected as possible choices, of which three balls were used in the competition. To deal with this challenge, we exploited augmentation by randomly substituting the ball in the images with the announced balls.

Our pipeline shows the ability to cover the entire field and detects objects that are far away from the robot. Furthermore, since the localization module depends on landmarks (lines, center circle, and goalposts), the improvements of the vision perception module allow us to localize more precisely in the new field, where penalty area lines were added, through minor modifications in the localization module, e.g., increasing the weights of the center circle and goalposts.

4 Robust Omnidirectional Gait with Diagonal Kick

4.1 Capture Step Walking

The walking of our robots is based on the Capture Step Framework [11]. This framework combines an open loop gait pattern [10], that generates leg-swinging motions with a sinusoid swing pattern, with a balance control layer that models the robot as a linear inverted pendulum (LIP) and computes the timing, and location, of the next step in order to absorb disturbances and return the robot to an open-loop stable limit cycle. The parameters of the LIP are fitted to center of mass data recorded from the robot while the robot is walking open loop. The footstep locations computed by the LIP are then met by modifying the amplitude of the sinusoid leg swings during a step. The timing of the footsteps modulates the frequency of the central pattern generator such that the step motion is slowed down, or sped up, in order to touch the foot down at the commanded time. This year, for the first time, all of our robots were equipped with a Capture Step-capable walk. The Capture Steps proved especially useful for regaining balance after a collision with an opponent while fighting for the ball, for regaining balance after moving the ball with our seamlessly integrated in-walk kicks, and, of course, for winning the Push Recovery Challenge. Figure 3 and Fig. 4 show plots of a lateral and a sagittal push recovery example, respectively.

4.2 Balance State Estimation

As with any real system, there is inherent noise in the IMU and joint sensors. Estimating balance-relevant state variables in presence of this noise is a critical task, as the control is strictly tied to the state estimation.

Before the 2022 competition [11], the Center of Mass (CoM) movement was estimated purely through a combination of joint encoders, kinematics, and a torso attitude estimator. The final output would be then smoothed with a Golay filter, also providing velocity and acceleration estimates. This solution, while working to an extent, was not ideal. Noise would then be further suppressed

---

[4] https://humanoid.robocup.org/hl-2022.

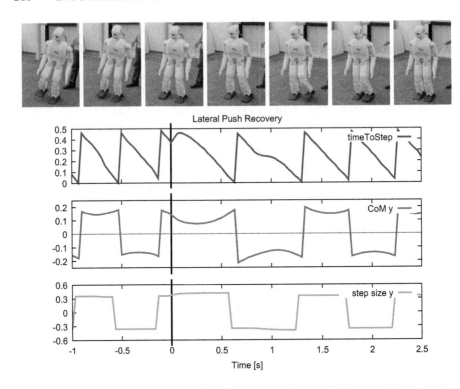

Fig. 3. Lateral push recovery example. From top to bottom, the remaining time of a step, the lateral center of mass (CoM) location, and the lateral step size are plotted. The push occurs at time 0. The timing of the step is immediately adjusted for a longer step duration. The center of mass approaches the stance foot and then returns to the support exchange location at approx. 0.6 s when the actual Capture Step is performed. The size of the Capture Step is adjusted from the default 30 cm to 35 cm. After the Capture Step at 1.25 s, the gait returns to nominal values.

in post-processing, along with balance-relevant dynamic effects, leading to the robot tending to walk open-loop at higher walking velocities.

We have adopted the idea of the Direct Centroidal Controller (DCC) from [5] to use a Kalman Filter to estimate the CoM state $\mathbf{c} = [c\ \dot{c}\ \ddot{c}]$ on the sagittal $\mathbf{c}_x$ and lateral $\mathbf{c}_y$ planes and supplementing the measurement model $\mathbf{z}_k$ with the unrotated and unbiased for gravity g trunk acceleration $^G\ddot{x}_t$ from the IMU:

$$\begin{bmatrix} ^G\ddot{x}_t \\ ^G\ddot{y}_t \\ ^G\ddot{z}_t \end{bmatrix} = \begin{bmatrix} ^T\ddot{x}_t \\ ^T\ddot{y}_t \\ ^T\ddot{z}_t \end{bmatrix} \mathbf{R_t} - \begin{bmatrix} 0 \\ 0 \\ g \end{bmatrix}, \quad \mathbf{z}_{k,x} = \begin{bmatrix} 1 & 0 \\ 0 & 0 \\ 0 & 1 \end{bmatrix} \begin{bmatrix} c_x \\ ^G\ddot{x}_t \end{bmatrix} + \mathbf{v}_{k,x}. \tag{1}$$

Unlike in the DCC, the mass position does not consider limb dynamics [3] and equates to a fixed point in the body frame. This simplification is an advantage in the sensing scheme, as the accelerations of the trunk are directly linked to the trunk-fixed CoM. As a result, Capture Steps operated on the estimated state directly, allowing for precise and uniform balance control, independent of the walking speed.

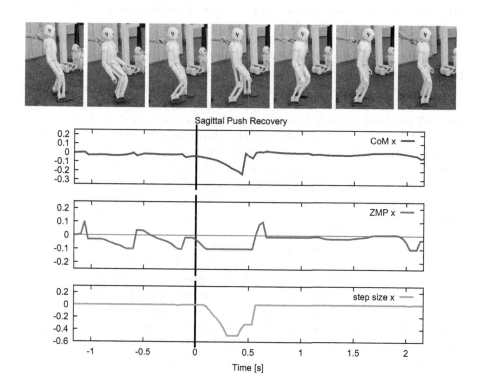

Fig. 4. Sagittal push recovery example. From top to bottom, the sagittal center of mass (CoM) location, the sagittal zero moment point (ZMP) coordinate, and the sagittal step size are plotted. At the time 0 the robot experiences a backwards push. The ZMP immediately moves into the physical limit in the heel, and the step size is adjusted to perform a backwards step. After the first Capture Step is finished at 0.5 s, the robot still has a small backwards momentum, which is nullified during the second recovery step and the gait returns to nominal values.

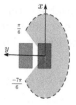

Fig. 5. The feasible kick direction for the right leg. The dotted boundary represents the optimal ball position relative to the foot, shown as an ellipse from the configured optimal distance for the side and front kick.

4.3   Phase-Based In-Walk Kick

We introduced the "in-walk kick" approach in [13] and implemented it on NimbRo-OP2(X) platforms to compete at RoboCup 2019. The "in-walk kick" approach eliminated unnecessary stops to execute kicking motions, resulting in an improvement in the overall pace of the game.

However, the previously introduced approach had limitations that needed to be addressed. The strength of the kick depended on the accuracy of the foot placement behind the ball; the closer or farther the ball was from the optimal distance, the weaker the kicks were. To shoot in the desired direction, the robot had to first rotate in that direction before executing the forward kick. For effective ball handling, the robot first had to align its foot behind the ball.

For RoboCup 2022, we addressed the above limitations to further improve the pace of the game. Our approach allows the robot to perform kicks in feasible directions, while adjusting the foot behind the ball before the ball is actually kicked. It also improves the strength of kicks where the ball is at a less than optimal distance from the foot. The kick direction is not feasible if the leg is physically limited to move in that direction. Figure 5 shows the feasible kick directions of the NimbRo-OP2 robot.

We define the kick coordinate frame such that the x axis is along the direction of the kick and the origin of the frame is at the center of the kicking foot. The formulation of the kick trajectories is represented in this frame and later transformed into the local frame of the robot and applied to the sagittal and lateral trajectories of the leg, allowing the robot to perform diagonal and side-kicks in addition to the forward kick. We create two swing trajectories: the kick swing s_{kick} and the adjust swing s_{adj}. The former applies a swing trajectory to the kick, the latter is a swing trajectory that adjusts the y offset of the foot to the ball. Figure 6 shows three examples of the generated swing trajectories[5]. The function of the swing trajectory s is defined as:

$$g\left(\phi, y_0, y_f, c\right) = y_0 + \left(y_f - y_0\right) \left(6\left(1 - \phi\right)^2 \phi^2 c + 4\left(1 - \phi\right)\phi^3 + \phi^4\right),$$

$$s\left(\phi, \alpha, \phi_p, c\right) = \begin{cases} g\left(\frac{\phi}{\phi_p}, 0, \alpha, c\right) & 0 \le \phi < \phi_p \\ g\left(\frac{\phi - \phi_p}{1 - \phi_p}, \alpha, 0, 1 - c\right) & \phi_p \le \phi \le 1 \end{cases}, \quad (2)$$

where y_0 and y_f are the initial and final domains for the quartic Bezier curve g. The function s for the phase variable ϕ represents a swing curve that reaches the peak amplitude α in the phase ϕ_p and returns to 0 at the end of the phase while following a curvature defined by c. The kick and swing trajectories are then formulated as follows:

$$\begin{aligned} s_{fw}(\phi) &= s\left(\phi, \alpha_{fw}, \phi_{fw}, c_{fw}\right), \\ s_{bw}(\phi) &= s\left(\phi, \alpha_{bw}, \phi_{bw}, c_{bw}\right), \\ s_{kick}(\phi) &= s_{fw}(\phi) + s_{bw}(\phi), \\ s_{adj}(x) &= s\left(\phi, \alpha_y, \phi_{adj}, c_{adj}\right), \end{aligned} \quad (3)$$

---

[5] Online in-walk kick graph: https://www.desmos.com/calculator/v7wlvjtchl.

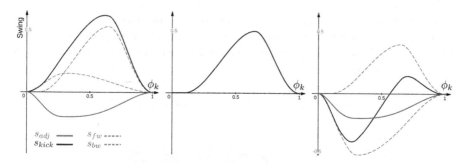

Fig. 6. In-walk kick swing trajectories. ϕ_k is the kick phase (like x_K in Fig. 6 of [13]). s_{fw}, s_{bw}, s_{kick}, and s_{adj} are the forward, backward, kick, and adjust swings, respectively, formulated in (3). Three examples are given, from left to right: the ball farther away than the optimal distance with an offset along the y-axis, the ball at the optimal distance without y-offset, and the ball close to the foot with y-offset.

where s_{kick} denotes a swing trajectory for kicking the ball and s_{adj} denotes an adjustment swing trajectory before kicking the ball. $\alpha_{fw} = \alpha_x + 0.8\,(\alpha_{opt} - \alpha_x)$ is the forward swing amplitude calculated with respect to the optimal swing amplitude α_{opt} and the ball position in the kick frame, $\alpha_{bw} = \alpha_{fw} - \alpha_{opt}$ is the back-swing amplitude, and c_{fw}, c_{bw}, and c_{adj} are the curvature gains of the forward, backward, and adjust swings, respectively.

With phase-based in-walk kicks, the behavior has more freedom to adjust the robot state before the kick and therefore makes faster decisions, which is especially beneficial in one-on-one fights where the opponent is also behind the ball and trying to kick it.

5   Behavior Control

For controlling the high-level behavior, which steers the robot behind the ball, avoids obstacles, and triggers inline kicks towards the opponent goal, we use a simple force field method. We are able to regard the robot as a holonomic point mass due to the omnidirectional control layer of our gait [10] and can control it with a directional input vector whose direction determines the walking direction and its length determines the walking velocity. The vector has a separate component for turning, which is independent of the 2D walking direction on the plane. Using this interface, we can simply sum up forces that pull the robot towards the behind ball position and push the robot away from obstacles to obtain a suitable gait control vector. Orthogonal forces to the robot-ball line and to the robot-obstacle line help with circumventing the ball and walking around obstacles, respectively. The orientation of the robot is controlled by its own set of forces that rotate the robot first towards the ball, and then towards the ball target when the robot is near the behind ball position. Tuning parameters that determine the strength of the individual forces allow us to quickly adapt the behavior controller to different types of robots. This year, where all our robots were of a very similar build, we were able to use the same parameters for all robots. Figure 7 shows visualizations of components involved with the behavior control. There is no special goalie behavior since we find it more efficient to have two active field players.

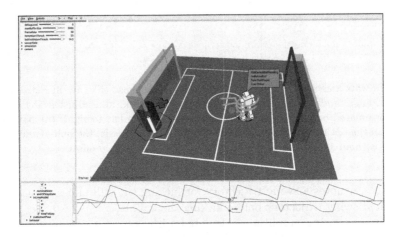

Fig. 7. The graphical debugging and diagnostics tool of team NimbRo. The 3D visualization area provides a large variety of information at one glimpse, such as: the whole-body state estimation of the robot, the localization, and the observed objects. The orange sphere is a ball. The red cross marks the spot towards which the robot should kick the ball. The small red circle is the behind ball position. The stack of arrows drawn around the robot model indicate the output of different control layers, where the top arrow is the force that controls the robot. The white vertical bar is the goalpost, and the gray cylinder around it represents an obstacle. The tool comes with an inbuilt plotter, and sliders for parameter tuning during operation. (Color figure online)

6   Debugging and Diagnostics

As the complexity of the soccer-playing software increases over time, it is very important to develop visualization tools that help with debugging and allow for comprehensive diagnostics of hardware failures. Figure 7 shows a screenshot of our graphical analyzer. A 3D OpenGL scene shows the whole-body state estimation of our robot, the localization on the soccer field, observations such as the ball and the goal posts, the game state according to the game controller, the role and task of the robot, and information about the behavior and gait controller output. Obstacles are marked with a black cylinder. Our tool also includes a plotter for variables that cannot be easily visualized, such as the estimated time until the next support exchange shown by the blue curve in the plot area at the bottom of the screen. The communication interface between the debugging tool and the robot seamlessly integrates into our ROS infrastructure.

7   Technical Challenges

In this section, we describe our approach to Technical Challenges, which is a separate competition within AdultSize League. In a 25 min long time slot, the robots are required to perform four individual tasks: Push Recovery, Parkour, High Kick, and Goal Kick from Moving Ball. The strict time restriction strongly advocates for robust and reliable solutions.

7.1   Push Recovery

In this challenge, the robot has to withstand three pushes in a row from the front and from the back while walking on the spot. The pushes are induced by a pendulum which hits the robot at the height of the CoM. The robots are ranked by the combination of pendulum weight and pendulum retraction distance, normalized by the robot mass. Our robot managed to withstand pushes from both 3 kg and 5 kg pendulums, thanks to the Capture Steps Framework, winning in this challenge.

7.2   Parkour

In this challenge, the robot has to go up a platform and then go back down. The robots are ranked by the height of the platform. We performed this challenge using a manually designed motion. Since our robots have parallel leg kinematics, it was advantageous for us to go on top of the platform with a side-step, where our robots have more mobility. The motion included a controlled fall on the platform, allowing the foot to land closer to the center of the platform, creating space for the other foot in the later stage of the ascent. During this phase, the gains of the actuators were reduced, providing a damping effect when landing on the platform. Then, a series of CoM shifts brought the other foot on the platform, achieving a stable stance (Fig. 8). We did not descend from the platform, because such motion imposed a significant load on the motors. Our robot managed to go up a 30 cm high platform, coming in second in this challenge.

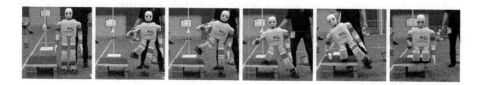

Fig. 8. Technical challenge: Parkour.

7.3   High Kick

In this challenge, the goal is to score a goal by kicking the ball over an obstacle. The teams are ranked by the height of the obstacle in a successful attempt.

The ball starts at the penalty mark and the obstacle is at the goal line. In order to reliably kick over the obstacle, we first perform a kick of low strength to move the ball closer to the obstacle. Then, we kick over it utilizing a kick motion with the foot making contact with the ball as low as possible and using a "scoop"-shaped foot. Our robot came in first, kicking over a 16 cm high obstacle.

7.4 Goal Kick from Moving Ball

In this challenge, a robot has to score a goal by kicking a moving ball. The robot is stationary at the penalty mark and the ball is at the corner. The ball is passed to the robot by a human. The teams are ranked by the number of goals scored in three successive attempts. In order to reliably score goals from a moving ball, we estimate the ball arrival time from velocity and acceleration estimates, which are calculated from the series of ball detections. This enables our robots to start kicking at the right moment, when the ball is approaching the foot (Fig. 9). Our robot came in first in this challenge, scoring in three successive attempts.

Fig. 9. Technical Challenge: Goal Kick from Moving Ball.

8 Soccer Game Performance

At the RoboCup 2022 AdultSize soccer competition, robots played 2 vs. 2 soccer games autonomously on a 9 × 14 m soccer field. In addition to the main tournament, there were Drop-in games, where robots from different teams formed temporary joint teams to compete in 2 vs. 2 games. Our robots performed well and won the AdultSize tournament, which included six round-robin games and the finals, with a total score of 43:2, winning the final 7:1. Our robots also won the Drop-In competition, accumulating 22 points compared to 5.5 points of the second-best team, finishing three games with a total score of 15:3. During the competition, our robots played 10 games with a total score of 58:5 and total duration of 200 min.

9 Conclusions

In this paper, we presented improvements of hardware and software components of our humanoid soccer robots which led us to winning all available competitions in the AdultSize league of RoboCup 2022 in Bangkok: main tournament, Drop-In games, and Technical Challenges, where our robots scored in each challenge. Consistently strong performance of our robots through the competition resulted in receiving the Best Humanoid Award. Improved components include: upgraded perception module, Capture Steps gait, and phase-based in-walk kicks. These innovations allowed for improved localization and ball perception, more robust walking, and more dynamic ball handling—contributing to winning the RoboCup 2022 AdultSize Humanoid League.

Acknowledgments. This work was partially funded by H2020 project EUROBENCH, GA 779963.

References

1. Amini, A., Farazi, H., Behnke, S.: Real-time pose estimation from images for multiple humanoid robots. In: Alami, R., Biswas, J., Cakmak, M., Obst, O. (eds.) RoboCup 2021. LNCS (LNAI), vol. 13132, pp. 91–102. Springer, Cham (2022). https://doi.org/10.1007/978-3-030-98682-7_8

2. Ficht, G., Allgeuer, P., Farazi, H., Behnke, S.: NimbRo-OP2: grown-up 3D printed open humanoid platform for research. In: 17th IEEE-RAS International Conference on Humanoid Robots (Humanoids) (2017)

3. Ficht, G., Behnke, S.: Fast whole-body motion control of humanoid robots with inertia constraints. In: IEEE International Conference on Robotics and Automation (ICRA), pp. 6597–6603 (2020)

4. Ficht, G., Behnke, S.: Bipedal humanoid hardware design: a technology review. Curr. Robot. Rep. **2**(2), 201–210 (2021)

5. Ficht, G., Behnke, S.: Direct centroidal control for balanced humanoid locomotion. In: Cascalho, J.M., Tokhi, M.O., Silva, M.F., Mendes, A., Goher, K., Funk, M. (eds.) CLAWAR 2022. LNNS, vol. 530, pp. 242–255. Springer, Cham (2022)

6. Ficht, G., et al.: NimbRo-OP2X: adult-sized open-source 3D printed humanoid robot. In: 18th IEEE-RAS International Conference on Humanoid Robots (Humanoids) (2018)

7. Ficht, G., et al.: Nimbro-OP2X: affordable adult-sized 3D-printed open-source humanoid robot for research. Int. J. of Humanoid Robot. **17**(05) 2050021 (2020)

8. He, K., Zhang, X., Ren, S., Sun, J.: Deep residual learning for image recognition. In: IEEE International Conference on Computer Vision and Pattern Recognition (CVPR), pp. 770–778 (2016)

9. Hirukawa, H., et al.: Humanoid robotics platforms developed in HRP. Robot. Auton. Syst. **48**, 165–175 (2004)

10. Missura, M., Behnke, S.: Self-stable omnidirectional walking with compliant joints. In: Workshop on Humanoid Soccer Robots, IEEE-RAS International Conference on Humanoid Robots (Humanoids). Atlanta, USA (2013)

11. Missura, M., Bennewitz, M., Behnke, S.: Capture steps: robust walking for humanoid robots. Int. J. Humanoid Robot. **16**(6), 1950032:1–1950032:28 (2019)

12. Park, I.W., Kim, J.Y., Lee, J., Oh, J.H.: Mechanical design of the humanoid robot platform, HUBO. Adv. Robot. **21**(11), 1305–1322 (2007)

13. Rodriguez, D., et al.: RoboCup 2019 adultsize winner nimbro: deep learning perception, in-walk kick, push recovery, and team play capabilities. In: Chalup,

S., Niemueller, T., Suthakorn, J., Williams, M.-A. (eds.) RoboCup 2019. LNCS (LNAI), vol. 11531, pp. 631–645. Springer, Cham (2019). https://doi.org/10.1007/978-3-030-35699-6_51

14. Ronneberger, O., Fischer, P., Brox, T.: U-Net: convolutional networks for biomedical image segmentation. In: Navab, N., Hornegger, J., Wells, W.M., Frangi, A.F. (eds.) MICCAI 2015. LNCS, vol. 9351, pp. 234–241. Springer, Cham (2015). https://doi.org/10.1007/978-3-319-24574-4_28

15. Shigemi, S., Goswami, A., Vadakkepat, P.: Asimo and humanoid robot research at honda. Humanoid Robot. Ref. 55–90 (2018)

HELIOS2022: RoboCup 2022 Soccer Simulation 2D Competition Champion

Hidehisa Akiyama[1]($\boxtimes$), Tomoharu Nakashima[2], Kyo Hatakeyama[2], and Takumi Fujikawa[2]

[1] Okayama University of Science, Okayama, Japan
hidehisa.akiyama@ous.ac.jp
[2] Osaka Metropolitan University, Osaka, Japan
tomoharu.nakashima@omu.ac.jp

Abstract. The RoboCup Soccer Simulation 2D Competition is the oldest of the RoboCup competitions. The 2D soccer simulator enables two teams of simulated autonomous agents to play a game of soccer with realistic rules and sophisticated game play. This paper introduces the RoboCup 2022 Soccer Simulation 2D Competition champion team, HELIOS2022, a united team from Okayama University of Science and Osaka Metropolitan University. The overview of the team's two recent approaches is also described. The first one is the method of online search of cooperative behavior for the setplay planning. The second is a performance evaluation system for efficient team development.

1 Introduction

This paper introduces the RoboCup 2022 Soccer Simulation 2D Competition champion team, HELIOS2022, a united team from Okayama University of Science and Osaka Metropolitan University. The team has been participating in the RoboCup competition since 2000, and won 2010, 2012, 2017, 2018 and 2022 competitions. The team released several open source software for developing a simulated soccer team using the RoboCup Soccer 2D simulator. A team base code, a visual debugger, and a formation editor are available now[1]. The details can be found in [1].

2 Soccer Simulation 2D Competition

The RoboCup Soccer Simulation 2D Competition is the oldest of the RoboCup competitions [6]. The simulation system[2] enables two teams of 11 autonomous player agents and an autonomous coach agent to play a game of soccer with realistic rules and game play. Figure 1 shows a scene from the final of RoboCup2022. Player agents receive a visual sensor message, an aural sensor message,

---

[1] Available at: https://github.com/helios-base.
[2] Available at: https://github.com/rcsoccersim.

© The Author(s), under exclusive license to Springer Nature Switzerland AG 2023
A. Eguchi et al. (Eds.): RoboCup 2022, LNAI 13561, pp. 253–263, 2023
https://doi.org/10.1007/978-3-031-28469-4_21

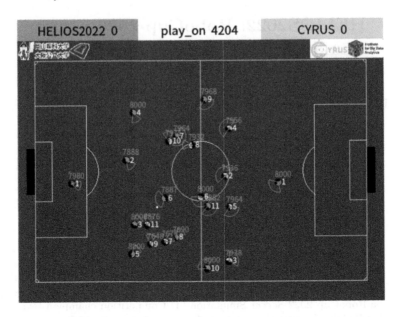

Fig. 1. A scene from the final of RoboCup2022, a match between HELIOS2022 (Okayama University of Science and Osaka Metropolitan University, Japan) and CYRUS (Dalhousie University, Canada).

and a body sensor message from the simulation server, and can send a few types of abstract action commands (kick, dash, turn, turn_neck, and so on). In the game of RoboCup Soccer Simulation 2D Competition, player agents make a decision at each simulation cycle in real time. In 2022, the player dash model was changed. This change made it more difficult for players to accelerate backward and required them to plan their movement actions more deliberately.

In RoboCup2022, 16 teams qualified and 12 teams participated in the competition. The competition consisted of five round-robin rounds and a final tournament. HELIOS2022 won the championship with 26 wins and 1 loss and 1 draw, scoring 92 goals and conceding 12 goals. CYRUS from Dalhousie University won second place, and YuShan2022 from Anhui University of Technology won third place. HELIOS2022 also won the Cooperation Challenge, a challenge competition that mixes players from two different teams to form one team.

3   Online Setplay Planning

This section shows an overview of online planning of cooperative behavior implemented in our team. First, the model of cooperative behavior and its planning process are described. Then, extensions to setplay planning by the coach agent are described.

3.1   Action Sequence Planning by Player Agents

In order to model the cooperative behavior among players as action sequence planning, we employed a tree search method for generating and evaluating action sequences performed by multiple players [3]. This method searches for the best sequence of ball kicking actions among several teammate players using a tree-structured candidate action generator and an evaluation function of the candidate actions. We assume the considered actions are abstracted ones, such as pass, dribble, and shoot. Best first search algorithm is used to traverse a tree and expand nodes. A lot of ball kicking action plans are generated during the search process and the best action plan is selected based on the evaluation value. A path from the root node to the branch represents a kick action sequence that defines a certain cooperative behavior planned by the ball holder player. If other players can generate the same tree, all players can cooperate using the same plan. However, due to the noise of observation, it isn't easy to generate the same tree. And, communication among players is limited. Therefore, the generated plan often is not completed as planned first.

An example of the planning process is depicted in Fig. 2. The kicker generates three candidates for the first action (i.e., pass, pass, and dribble). Each of the three actions has an evaluation value in the corresponding node. That is, the evaluation value of the first pass is 30, the second pass is 20, and the dribble is 15. In this case, the first pass with the highest evaluation value is employed as the first action. Further candidate actions are generated from the selected pass action. We call the level of the tree the depth of the action sequence. Two actions (pass and dribble) in Depth 2 are added as the candidate action with the corresponding evaluation values. The action sequence is updated as the one with the highest evaluation value among the candidate. In this case, the pass in Depth 2 is selected as it has the highest evaluation of 35. Thus the resultant action sequence is "pass–pass".

The decision of players highly depends on the evaluation function, which computes the evaluation value of each action plan. We have to design an appropriate evaluation function in order to select the action plan corresponding to the team strategy and tactics. We have proposed several methods to acquire the evaluation function and are obtaining promising results [2,5,7]. However, because it is still difficult to acquire evaluation functions that can be used in competitions by machine learning, our team used a manually designed evaluation function in RoboCup2022.

3.2   Setplay Planning by Coach Agent

This year, we extend the above action sequence planning method to the coach agent. The coach agent in the Soccer Simulator is supposed to analyze the game and advice the players. The coach can observe the field conditions without error, whereas the player's field of view is limited and much noise is introduced into the visual information. Therefore, under the current rules, communication from the coach agent to player agents is strongly restricted during play on in order to

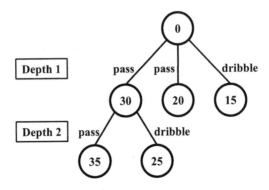

Fig. 2. An example search tree of action sequence planning.

restrict the central control. On the other hand, during setplays, the coach agent is relatively free to send advice to players.

Our coach agent tries to plan two sequential actions among teammate players during our freekick period. In the current implementation, only pass actions are subject to search. However, in the case of setplays, the optimal position of the player's movement can be calculated according to the simulator's physics model because the coach agent can observe noiseless information and players can stand still until it is time to move. If properly communicated, the timing of the players' movements can be synchronized so that the fastest possible pass actions can be executed.

Figure 3 shows an example of setplay planning. The first set play kicker is the number 7 player, slightly above center in the image. That player has the ball in a controllable position. The small squares indicate candidate receiving positions for the first pass. The line segments radiating from players indicate the paths of movement along which each player can receive the first pass without being intercepted by an opponent player. The small circles indicate candidate receiving positions for the second pass. The double circle indicates the positions where the second pass can be received without being intercepted by an opponent player. These combinations are searched and evaluated with an evaluation function to find the best combination. Once the best combination is determined, the coach informs the players of their receiving positions and when each player will move. The result is a chain of the fastest passes, often breaking through a wall of the opposing player's defense.

4 Performance Evaluation System

When developing a team, it is important to be efficient in team performance evaluation. This section describes our performance evaluation system in order to facilitate team development.

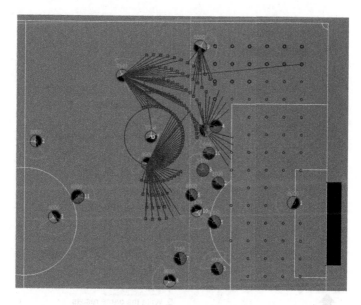

Fig. 3. An example of setplan plan and candidate receiver actions generated by the coach agent.

4.1 Overview of the System

Figure 4 shows the overview of our performance evaluation system. This system reduces the burden of the developers and promotes efficient performance evaluation of the developed teams. The performance evaluation system is available at our code repository[3]. The procedure of the system is as follows:

1. Select game settings
2. Assign computers
3. Execute games respectively
4. Analyze game log og files
5. Write game results

- Select game settings
  First, the user sends game settings (the branch name in Git repository, the name of the opponent team, the number of games, and so on) to the server computer through a chat bot application as a request. We used Slack as the chat system and developed a chat bot that runs on it. A chat bot is pragmatically controlled via a bot user token that can access one or more APIs. One advantage of using a chat bot as UI is that it can be used on multiple devices, such as PCs, smartphones, and tablets. Another advantage is that multiple users can execute the same procedure at the same time. Figure 5 shows an example of a dialog screen between a user and the chat bot.

---

[3] Available at: https://github.com/opusymcomp/autogame.

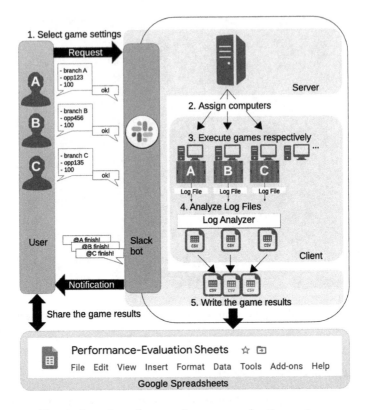

Fig. 4. Overview of our performance evaluation system.

- Assign Computers
  Second, in response to the request in Step 1, the server assigns the requested games to the host computers according to the availability of computational resources. The server searches for available host computers by checking the CPU usage. If a host computer returns a busy status, the server does not assign any task to that host computer to avoid a resource conflict.
- Execute games respectively
  Third, each assigned host runs the soccer simulator according to the specified game settings. Note that the assigned host computers keep the game log files after the games are finished.
- Analyze game log files
  After finishing the game, the host computer analyzes the game log files. In the developed system, LogAnalyzer3[4] is used to analyze game log files. The analyzed game results are saved in CSV format on the assigned host computers. Then the CSV file is transferred to the server.
- Write game results

---

[4] Available at: https://github.com/opusymcomp/loganalyzer3.

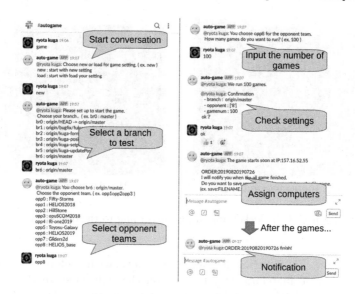

Fig. 5. An example of a dialogue between a user and the chat bot.

			AQ	AR	AS	AT	AU
1	A	B			HELIOS_base		
2	ORDER	branch	win	lose	draw	our-score	opp-score
82	20190806003635	origin/yama-6-press	60	22	18	2.59	1.56
83	20190806110700	origin/feature/fuku-switch-evaluator	77	6	17	2.75	0.8
84	20190806110233	origin/kuga-updatePosition	88	7	5	2.99	0.65
85	20190806103050	origin/omori_penalty_kick_adjust	82	5	13	3.15	0.87
86	20190806111419	origin/feature/fuku-switch-evaluator	77	11	12	2.74	0.86
87	20190806173948	origin/yama-6-press	79	13	8	2.72	0.93
88	20190806112930	origin/kuga-setplay-bugfix	95	1	4	4.14	0.79
89	20190806190644	origin/master	81	6	13	2.73	0.81
90	20190807114731	origin/omori_penalty_kick_adjust	85	6	7	3.102040816	0.7551020408
91	20190807144303	origin/master	83	9	8	2.96	0.67
92	20190808002714	origin/yama-6-press	61	17	21	2.373737374	1.111111111

Fig. 6. An example of game results shared by Google Spreadsheet.

Finally, the game results are summarized in Google Spreadsheets in order to share the results with all team members. The system uses SheetsAPI[5], which is an API to read and write data in Google Spreadsheets, to write the analysis results. Figure 6 shows an example of the game results shared by Google Spreadsheets on the Web.

4.2 Case Study: Effect of Team Names on the Team Strategy

Changing Team Strategy According to Opponent Team Names. In the 2D competition, almost all teams have developed their own strategies while there seem to be a few teams that have specialized strategies to some particular teams. Our team has also been trying to adopt the strategy according to the specific opponent strategy [4].

---

[5] https://developers.google.com/sheets/api.

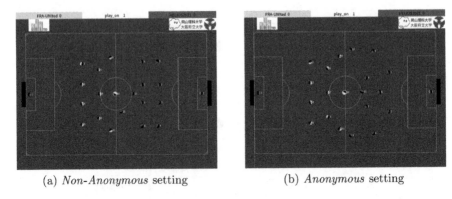

(a) *Non-Anonymous* setting   (b) *Anonymous* setting

Fig. 7. Difference of the kickoff formations according to the opponent team name

For example, Fig. 7 shows the difference in formations before the start of the game in the same matchup. Our team is on the right side. In the left of the figure (Fig. 7(a)), both teams know their opponent teams. That is, the information about the opponent's team name is known before the game starts. On the other hand, the right figure (Fig. 7(b)) shows the kickoff formation when the information on their opponent team (i.e., team names of each other's opponent) was not allowed to be sent to both teams. As shown in the figures, our team obviously changes the formation only by the name of the opponent team. This indicates that our team has a specialized strategy for a particular team. Because the specialized team strategies indicate that the phase of team development is shifting to the second way, the investigation into this will give us some information on the progress in this league in terms of team development.

Numerical Experiments. In order to assess the effect of team names on the team strategy, we conducted numerical experiments using our performance evaluation system. We investigate the difference in team performance between *Anonymous* and *Non-Anonymous* settings. We have collected the binaries of the top 13 teams in RoboCup 2021. The teams played round-robin games 1000 times. This process was applied to both cases of *Anonymous* mode and *Non-Anonymous* mode.

In order to see whether teams change strategy according to the opponent team name or not, the difference in winning rates between *Anonymous* and *Non-Anonymous* modes are calculated for each team. Table 1 shows the total average point and the rank of the team in this experiment. The points of some teams have increased or decreased by 1.0 or more. Although the fluctuation of the ranking is small, it is considered that there is a difference in terms of points.

The significance of the differences in the team strategies was checked by using the Chi-squared test. The test was conducted in two rounds. In the first round of the test, we used two indices the number of winning and the others. Second, we used three indices: The number of wins, draws, and losses. If both tests proved

Table 1. Total average point (rank) in 1000 games.

Team	Non-anonymous	Anonymous
CYRUS	26.047 (3)	26.219 (3)
HELIOS2021	32.188 (1)	32.281 (1)
YuShan2021	27.812 (2)	26.781 (2)
HfutEngine2021	16.656 (7)	16.281 (7)
Alice2021	24.391 (4)	25.500 (4)
Oxsy	21.688 (5)	18.656 (6)
RoboCIn	12.938 (9)	12.977 (9)
FRA-UNIted	17.375 (6)	19.219 (5)
Jyo_sen2021	9.102 (10)	9.039 (10)
MT2021	14.023 (8)	14.117 (8)
ITAndroids	4.820 (13)	5.207 (13)
Persepolis	7.453 (11)	7.695 (11)
ARAS	6.227 (12)	6.188 (12)

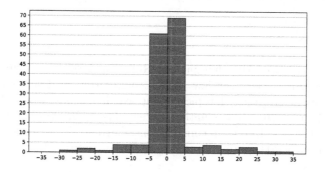

Fig. 8. Distribution of difference in winning rate

that the difference is significant, Point 1 is given to the team. If the significant difference is proved only from one of the two tests, point 0.5 is given. If neither of the two tests recognizes any significant difference, no point is given (i.e., the point is zero).

Figure 8 shows the difference in the winning rates. Most of the differences in the winning rates are less than 5%. However, there are some matches where the difference was tested significantly.

Figure 9 shows heat maps of the winning rate and the result of the Chi-squared test. The value represents the winning rates of the teams in a row against the teams in a column. This means that the larger the positive value is, the stronger the corresponding team is, and the smaller the negative value is, the weaker the corresponding team is in the *Non-Anonymous* mode. On the other hand, if the value is near 0, there is not much difference between *Anonymous* and

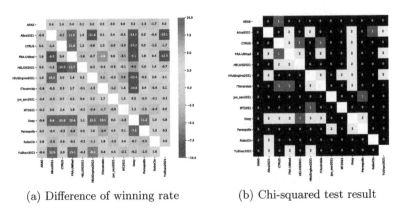

(a) Difference of winning rate   (b) Chi-squared test result

Fig. 9. Heat map representations.

Non-Anonymous . Regarding the same matchups whose difference of winning rate is more than several percent: We found that the Chi-squared test decided that there is a difference in such frequency of result.

5   Conclusion

This paper introduced the champion of RoboCup 2022 Soccer Simulation 2D Competition. First, the overview of the competition is described. Then, we described our current research topics, online search of cooperative behavior planning, and the performance evaluation system. The HELIOS team won 4 championships in the past RoboCup competitions. Currently, our released software is widely used in the 2D community not only for competition but for research.

References

1. Akiyama, H., Nakashima, T.: HELIOS base: an open source package for the RoboCup soccer 2D simulation. In: Behnke, S., Veloso, M., Visser, A., Xiong, R. (eds.) RoboCup 2013. LNCS (LNAI), vol. 8371, pp. 528–535. Springer, Heidelberg (2014). https://doi.org/10.1007/978-3-662-44468-9_46
2. Akiyama, H., Fukuyado, M., Gochou, T., Aramaki, S.: Learning evaluation function for RoboCup soccer simulation using humans' choice. In: Proceedings of SCIS & ISIS 2018 (2018)
3. Akiyama, H., Nakashima, T., Igarashi, H.: Representation and learning methods for situation evaluation in RoboCup soccer simulation. J. Jpn. Soc. Fuzzy Theory Intell. Inform. **32**(2), 691–703 (2020)
4. Fukushima, T., Nakashima, T., Akiyama, H.: Online opponent formation identification based on position information. In: Akiyama, H., Obst, O., Sammut, C., Tonidandel, F. (eds.) RoboCup 2017. LNCS (LNAI), vol. 11175, pp. 241–251. Springer, Cham (2018). https://doi.org/10.1007/978-3-030-00308-1_20

5. Fukushima, T., Nakashima, T., Akiyama, H.: Evaluation-function modeling with multi-layered perceptron for RoboCup soccer 2D simulation. Artif. Life Robot. **25**(3), 440–445 (2020). https://doi.org/10.1007/s10015-020-00602-w
6. Noda, I., Matsubara, H.: Soccer server and researches on multi-agent systems. In: Proceedings of IROS-96 Workshop on RoboCup, pp. 1–7 (1996)
7. Suzuki, Y., Fukushima, T., Thibout, L., Nakashima, T., Akiyama, H.: Game-watching should be more entertaining: real-time application of field-situation prediction to a soccer monitor. In: Chalup, S., Niemueller, T., Suthakorn, J., Williams, M.-A. (eds.) RoboCup 2019. LNCS (LNAI), vol. 11531, pp. 439–447. Springer, Cham (2019). https://doi.org/10.1007/978-3-030-35699-6_35

Tech United Eindhoven @Home 2022 Champions Paper

Arpit Aggarwal, Mathijs. F. B van der Burgh, Janno. J. M Lunenburg, Rein. P. W Appeldoorn, Loy. L. A. M van Beek, Josja Geijsberts, Lars. G. L Janssen, Peter van Dooren[✉], Lotte Messing, Rodrigo Martin Núñez, and M. J. G. van de Molengraft

Eindhoven University of Technology, Den Dolech 2, P.O. Box 513, 5600, MB Eindhoven, The Netherlands
techunited@tue.nl
http://www.techunited.nl, https://github.com/tue-robotics

Abstract. This paper provides an overview of the main developments of the Tech United Eindhoven RoboCup@Home team. Tech United uses an advanced world modeling system called the Environment Descriptor. It allows for straightforward implementation of localization, navigation, exploration, object detection & recognition, object manipulation and human-robot cooperation skills based on the most recent state of the world. Other important features include object and people detection via deep learning methods, a GUI, speech recognition, natural language interpretation and a chat interface combined with a conversation engine. Recent developments that aided with obtaining the victory during RoboCup 2022 include people and pose recognition, usage of HSR's display and a new speech recognition system.

1 Introduction

Tech United Eindhoven[1] (established 2005) is the RoboCup student team of Eindhoven University of Technology[2] (TU/e), which joined the ambitious @Home League in 2011. The RoboCup@Home competition aims to develop service robots that can perform everyday tasks in dynamic and cluttered 'home' environments. The team has been awarded multiple world vice-champion titles in the Open Platform League (OPL) of the RoboCup@Home competition during previous years, and two world champion titles in 2019 and 2022[3] [4] [5] in the Domestic Standard Platform League (DSPL). In the DSPL, all teams compete with the same hardware; all teams compete with a Human Support Robot (HSR), and use the same external devices. Therefore, all differences between the teams regard only the software used and implemented by the teams.

---

[1] http://www.techunited.nl.
[2] http://www.tue.nl.
[3] https://tinyurl.com/DSPLBangkok2022Stage1Score.
[4] https://tinyurl.com/DSPLBangkok2022Stage2Score.
[5] https://tinyurl.com/DSPLBangkok2022FinalScore.

© The Author(s), under exclusive license to Springer Nature Switzerland AG 2023
A. Eguchi et al. (Eds.): RoboCup 2022, LNAI 13561, pp. 264–275, 2023
https://doi.org/10.1007/978-3-031-28469-4_22

Tech United Eindhoven consists of (former) PhD and MSc. students and staff members from different departments within the TU/e. This year, these team members successfully migrated the software from our TU/e built robots, AMIGO and SERGIO, to HERO, our Toyota HSR. This software base is developed to be robot independent, which means that the years of development on AMIGO and SERGIO are currently being used by HERO. Thus, a large part of the developments discussed in this paper have been optimized for years, whilst the DSPL competition has only existed since 2017[6]. All the software discussed in this paper is available open-source at GitHub[7], as well as various tutorials to assist with implementation. The main developments that resulted in the large lead at RoboCup 2022, and eventually the championship, are our central world model, discussed in Sect. 2, the generalized people recognition, discussed in Sect. 4, the head display, discussed in Sect. 5.3 and the new speech recognition system in Sect. 5.4.

2   Environment Descriptor (ED)

The TU/e Environment Descriptor (ED) is a Robot Operating System (ROS) based 3D geometric, object-based world representation system for robots. ED is a database system that structures multi-modal sensor information and represents this such that it can be utilized for robot localization, navigation, manipulation and interaction. Figure 1 shows a schematic overview of ED.

ED has been used on our robots in the OPL since 2012 and was also used this year in the DSPL. Previous developments have focused on making ED platform independent, as a result ED has been used on the PR2, Turtlebot, Dr. Robot systems (X80), as well as on multiple other @Home robots.

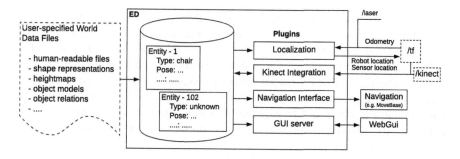

Fig. 1. Schematic overview of TU/e Environment Descriptor. Double sided arrows indicate that the information is shared both ways, one sided arrows indicate that the information is only shared in one direction.

---

[6] https://athome.robocup.org/robocuphome-spl.
[7] https://github.com/tue-robotics.

ED is a single re-usable environment description that can be used for a multitude of desired functionalities such as object detection, navigation and human machine interaction. Improvements in ED reflect in the performances of the separate robot skills, as these skills are closely integrated in ED. This single world model allows for all data to be current and accurate without requiring updating and synchronization of multiple world models. Currently, different ED plug-ins exist that enable robots to localize themselves, update positions of known objects based on recent sensor data, segment and store newly encountered objects and visualize all this in RViz and through a web-based GUI, as illustrated in Fig. 9. ED allows for all the different subsystems that are required to perform challenges to work together robustly. These various subsystems are shown in Fig. 2, and are individually elaborated upon in this paper.

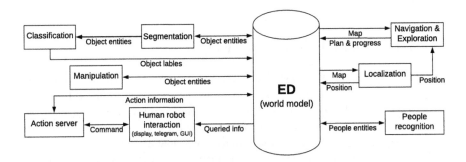

Fig. 2. A view of the data interaction with robot skills that ED is responsible for.

2.1   Localization, Navigation and Exploration

The *ed_localization*[8] plugin implements AMCL based on a 2D render of the central world model.With use of the *ed_navigation* plugin[9], an occupancy grid is derived from the world model and published. With the use of the *cb_base_navigation* package[10] the robots are able to deal with end goal constraints. The *ed_navigation* plugin allows to construct such a constraint w.r.t. a world model entity in ED. This enables the robot to navigate not only to areas or entities in the scene, but to waypoints as well. Figure 3 also shows the navigation to an area. Modified versions of the local and global ROS planners available within *move_base* are used.

---

[8] https://github.com/tue-robotics/ed_localization.
[9] https://github.com/tue-robotics/ed_navigation.
[10] https://github.com/tue-robotics/cb_base_navigation.

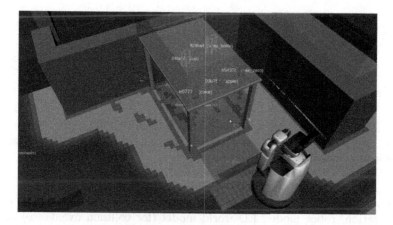

Fig. 3. A view of the world model created with ED. The figure shows the occupancy grid as well as classified objects recognized on top of the cabinet.

2.2 Detection and Segmentation

ED enables integrating sensors through the use of the plugins present in the *ed_sensor_integration* package. Two different plugins exist:

1. *laser_plugin*: Enables tracking of 2D laser clusters. This plugin can be used to track dynamic obstacles such as humans.
2. *kinect_plugin*: Enables world model updates with use of data from a RGBD camera. This plugin exposes several ROS services that realize different functionalities:
   (a) *Segment*: A service that segments sensor data that is not associated with other world model entities. Segmentation areas can be specified per entity in the scene. This allows to segment object 'on-top-of' or 'in' a cabinet. All points outside the segmented area are ignore for segmentation.
   (b) *FitModel*: A service that fits the specified model in the sensor data of a RGBD camera. This allows updating semi-static obstacles such as tables and chairs.

The *ed_sensor_integration* plugins enable updating and creating entities. However, new entities are classified as unknown entities. Classification is done in *ed_perception* plugin[11] package.

2.3 Object Grasping, Moving and Placing

The system architecture developed for object manipulation is focused on grasping. In the implementation, its input is a specific target entity in ED, selected by a Python executive and the output is the grasp motion joint trajectory. Figure 4 shows the grasping pipeline.

---

[11] https://github.com/tue-robotics/ed_perception.

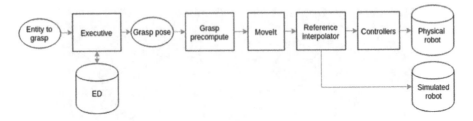

Fig. 4. Custom grasping pipeline base on ED, MoveIt and a separate grasp point determination and approach vector node.

MoveIt! is used to produce joint trajectories over time, given the current configuration, robot model, ED world model (for collision avoidance) and the final configuration.

The grasp pose determination uses the information about the position and shape of the object in ED to determine the best grasping pose. The grasping pose is a vector relative to the robot. An example of the determined grasping pose is shown in Fig. 5. Placing an object is approached in a similar manner to grasping, except for that when placing an object, ED is queried to find an empty placement pose.

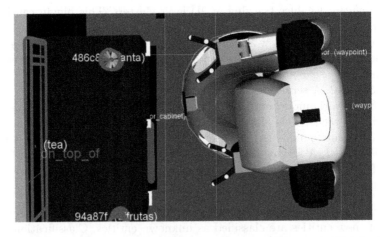

Fig. 5. Grasping pose determination result for a cylindric object with TU/e built robot AMIGO. It is unpreferred to grasp the object from behind.

2.4   World Model Creation

A world model is described in a SDFormat file for compatibility with the Gazebo simulation engine. Currently, world models in ED are generated in a semi-automated way by manually composing and updating multiple object models templates[12] built over the years. New ED plugins are in works that would fully automate the updation of these templates in the future.

3   Image Recognition

The *image_recognition* packages apply state of the art image classification techniques based on Convolutional Neural Networks (CNN).

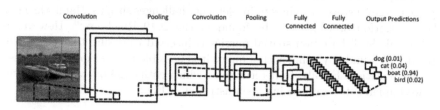

Fig. 6. Illustration of Convolutional Neural Networks (CNN) used in our object recognition nodes with use of Tensorflow.

1. **Object recognition:** Tensorflow™ with retrained top-layer of a Inception V3 neural network, as illustrated in Fig. 6.
2. **Face recognition:** OpenFace[13], based on Torch.
3. **Pose detection:** OpenPose[14].

Our image recognition ROS packages are available on GitHub[15] and as Debian packages: *ros-kinetic-image-recognition*.

4   People Recognition

As our robots need to operate and interact with people in a dynamic environment, our robots' people detection skills have been upgraded to a generalized system capable of recognizing people in 3D. In the people recognition stack, an RGB-D camera is used as the sensor to capture the scene information. A recognition sequence is completed in four steps. First, people are detected in the scene using OpenPose and if their faces are recognized, using OpenFace, as one of the learned faces in the robot's database, they are labeled using their known name. The detections from OpenPose are associated with the recognitions from OpenFace by maximizing the IoUs of the face ROIs. Then, for each of the recognized people, additional properties such as age, gender and the shirt color are identified. Furthermore, the pose keypoints of these recognitions are coupled with the depth information of the scene to re-project the recognized people to 3D as skeletons. Finally, information about the posture of each 3D skeleton is calculated using geometrical heuristics. This allows for the addition of properties such as "pointing pose" and additional flags such as 'is_waving', 'is_sitting', etc.

---

[12] https://github.com/tue-robotics/ed_object_models.
[13] https://cmusatyalab.github.io/openface/.
[14] https://github.com/CMU-Perceptual-Computing-Lab/openpose.
[15] https://github.com/tue-robotics/image_recognition.

4.1   Pointing Detection

Similar to the previous years, this year's tournament challenges too involved various non-verbal user interactions such as detecting an object the user was pointing to. In the previous section, we explained our approach to recognizing people in 3D. Once the recognition results are inserted into the world model, additional properties can be added to the people taking other entities in the world model into account, e.g. *"is_pointing_at_entity"*. This information is used by the top-level state machines to implement challenges such as 'Hand Me That', the description of which can be found in the 2019 Rulebook[16]. However an additional check based on spatial queries is inserted to ensure that the correct operator is found. By using such a query it is possible to filter out people based on their location. Finally, to determine at which entity the operator is pointing to, we implemented ray-tracing, as illustrated in Fig. 7.

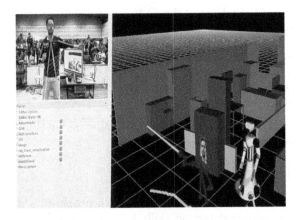

Fig. 7. Ray-tracing based on pose detection with AMIGO.

5   Human-Robot Interface

We provide multiple ways of interacting with the robot in an intuitive manner: WebGUI, Subsect. 5.1, and Telegram™ interface, Subsect. 5.2, which uses our *conversation_engine*, Subsect. 5.2.

5.1   Web GUI

In order to interact with the robot, apart from speech, we have designed a web-based Graphical User Interface (GUI). This interface uses HTML5[17] with the Robot API written in Javascript and we host it on the robot itself.

---

[16] http://www.robocupathome.org/rules.

[17] https://github.com/tue-robotics/tue_mobile_ui.

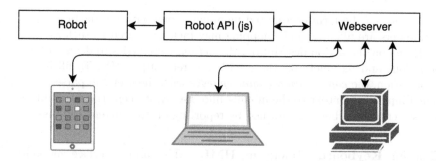

Fig. 8. Overview of the WebGUI architecture. A webserver that is hosting the GUI connects this Robot API to a graphical interface that is offered to multiple clients on different platforms.

Fig. 9. Illustration of the 3D scene of the WebGUI with AMIGO. User can long-press objects to open a menu from which actions on the object can be triggered

Figure 8 gives an overview of the connections between these components and Fig. 9 represents an instance of the various interactions that are possible with the Robot API.

5.2 Telegram™

The Telegram interface[18] to our robots is a ROS wrapper around the *python-telegram-bot* library. The software exposes four topics, for images and text resp. from and to the robot. The interface allows only one master of the robot at a time. The interface itself does not contain any reasoning. This is all done by the *conversation_engine*, which is described in the following subsection.

Conversation Engine. The *conversation_engine*[19] bridges the gap between text input and an action planner (called action_server). Text can be received from either Speech-to-Text or from a chat interface, like Telegram™. It is then parsed according to a (Feature) Context Free Grammar, resulting in an action

---

[18] https://github.com/tue-robotics/telegram_ros.
[19] https://github.com/tue-robotics/conversation_engine.

description in the form of a nested mapping, along with (sub)actions and their parameters are filled in. This mapping may include references such as "it".

Based on the action description, the *action_server* tries to devise a sequence of actions and parameterize those with concrete object IDs. To fill in missing information, the *conversation_engine* engages with the user and parses any additional inputs in context to the missing info. Lastly, it keeps the user "informed" whilst actions are being performed by reporting on the current sub-task.

Custom Keyboard, Telegram HMI.   The user interface modality as explained above has been extended to reduce the room for operator error by only presenting the user with a limited number of buttons in the Telegram app. This has been realized through Telegram's *custom_keyboards*[20] feature. This feature is especially useful when there are only a few options, like a predetermined selection of drinks, as shown in our RoboCup@Home 2019 Finals.

We have employed this custom keyboard to compose commands word-for-word (*hmi_telegram*[21]). After a user input has been received, either via text or previous buttons, for example "Bring me the ...", the user is presented with only those words as options that might follow the input according to the grammar, eg. "apple", "orange" etc. This process iterates until a full command has been composed.

5.3   Head Display

Most people find interacting with robots a very challenging task, especially when they do not deal with them on a regular basis. It is often difficult for people to hear what the robot is saying and not always intuitive to know when to respond to it. As a solution, we use the integrated screen on the Toyota HSRs' 'head' to display useful information. Through the *hero_ display*[22] we have integrated a few different functionalities. As a default, our Tech United @Home logo with a dynamic background is shown on the screen, as depicted in Fig. 10. When the robot is speaking, the spoken text is displayed, and when it is listening, a spinner along with an image of a microphone is shown. It is also possible to display custom images on this screen.

---

[20] https://github.com/tue-robotics/telegram_ros.
[21] https://github.com/tue-robotics/hmi_telegram.
[22] https://github.com/tue-robotics/hero-display.

Fig. 10. The default status of HERO's head display.

5.4 Speech Recognition

Over the years, with the change of base hardware and operating systems, we implemented and experimented with multiple speech recognition systems to allow our robots to hear and understand the operator in noisy environments. We started with the Dragonfly speech recognition framework[23] using Windows Speech Recognition engine as the backend on a Windows 10 virtual machine. This system proved to not be robust against noisy environments primarily due to the default microphone of HERO. As an alternative, we investigated Kaldi-ASR [4], but finally settled with Picovoice[24], alongside the existing Dragonfly system, as it provided seamless support for our custom context-free grammars.

6 Task Execution

In the previous sections, we have described the various modules that contribute towards the functioning of our robots. However, for a challenge to be successfully performed by HERO, these modules need to be strategically integrated together in context to the said challenge. We do this by creating hierarchical state machines using SMACH[25]. Over the years, we have extracted all the commonly used integrations into the modules *"robot_smach_states"* and *"robot_skills"*, and started retaining only challenge specific behaviors within the challenges.

---

[23] https://dragonfly2.readthedocs.io/en/latest/.
[24] https://picovoice.ai/.
[25] https://github.com/tue-robotics/tue_robocup.

7   Re-usability of the System for Other Research Groups

Tech United takes great pride in creating and maintaining open-source software and hardware to accelerate innovation. Tech United initiated the Robotic Open Platform website[26], to share hardware designs. All our software is available on GitHub[27]. All packages include documentation and tutorials. The finals of Robocup@Home 2022[28] demonstrates all the capabilities of HERO, as described in the previous sections. Tech United and its scientific staff have the capacity to co-develop (15+ people), maintain and assist in resolving questions.

8   Community Outreach and Media

Tech United has organised 3 tournaments: Dutch Open 2012, RoboCup 2013 and the European Open 2016. Our team member Loy van Beek was a member of the Technical Committee between 2014-2017 and Peter van Dooren has been since 2022. We also carry out many promotional activities for children to promote technology and innovation. Tech United often visits primary and secondary schools, public events, trade fairs and has regular TV appearances. Each year, around 50 demos are given and 25k people are reached through live interaction. Tech United also has a very active website[29], and interacts on many social media like: Facebook[30], Instagram[31], YouTube[32], Twitter[33] and Flickr[34]. Our robotics videos are often shared on the IEEE video Friday website.

A   HSR's Software and External Devices

A standard Toyota™ HSR robot is used. To differentiate our unit, it has been named HERO. This name also links it to our AMIGO and SERGIO domestic service robots.

HERO's Software Description. An overview of the software used by the Tech United Eindhoven @Home robots can be found in Table 1.

---

[26] http://www.roboticopenplatform.org.

[27] https://github.com/tue-robotics.

[28] https://tinyurl.com/TechUnited2022AtHomeFinals.

[29] http://www.techunited.nl.

[30] https://www.facebook.com/techunited.

[31] https://www.instagram.com/techunitedeindhoven.

[32] https://www.youtube.com/user/TechUnited.

[33] https://www.twitter.com/TechUnited.

[34] https://www.flickr.com/photos/techunited.

Table 1. Software overview

Operating system	Ubuntu 20.04 LTS Server
Middleware	ROS Noetic [5]
Simulation	Gazebo
World model	Environment Descriptor (ED), custom
	https://github.com/tue-robotics/ed
Localization	Monte Carlo [2] using Environment Descriptor (ED), custom
	https://github.com/tue-robotics/ed_localization
SLAM	Gmapping
Navigation	CB Base navigation
	https://github.com/tue-robotics/cb_base_navigation
	Global: custom A* planner
	Local: modified ROS DWA [3]
Arm navigation	MoveIt!
Object recognition	Inception based custom DNN [6]
	https://github.com/tue-robotics/image_recognition/image_recognition_tensorflow
People detection	Custom implementation using contour matching
	https://github.com/tue-robotics/people_recognition
Face detection & recognition	OpenFace [1]
	https://github.com/tue-robotics/image_recognition/image_recognition_openface
Speech recognition	Windows Speech Recognition, Picovoice
	https://github.com/reinzor/picovoice_ros.git
Speech synthesis	Toyota™ Text-to-Speech
Task executors	SMACH
	https://github.com/tue-robotics/tue_robocup

External Devices. *HERO relies on the following external hardware:*

- Official Standard Laptop
- Gigabit Ethernet Switch
- Wi-Fi adapter

References

1. Amos, B., Ludwiczuk, B., Satyanarayanan, M.: Openface: a general-purpose face recognition library with mobile applications. Technical report, CMU-CS-16-118, CMU School of Computer Science (2016)
2. Fox, D.: Adapting the sample size in particle filters through KLD-sampling. Int. J. Robot. Res. **22**(12), 985–1003 (2003)
3. Fox, D., Burgard, W., Thrun, S.: The dynamic window approach to collision avoidance. IEEE Mag. Robot. Autom. **4**(1), 23–33 (1997)
4. Povey, D., et al.: The kaldi speech recognition toolkit. In: IEEE Signal Processing Society (2011). http://infoscience.epfl.ch/record/192584. IEEE Catalog No.: CFP11SRW-USB
5. Quigley, M., et al.: ROS: an open-source robot operating system. In: ICRA Workshop on Open Source Software (2009)
6. Szegedy, C., et al.: Going deeper with convolutions. In: 2015 IEEE Conference on Computer Vision and Pattern Recognition (CVPR), pp. 1–9 (2015). https://doi.org/10.1109/CVPR.2015.7298594

RoboCup 2022 SSL Champion TIGERs Mannheim - Ball-Centric Dynamic Pass-and-Score Patterns

Mark Geiger(✉), Nicolai Ommer, and Andre Ryll

Department of Information Technology,
Baden-Wrttemberg Cooperative State University,
Coblitzallee 1-9, 68163 Mannheim, Germany
info@tigers-mannheim.de
https://tigers-mannheim.de

Abstract. In 2022, TIGERs Mannheim won the RoboCup Small Size League competition with individual success in the division A tournament, the blackout technical challenge and the dribbling technical challenge. The paper starts with an outline of the robot's dribbling hardware and ball catching computations, followed by a high level summary of the AI used in the tournament. Given 62 scored goals and no conceded goals at RoboCup 2022, the focus is on describing the used attack and support behaviors and how they are selected. The paper concludes with a statistic of the tournament backing the efficiency of our employed strategies.

1 Robot Dribbling Hardware and Ball Interaction

As in all other RoboCup soccer leagues ball handing and control is a key factor to success. It has gained more importance recently as our offensive employs an increasing number of actions to steal the ball from opponents, to move with the ball, or to protect it from opponents (see Sect. 2). Section 1.1 gives an overview of the robot hardware which is in direct contact with the ball and recent updates applied to it. Section 1.2 describes how to approach the ball to actually make use of the hardware.

1.1 Dribbling Device

In the SSL a golf ball is used, which is the most rigid game ball of all leagues. Hence, the ball itself provides only very little damping during reception and dribbling. It also has a low friction coefficient, complicating ball control even further.

Consequently, damping and a high friction coefficient must be provided by the robots controlling the ball. This is done by a unit which is called the dribbling device. It is depicted in Fig. 1 for our v2022 robot generation.

We decided to use a design with two degrees of freedom, as we did in our v2016 robots [1]. We combined the v2016 2-DoF dribbler with ZJUNlict's additional

© The Author(s), under exclusive license to Springer Nature Switzerland AG 2023
A. Eguchi et al. (Eds.): RoboCup 2022, LNAI 13561, pp. 276–286, 2023
https://doi.org/10.1007/978-3-031-28469-4_23

dampers [2]. The top damper is mainly used to absorb impact energy of incoming passes. As soon as the ball is actively controlled the exerted backspin on the ball can push the whole dribbler upwards on the sideward sliders. The additional load is absorbed by flexible elements and the motor is current and temperature controlled to prevent overstress. If the dribbler drops during the dribbling process (either due to a skirmish or an uneven ground) it is damped via the small bottom dampers. All dampers are 3D printed from a flexible TPE material with a 70A shore hardness. The damping properties of the top damper can be adjusted by changing its shape (mainly by varying the branch thickness).

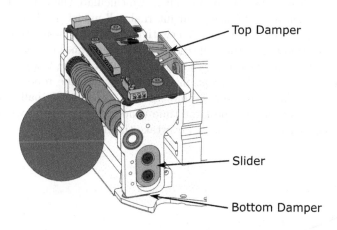

Top Damper

Slider

Bottom Damper

Fig. 1. Dribbling device with highlighted damping elements.

Compared to the version used in the 2021 hardware challenges some additional improvements were made [3]. The material of the dribbling bar has been changed to a soft silicone, which is much less abrasive than the previously used polyurethane. Due to the complex shape of the dribbling bar it is molded by using two 3D-printed half shells as a mold and pouring in the liquid silicone from the top. Furthermore, the gear modulus has been changed from 0.5 to 0.7 as the small gears tended to break under heavy load.

With the updated dribbling device, we achieve excellent damping properties and can stop an incoming pass directly at the robot. This was tested with another robot kicking the ball so that it reaches the dribbling device with 5 m/s. The rebound was assessed visually and no separation of the ball from the dribbling device could be identified. To retain ball control the dribbler can run at up to 25000 rpm. Depending on carpet friction it consumes between 2 A and 8 A of current. A higher current corresponds to a better grip of the ball. This current is also reported back to our central AI which uses it to asses if a difficult move with the ball can be executed. A detailed description of our v2020 hardware can be found in [4,5].

1.2   Catching a Rolling Ball

When a ball is rolling on the field, the robots have to stop or catch the ball. We use two different approaches to catch such a ball. If it is possible to intercept the ball by moving onto the ball travel line just in time, we try to intercept. Otherwise, we try to approach the ball from behind and stop it with the dribbler.

Intercepting the Ball. The approach of intercepting the ball is based on a method from CMDragons [6]. It samples multiple points along the ball travel line. A robot trajectory is then planned to each point and the resulting travel time is associated with the respective point. Then, we calculate the time that the ball needs to reach each point by using our internal ball model [7]. This gives us the slack time that a given point has. A negative slack time means that the robot reaches the position before the ball. Plotting the slack times results in the graph shown in Fig. 2. In most situations in which a ball is rolling towards the robot, there are two time slots (interception corridors) where the robot can actually catch the ball. Usually one small time window to catch the ball close to the robots current position and one large time window far in future, when the ball is getting so slow that the robot can overtake the ball again.

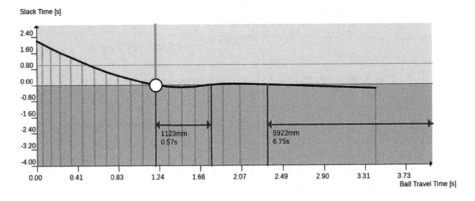

Fig. 2. Ball interception calculation.

The robot will try to move towards the first reachable interception corridor (negative slack time) that meets some requirements (corridor width > 0.2 s and min slack-time < −0.2 s). The selected corridor begins at a given ball travel time (x-axis), which we can use to feed the ball model to calculate a target position where to actually catch the ball. This will be done for each robot on the field. The robot that can catch the ball most rapidly will be selected as the primary offensive robot. If it is uncertain that the primary robot is able to catch the ball, then multiple robots may try to intercept the ball.

Approach and Stop Ball. The fallback, when intercepting the ball is not feasible is to approach the ball from behind by moving onto the ball travel line and then approaching the ball until it hits the spinning dribbler. As soon as the ball is on the dribbler, the robot brakes as quickly as possible without loosing the ball. If tuned well, this is quite an effective approach to quickly gain back ball control.

2 Offensive Strategies

This section introduces the basic foundation of the offensive decision making. One key aspect of the offensive strategy are the *OffensiveActionMoves*. An *OffensiveActionMove* represents a specific action a robot can execute. An *OffensiveActionMove* can be a simple pass, a kick on the opponents goal, or a special behavior in close engagements with robots from the opponent team. Currently, we have ten *OffensiveActionMoves*. There are three methods that each *OffensiveActionMove* has to implement. The method *isActionViable* determines the viability of an *ActionMove*. The viability can either be *TRUE*, *PARTIALLY* or *FALSE*. The method *activateAction* controls the actual execution of the move. The method *calcViabilityScore* will determine a score between 0 and 1 for the current situation. This score should be connected to the likelihood, that this action can be executed successfully. The viability and its score are calculated in a unique way for each OffensiveActionMove. For example, the viability of a GOAL_SHOT is determined mainly by the open angle through which the ball can enter the opponent's goal. The viability of a PASS is mainly determined by the pass target rating (see Sect. 3.2). The different scores are made comparable by additional weights set by hand, based on an educated guess. In addition, a self-learning algorithm is used that takes into account the successes and failures of past strategies to fine-tune these weights during a match. This algorithm was first presented in our 2018 TDP [1].

Algorithm 1 shows how the best *OffensiveActionMove* out of one given *OffensiveActionMoveSet* is determined. It is important to note that the *OffensiveActionsMoves* inside a given set have a specific ordering, which represents the priority. The *OffensiveActionMove* in the first position of the set has the highest priority. An *OffensiveActionMove* will be activated if its viability returns *TRUE* and it has a higher priority than all other *OffensiveActionMoves* that return a *TRUE* viability. Actions that return the viability *FALSE* will be ignored in any further processing. All actions that are *PARTIALLY* viable are sorted by their *viabilityScore* and if there is no action that has a *TRUE* viability, then the action with the highest *viabilityScore* will be activated. In case all actions have a *FALSE* viability then a default strategy will be executed.

Algorithm 1. Pseudocode - Find the best OffensiveActionMove

```
for (var action : actionsSet) {
  var viability = action.isActionViable();
  if (viability == TRUE) {
    // activate first move that got declared as viable
    return action.activateAction();
  } else if (viability == PARTIALLY)
    partiallyMoves.add(action);
}

partiallyMoves.sort(); // sort by viabilityScore
if (!partiallyMoves.isEmpty()) {
  // choose best partially viable move to be activated
  return partiallyMoves[0].activateAction();
}
return defaultMove.activateAction()
```

The separation into viable and partially viable actions, combined with priorities leads to a very stable and easily modifiable/extendable algorithm for the offensive strategy. For example, the *OffensiveActionMove* that controls direct kicks on the opponent goal will return a *TRUE* viability if there is a high chance to score a goal. If there is a extremely low chance to score a goal it will return *FALSE*. Otherwise, if the hit chance is reasonable but not really high, it will return *PARTIALLY*. Additionally, this action has a high priority. Thus, the robot will surely shoot on the goal if there is a good opportunity to score a goal. However, if the viability is *PARTIALLY* the action will be compared with the other actions and based on the *viabilityScores* the robot will decide whether it should shoot on the goal or execute another action, e.g. a pass to another robot.

2.1 Offensive Dribbling

Another offensive action that the robot may choose is the so called *DribbleKick*, which is one of the dribbling actions the robot can do. Figure 3a shows a typical scenario of a ball located in front of the opponent goal. In this case the robot chooses to do a DribbleKick. The robot approaches the ball and tries to bring it onto its dribbler. The strength of the dribble contact can be estimated from the power drawn by the dribble motor (see Sect. 1.1). The robot will wait until the ball has a strong contact and also checks if it is possible to score a goal from another position on a curve around the opponents penalty area. Multiple points on the curve are sampled and evaluated for their chance to score a goal (white = high chance to score, gray = low chance to score). The robot will drive along the curve towards the best point, while keeping the ball on the dribbler. As soon as the target is not blocked anymore the robot will kick the ball as shown in Fig. 3b.

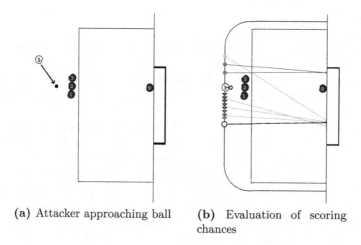

(a) Attacker approaching ball   (b) Evaluation of scoring chances

Fig. 3. Execution of a DribbleKick

The entire sample-curve can move closer or further away from the opponent goal, depending on the behavior of the defending robots. In general the robot will try to avoid coming to close to opponent robots. The robot must also adhere to the maximum dribble distances allowed. Therefore, it does not sample positions farther away than the maximum allowable dribble distance (1 m) to avoid dribble rule violations. The robot tries to move laterally and shoot the ball while it is still in the acceleration phase. Since the opposing robot only reacts to the measured position of our robot, it will always have a disadvantage due to overall system latency. As the robot tries to shoot during acceleration it may not be possible to change movement anymore if a dribbling violation is imminent. In such a case, the robot will simply shoot the ball to avoid a violation, even if there is no good chance to score a goal.

The calculations are done on every AI frame. Meaning that there is no *plan* that the robot follows. Each frame the destination or the kick target can change. This is important, because we need to react fast to the opponents movement and re-evaluate our strategy constantly. In order not to lose the ball while dribbling, the robot balances its orientation so that the rotational force of the ball points in the direction of our robot. When a dribbling robot changes its orientation, the force vector of the rotating ball also changes. However, it lags behind the robots movement. If the orientation or the direction of movement is changed too quickly, ball control may be lost. The robot will give priority to ball control during the movement. However, if the robot sees that it could score a goal, it will quickly align itself towards the target and shoot. For the final shot, the robot will take into account its current velocity to calculate the final alignment towards the target to make an accurate shot.

2.2   Defensive Dribbling

Our AI distinguishes between defensive and offensive dribbling. Defensive dribbling is concerned with getting the ball and protecting it from the opponent robots while always adhering to the dribbling rule constraints. Our attacking robot will remain in the defensive dribbling state until a good enough offensive strategy has been found.

Figure 4 shows a common situation. The ball is located in front of the translucent robot and an opponent robot is about to attack us. Our robot has ball control, but no offensive action with a good enough viability score. Thus, the robot will enter the defensive dribbling mode. The robot will then try to protect the ball from the opponent robots. Multiple points within the allowed dribbling radius are sampled and evaluated. The robot will then dribble the ball towards the position that is rated to be the safest from opponent robots. At the same time the robot will try to turn the ball away from the opponent robots.

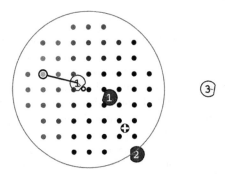

Fig. 4. Defensive dribbling calculations.

3   Support Strategies

Robots which are not assigned to any attacking or defending role become supporting robots. They are supposed to run free, look for good positions on the field from where they can safely receive a pass and ideally also have a good chance to score a goal. Section 3.1 gives an overview of the high-level behaviors a supporting robot may get assigned. They define where a robot should go. Section 3.2 outlines where our robots may receive passes and forms the connection between support and attack strategies.

3.1   Supporting Robots

Given the fast-paced nature of the Small Size League, planning too far in the future is not advisable. Situations change within fractions of a second. So instead of finding good positions globally on the field, we focus on optimizing current robot positions first, while still observing the global supporter distribution. With the increasing number of robots in the league (2012–2017: 6, 2018–2019: 8, 2021–2022: 11) more and more robots take over the supporting roles. During a free kick in the opponent half, we may use up to 9 supporters, while 5 years ago, it were 4 at most.

Over the previous years, we developed different supporting behaviors. Each supporter is assigned a behavior. There can be limits on the number of robots having a certain behavior and behaviors may be disabled based on situation, tactics or game state. For each robot, the viability and a score between 0 and 1 of each behavior is determined and the best rated behavior is assigned. The viability algorithm is similar to the one described in Sect. 2. The following sections describe some of the most important behaviors.

Direct Goal Redirector. Find a position from where a goal can be scored, optimizing for the redirect angle, namely the angle between the current ball position, the desired supporter position and the goal center. A small redirect angle is better, because receiving it is more reliable and precise.

Fake Pass Receiver. If a supporter is near an ongoing or planned pass, it pretends to receive this pass by standing close to the passing line, but without actually receiving the ball. Opponents will need to figure out which is the right receiver or need to defend all potential receivers. This behavior could often be observed quite clearly in matches[1].

Penalty Area Attacker. Position the robot as close as possible to the opponent penalty area to prepare it for a goal kick. Passing through the penalty area and scoring from that position will leave the defense few chances to block the goal kick.

Repulsive Attacker. Bring the supporter to a good attacking position without interfering with other supporters using a force field with several force emitters. For example field boundaries, other robots and a general trend towards the opponent goal. The desired position is determined by following the forces in the field a fixed number of iterations. Figure 5 shows such a force field fully visualized for the team playing towards the left goal. In the own half, forces are directly towards the opponent half, while in the opponent half, forces are directly towards the middle of the left or right side of the opponent half and away from opponents and the ball.

Repulsive Pass Receiver. Based on the same repulsive principal as the Repulsive Attacker behavior, a position with a certain distance to the ball that is not covered by opponents is targeted.

---

[1] https://youtu.be/W8Z_2a2Ieak?t=80.

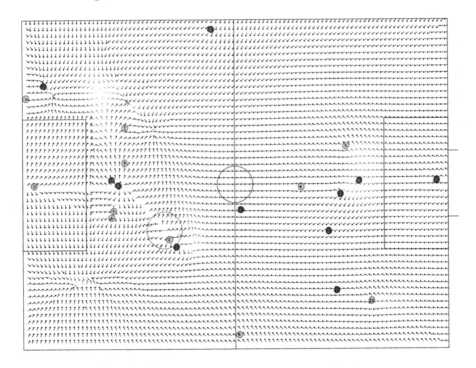

Fig. 5. Repulsive force field for Repulsive Attacker behavior

3.2   Pass Targets

Pass targets are potential positions where the ball can be received and are calculated for each friendly robot on the field, except for the keeper. Multiple pass
targets exist for each robot. Each one having a rating which is based on its pass-
ability, chance to score a goal from the pass targets location and the chance of an
opponent intercepting the ball on its way to the pass target. The pass targets are
calculated in a circle around the robot, where the circle is additionally shifted
in the current motion path of the robot. The radius of the generation circle is
determined by the current velocity of the robot. A fixed number of unfiltered targets is generated in each frame. The list of targets is then filtered to ensure that
all targets can be reached in time by the designated robot until the scheduled
pass arrives. Different times are taken into account: The time until the attacking
robot is able to shoot the ball, the travel time of the ball and the time needed
for the pass receiving robot to reach its passing target. The best pass targets
from the past are reused to efficiently optimize the targets over time. Figure 6a
shows the calculated pass targets for a single robot.

Once an offensive robot gets close to the ball, it will choose the currently
calculated offensive action. In case of a pass, the offensive strategy will take over
the robot with the best rated pass target as the pass receiver shortly before the
pass is executed. This allows a tighter and more stable coordination between the

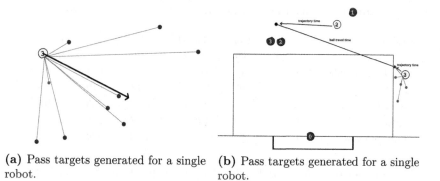

(a) Pass targets generated for a single robot.   **(b)** Pass targets generated for a single robot.

Fig. 6. Illustration of pass target generation and selection.eps

robots. Figure 6b shows a typical pass situation generated by our AI. Robot 2 plans to pass the ball to robot 3. Rather than passing to the robot's position, it passes to a pass target near robot 3. Since we use trajectories to control our robot movement, we can calculate the time at which robot 2 needs to kick the ball. Furthermore, we can use the initial ball velocity and our ball model [7] to calculate the time at which the ball needs to reach the pass target. Also, we can calculate the time robot 3 needs to reach his pass target. By combining these numbers we can synchronize the time in which the ball and robot 3 will reach the pass target.

4   Conclusion

At RoboCup 2022, TIGERs Mannheim played 10 official matches during the group and elimination phase of the division A tournament and scored 62 goals in total, while not conceding any goal throughout the tournament. Every second attempted goal shot was successful and two thirds of passes between TIGERs robots succeeded on average. These numbers support the focus and strength of the team: A fast paced and dynamic attack strategy.

The numbers were extracted from the official log files[2] using the TIGERs log analyzer from the technical challenge 2019[3] and the leagues match statistics[4]. Table 1 shows the full set of gathered statistics. Only shots with a duration of at least 300 ms were considered. The ball possession specifies the amount of time that the team uniquely possessed the ball relative to the time that either team uniquely possessed the ball. So a value larger than 50% means that this team possessed the ball more often than the other team.

---

[2] https://ssl.robocup.org/game-logs/.
[3] https://ssl.robocup.org/robocup-2019-technical-challenges/.
[4] https://ssl.robocup.org/match-statistics/.

Table 1. Tournament statistics from RoboCup 2022 (Division A)

Team	Goals scored	Goal shots	Goal shot success	Ball possession	Passes	Pass success
TIGERs Mannheim	62	113	54.9%	63.2%	648	66.6%
ER-Force	13	95	13.7%	59.3%	619	53.3%
RoboTeam Twente	3	28	10.7%	37.6%	255	34.9%
KIKS	1	29	3.4%	52.1%	268	34.3%
RoboDragons	1	127	0.8%	35.5%	270	29.3%

The number of goals scored indicates a good positioning by the supporters and also a reliable and precise execution of kicks by the actual robots on the goal. The outstanding goal shot success ratio underlines very well the offensive action selection based on viabilities. Our robots do not blindly force kick towards the opponent goal on every opportunity but carefully decide if this action would have a chance to score at all. Alternative actions like the defensive dribbling ensure a high ball possession rate in case no other reasonable offensive strategy is available. The high number of passes and the pass success ratio show that our supporters are in good positions to receive passes and the offense often selects a passing action to get in a good position to score.

5   Publication

Our team publishes all their resources, including software, electronics/schematics and mechanical drawings, after each RoboCup. They can be found on our website[5]. The website also contains several publications with reference to the RoboCup, though some are only available in German.

References

1. Ryll, A., Geiger, M., Carstensen, C., Ommer, N.: TIGERs Mannheim - Extended Team Description for RoboCup 2018 (2018)
2. Huang, Z., et al.: ZJUNlict Extended Team Description Paper for RoboCup 2019 (2019)
3. Ommer, N., Ryll, A., Geiger, M.: TIGERs Mannheim - Extended Team Description for RoboCup 2022 (2022)
4. Ryll, A., Jut, S.: TIGERs Mannheim - Extended Team Description for RoboCup 2020 (2020)
5. Ryll, A., Ommer, N., Geiger, M.: RoboCup 2021 SSL champion TIGERs Mannheim - a decade of open-source robot evolution. In: Alami, R., Biswas, J., Cakmak, M., Obst, O. (eds.) RoboCup 2021. LNCS (LNAI), vol. 13132, pp. 241–257. Springer, Cham (2022). https://doi.org/10.1007/978-3-030-98682-7_20
6. Biswas, J., Mendoza, J.P., Zhu, D., Klee, M., Veloso, S.: CMDragons 2014 Extended Team Description (2014)
7. Ryll, A., et al.: TIGERs Mannheim - Extended Team Description for RoboCup 2015 (2015)

---

[5] Open source/hardware: https://tigers-mannheim.de/publications.

B-Human 2022 – More Team Play with Less Communication

Thomas Röfer[1,2]([✉]), Tim Laue[2], Arne Hasselbring[1], Jo Lienhoop[2], Yannik Meinken[2], and Philip Reichenberg[3]

[1] Deutsches Forschungszentrum für Künstliche Intelligenz, Cyber-Physical Systems, Enrique-Schmidt-Str. 5, 28359 Bremen, Germany
{thomas.roefer,arne.hasselbring}@dfki.de
[2] Fachbereich 3 – Mathematik und Informatik, Universität Bremen, Postfach 330 440, 28334 Bremen, Germany
{tlaue,joli,ymeinken}@uni-bremen.de
[3] JUST ADD AI GmbH, Konsul-Smidt-Straße 8p, 28217 Bremen, Germany
philip.reichenberg@justadd.ai

Abstract. The B-Human team won all of its seven games at the RoboCup 2022 competition in the Standard Platform League (SPL), scoring a total of 48 goals and conceding 0. B-Human achieved this high level of performance with a new behavior architecture that enables more cooperative game play while sending fewer team communication messages. This paper presents the parts of the behavior that we consider crucial for this year's success. We describe the strategic evaluation, action selection, and execution aspects of our new pass-oriented play style. The effectiveness of our algorithms is supported by statistics from competition games and extensive testing in our simulator. Empirically, our approach outperforms the previous behavior with a significant improvement of the average goal difference.

1 Introduction

Soccer is a team sport. Thus, the overall success always depends on the proper coordination among all players, which is, for instance, reflected in their positioning and the execution of cooperative actions such as passes. In recent years, the implementation of passing skills has been encouraged in the SPL by holding different technical challenges as well as by introducing (indirect) set pieces. However, in the normal course of most games, only very few actual passes were seen. Such a complex cooperation requires a common understanding of the current game situation as well as of the intentions of the teammates. Although cooperation has already been complicated in itself, in 2022, the SPL decided to significantly reduce the number of wireless messages that the robots can exchange within their team.

In this paper, we present our contributions that we assume had the most impact on winning RoboCup 2022 as well as the RoboCup German Open 2022: a new behavior architecture along with new skills, such as passing, as well as a new

© The Author(s), under exclusive license to Springer Nature Switzerland AG 2023
A. Eguchi et al. (Eds.): RoboCup 2022, LNAI 13561, pp. 287–299, 2023
https://doi.org/10.1007/978-3-031-28469-4_24

communication approach. Overall, we aimed for more sophisticated cooperation despite lower communication capabilities.

In our 2017 champion paper [11], we already argued that in the current state of the RoboCup Standard Platform League, the development of flexible robot behaviors makes a significant difference, while other tasks such as perception, modeling and motion have mostly been solved. Therefore, we developed a new behavior architecture this year, which is described in Sect. 2. It decouples different layers of complexity, along with the new cooperative skills for passing and positioning. Furthermore, we enhanced our behavior in one vs. one situations.

To cope with the SPL's new team-wide message budget, which allows way less communication than in previous years, the implementation of an approach that is more intelligent than just sending messages in static intervals became necessary to allow the intended high level of team play. In Sect. 3, we describe our new techniques to remain within the given budget whilst timely communicating important information to teammates.

2   Behavior

With the advancement of RoboCup, team play turns into a key feature of success [5]. It appears that individual robots dribbling and shooting at the goal is an approach that fails to fully explore the tactical possibilities of the game. This is why we introduce a more progressive approach to the SPL: coordinated passing of the ball between teammates. This involves planning and evaluating possible passes and goal shot opportunities as well as appropriate positioning of the cooperating robots. Furthermore, alternative actions for situations with opponents in the vicinity of the ball need to be taken into account.

2.1   Strategy

This year, B-Human extended the behavior architecture from 2019 [10] with a new strategy component. It encapsulates team and long- to mid-term decisions and lowers them to high-level commands for an individual robot. Team decisions include, e.g., which tactic to use, which position (i.e. defender/midfielder/forward) to fill, whether to play the ball, etc. Different tactics can be defined in configuration files by specifying a set of base positions, including their prioritization, which influence dynamic task allocation and positioning of the individual players. A state machine switches between tactics depending on the number of active field players and the ball position, which is stabilized by a team majority vote so all players use the same tactic. The strategy component is not interrupted by short-term events, such as falling down or one vs. one situations, such that long-term desires of the team are kept. The output of the strategy is rather abstract, e.g. *shoot at the goal, pass to player X, position at Y*, without defining which kick type, angle or path to use. These decisions are up to the skill layer, which can also decide for a completely different action if short-term circumstances require this (e.g. one vs. one situations), the action is currently impossible (e.g. if the goal is out of range, dribbling is preferred), or the resulting behavior would be illegal.

2.2   Passing

Passing has been dominant in other RoboCup leagues for many years, especially with the wheeled robots of the Small Size League (SSL) and Middle Size League (MSL). The objective of passing is getting closer to the opponent's goal while keeping possession of the ball to create scoring opportunities in a controlled way [3]. This creates tactical opportunities as the opponent team is often unable to cover multiple angles on their goal at the same time. This stems from the fact that the ball can be kicked and rolls faster than the robots are able walk or react. Therefore, if the ball is moved over a large distance with a pass, as long as a teammate is able to play the ball within a given time, an opening is guaranteed.

Planning. At the strategic level, a player wanting to manipulate the ball must decide between shooting at the goal or which teammate to pass the ball to using our so called "SmashOrPass" algorithm. We opted for an approach where the action is selected based on the probability that it will lead to a goal in the short term. This must consider the fact that the outcome of all actions is uncertain. For passing in particular, a kicking robot might hit the ball imprecisely or even fall over in the process [4]. In addition, the inaccuracies in the perception and the resulting world model make it unrealistic to compute accurate probabilities. Therefore, the chance of success of these actions is estimated by rating functions with hand-crafted features [1] that attempt to approximate the probability of succeeding in a computationally feasible way [2].

Goal Evaluation. The goal rating is the estimated probability of scoring a goal from a hypothetical ball position. This means, there is a wide enough open angle on the opponent's goal that is not blocked by obstacles currently present in the world model [8] and the goal is within kicking range [1]. These criteria are combined in the goal rating function $r_{\mathrm{goal}}(\boldsymbol{x})$, where $\boldsymbol{x}$ is the hypothetical ball position. The resulting rating is shown in Fig. 1a.

Pass Evaluation. The pass rating is the estimated probability that a teammate can receive the ball at a hypothetical target position when kicked from its current position [2]. Several criteria take into account the positions of the opponents currently present in the world model of the robot to assess the risk of an interception of the rolling ball before it arrives at the target [3]. Figure 1b shows the pass rating function $r_{\mathrm{pass}}(\boldsymbol{x})$ where $\boldsymbol{x}$ is the hypothetical target position.

Action Selection. The "SmashOrPass" algorithm iterates over the positions of the teammates **T** and compares their combined rating (see Fig. 1c) for a pass

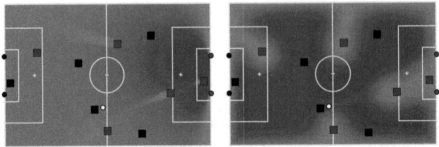

(a) The rating r_{goal} for goal shots from all positions on the field

(b) The rating r_{pass} for passes from the ball position to all positions on the field

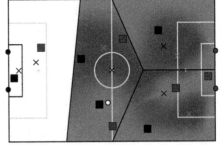

(c) The combined rating of (a) and (b) for passes to all positions on the field and following goal shots

(d) Normalized position ratings of a midfielder and two forwards with their corresponding Voronoi regions (black lines), base positions (black crosses), and the found local maxima (cyan crosses)

Fig. 1. Visualization of the rating functions. The world state of the robot shows the teammates (black squares), opponents (blue squares), ball (white circle), and goalposts (colored circles). The black team is playing from left to right. (Color figure online)

to the player and a following goal shot by that player. The best pass target $\boldsymbol{x}^*$ is selected as:

$$\boldsymbol{x}^* = \arg\max_{\boldsymbol{x} \in \mathbf{T}} r_{\text{pass}}(\boldsymbol{x}) \cdot r_{\text{goal}}(\boldsymbol{x}) \tag{1}$$

Equation (1) maximizes the estimated probability of success for a pass to the player at position $\boldsymbol{x}$ directly followed by a goal shot. This is compared to the rating of a direct goal shot from the current ball position $\boldsymbol{x}_{\text{ball}}$, to ensure that a pass is only played if its combined rating is higher, i.e. the following condition is true:

$$r_{\text{pass}}(\boldsymbol{x}^*) \cdot r_{\text{goal}}(\boldsymbol{x}^*) > r_{\text{goal}}(\boldsymbol{x}_{\text{ball}}) \tag{2}$$

Skill Execution. The pass skill interprets the pass request from the strategy layer and calculates the optimal parameters for the kick towards the receiver with the specified player number [4]. This process involves planning a trajectory for the pass that avoids interception by opponents and maximizes the probability of the teammate successfully receiving the ball [2].

2.3   Positioning

One of the most important aspects of the strategy, which makes a soccer game dynamic, is the positioning of field players, who are not currently playing the ball [3]. When the pass receiver is in the shadow area behind an opponent, the pass will most likely end in a loss of ball possession [3]. Therefore, we developed a procedure according to which players can create space and make themselves available for a pass. Initially, the entire field was considered for the recognition of open space. This is both inefficient, because distant areas cannot be reached quickly, and problematic, because role assignment could oscillate, when approaching a teammate's base position from the tactic [4]. As a solution, the search space can be reduced to the area that this player of the team is currently assigned to by the strategy. For this purpose, Voronoi diagrams have been integrated into the B-Human system, to be utilized for different tasks, such as the ball search and the positioning for passes or field coverage.

As mentioned before, the B-Human behavior architecture provides multiple tactics for a given number of players, each describing exactly one base position for each player and thus the general formation of the team [4]. A Voronoi diagram can be generated from these base positions of the currently active tactic. Since all possible combinations of base positions are known in advance, the Voronoi diagrams can be precomputed. The behavior can then utilize the corresponding Voronoi regions for any situation that can arise in the game. To find a specific position inside the Voronoi region, we utilize rating functions that differ between the roles to express their duty [6]. In general, the task of a position role is to calculate a position, where the player is of great strategic use [4]. Examples of the resulting rating functions for the position roles in their corresponding Voronoi region are shown in Fig. 1d.

For the forward, the rating function consists of a combination of the pass and goal ratings, which are described in Sect. 2.2, as well as the distance to the base position b from the tactic. They are combined using the following formula:

$$r_{\text{forward}}(x) = r_{\text{pass}}(x) \cdot r_{\text{goal}}(x) \cdot \underbrace{e^{-\frac{\|x-b\|^2}{2\sigma^2}}}_{\text{distance factor}} \tag{3}$$

This represents the task of the forward position to position itself in such a way that it can receive a pass and then shoot directly at the opponent's goal.

The midfielder uses a different rating function that focuses on field coverage, as it is beneficial to position away from teammates and the field border. It also takes the position's distance to the base position and a preferred distance to the ball into account.

Using gradient ascent, a local optimum of the rating function inside the Voronoi region is found. This makes it possible to react dynamically to changes in the world model [5]. As we start each cycle at the position that was found in the last frame, we tend to stick to the same local optimum, which provides a good resistance to noise. The limitation to find only local optima due to gradient ascent is not drastic as in most situations there are only a few that do not differ much.

2.4 One vs. One Situations

Although passing is supposed to prevent one vs. one situations with other robots in the first place, they are still a major part of the game, so it is clearly advantageous to win as many of them as possible. An example of such a situation can be seen in Fig. 2.

Falling or executing kicks that let the ball bounce off the opponent will always result in losing ball possession in the short term. Walking faster than the opponent and wasting less time also gives an advantage to gain or keep ball possession. In previous competitions the behavior ignored that the robot could fall while walking to the ball or while kicking, but also that kicking the ball far away without a teammate in sight would result in losing ball possession. This resulted in many situations, in which the robots walked or kicked directly into the opponent and fell as a result of the collision, or they kicked the ball far away, so no teammate could gain ball possession.

We now use two new approaches: On the one hand, as long as no pass is desired, the ball is kept close and only dribbled a short distance forward. This allows to react quickly to changes of the situation, for instance a pass becomes possible or an opponent walks into our path. On the other hand, we explicitly determine whether a kick would lead to a collision with another robot and only use such kicks when no other one is possible. Even then, the relative position behind the ball is adjusted to prevent walking into the opponent. The kick itself will adjust automatically, as the movement is calculated based on relative ball position offsets. If no kick is possible, e.g. if an opponent stands directly behind the ball, the robot places itself behind the ball and forms a V-shape with its feet. Afterwards, the swing leg will be rapidly moved to the ball and back to the starting position to move the ball sideways away from the opposing player. This gives us a head start to follow the ball and can generate an opening to outmaneuver the opponent.

Although passes are hard to execute in such situations, the high-level pass request is used to determine passes to a specific teammate. A bad pass is preferred over a one vs. one situation and losing the ball, mainly because the ball would be moved into a better position for scoring a goal. Therefore, instead of positioning exactly behind the ball to kick it as close to the target direction as possible, we willingly allow high deviations in the direction and range. To reduce the execution time as much as possible, based on the distance to the next robot, a handful of restrictions for the kicks are deactivated and thresholds increased. For example, most kicks require the robot to stop or have executed only a small walking step before. Moreover, the relative ball position to the kicking foot must lie within a given threshold range in order to start a kick. But given a close

Fig. 2. A one vs. one situation. A B-Human robot (black) kicking the ball past the opponent (blue). Photos by JT Genter (2022, CC BY-SA 4.0) (Color figure online)

opponent, we prefer the risk of a failed kick or higher deviations to not kicking at all, as positioning behind the ball takes too much time. Note that a failed kick means the ball was not touched or the ball moved significant less than expected.

This combination of short dribble kicks, dynamic ball adjustments, and far passes, whilst maintaining a stable posture without the need for accurate positioning, is possible thanks to our walk step adjustment [9]. It ensures no matter the situation, like bumping into other robots, switching the target multiple times or an executed kick, the robots will fall as little as possible but still maintain a high walking speed and can react to changes within one walk step duration, which is currently 250 ms. A more detailed description how the behavior works in one vs. one situations can be found in the wiki[1] of this year's code release.

2.5 Results

Simulation. In order to test the effectiveness of our new passing behavior, an experiment in our simulator "SimRobot" [7] was designed based on an *a priori* power analysis. A one-sample t-test was conducted to compare the average goal difference of a passing team, that scored 234 goals and conceded 100 goals against a team that cannot pass. There was a very significant difference in the mean value of the goal difference ($M = 0.67$, 99% CI $[0.43, 0.91]$, $SD = 1.31$) for the passing team; $t(199) = 7.225$, $p < .0001$. These results suggest that passing leads to more goals, based on a sample of $N = 200$ simulated halves. Specifically, the passing team scored 70% more goals and conceded 30% fewer goals against a team that cannot pass, compared to the results of two teams that both cannot pass. Therefore, the detected effect size of this analysis ($d = 0.51$) is considered practically significant.

Competitions. The new behavior on real robots enabled more frequent passing to teammates during the competition games at RoboCup 2022. Compared to RoboCup 2019 the average number of passes per game in normal play increased sevenfold, while remaining at a comparable success rate, as seen in Table 1. Fewer passes have been attempted at the German Open 2022. This could be due to

---

[1] https://wiki.b-human.de/coderelease2022/behavior/#zweikampf.

Table 1. B-Human's pass statistics for the last three regular competitions

Competition	#Games	Normal play		Set play	
		#Passes	Success	#Passes	Success
RoboCup 2019	7	11	81.8%	18	83.3%
German Open 2022	8	29	75.9%	7	85.7%
RoboCup 2022	6[a]	68	72.1%	15	73.3%

[a]Due to missing video footage, one game could not be evaluated.

Table 2. B-Human's duel statistics at RoboCup 2019 and 2022

Games	RoboCup	#Duels	Success
All	2019	129	52.7%
	2022	149	70.5%
Final	2019	45	42.2%
	2022	42	59.5%

Table 3. Fall statistics of the top 4 teams at RoboCup 2019 and 2022 (Due to missing video footage, the number of games per team varies. The numbers in parentheses include falls due to already broken gears, the regular numbers do not.)

Team	RoboCup	Avg. #falls per game		
		Overall	Collision	Walking
B-Human	2019	10.1	9.4	0.7
	2022	9.3 (14.3)	6.0	3.3 (8.3)
HTWK Robots	2019	18.5	14.1	4.4
	2022	21.9	16.0	5.9
rUNSWift	2019	16.6	14.0	2.6
	2022	22.2	14.7	7.5
Nao Devils	2019	18.3	12.6	5.7
	2022	31.5	21.2	10.3

the fact, that the dynamic positioning based on rating functions, as presented in this paper, was not yet developed. Additionally, the early iteration of the passing algorithm had an inaccuracy in the implementation of the rating functions that led to systematically lower pass ratings.

For one vs. one situations, the number of won encounters increased significantly, as shown in Table 2; on average by about 33.8% and about 41% against the runner-up team HTWK Robots. Meanwhile, the number of falls through collisions (Table 3) decreased by about 36.2%, while it increased for the other top 4 teams by between 5% to 69%. Overall, it must be noted that the fall rate for the other teams increased dramatically, by between 18% to 73%, while B-Human decreased by about 9%, despite a higher walking speed.

3 Limited Team Communication

In this year, the major rule change was to limit the maximum number of messages robots can broadcast to their teammates to 1200 per game for the whole team. This is significantly less than the single message each robot was allowed to send at RoboCup 2019 per second. For instance, the 2022 final lasted around 25 min.

According to the 2019 rules, the robots would have been allowed to send 7500 messages, but now have to make do with only 16% of that amount. The league's referee application keeps track of the number of messages each team sent. It broadcasts information about the state of the game to the robots twice per second. The remaining message budget and the remaining game time are part of that information, allowing the teams to monitor their budget use.

Although the intention of this rule change is that robots should rely more on their own perceptions rather than sharing them regularly, our general approach is to exchange mostly the same information as before, but limit the communication based on relevance. There is a multi-step approach to sending messages. Firstly, each robot waits at least one second after it sent the previous message before it sends the next one. On the one hand, this limits the maximum sending frequency to its 2019 counterpart. On the other hand, it ensures that the message sent was already counted by the referee application and is reflected in its latest broadcast. The second step is to make sure that the message budget is never exceeded, because this would nullify any goals scored according to the rules, which would prevent our robots from winning the game. There are two different approaches based on the importance of the information that should be sent.

3.1   Priority Messages

If the whistle was just detected, the message should be sent as soon as possible, because teammates would interpret a missing message as the whistle not being heard, impeding the majority decision of the team whether the referee actually whistled. Therefore, such messages are sent immediately as long as there is still a budget left (keeping a small reserve), i.e. if $\beta_{\text{remaining}} > \beta_{\text{reserve}}$, where $\beta_{\text{remaining}}$ is the message budget remaining as broadcast by the referee application and β_{reserve} is the number of messages we want to keep at the end of the game. This must be at least $\#robots - 1$ messages, because all robots could send a message at the same time and thereby reduce the budget by $\#robots$ at once.

3.2   Normal Messages

For all other information, a sliding budget is used based on the remaining game time and the remaining message budget using the following condition (ignoring some edge cases here):

$$\frac{\beta_{\text{remaining}} - \beta_{\text{reserve}}}{t_{\text{remaining}} - t_{\text{lookahead}}} > \frac{\beta_{\text{game}} - \beta_{\text{reserve}}}{t_{\text{game}}} \tag{4}$$

β_{game} is the overall message budget for a game, i.e. 1200. $t_{\text{remaining}}$ is the remaining time in the game. $t_{\text{lookahead}}$ is a time period into the future, of which the message budget can already be used. It allows to send a certain number of messages at the beginning of the game that are basically borrowed from the end of the game. This was set to 5 s. t_{game} is the overall time of a game. However, the overall duration of a game is unknown, because there are certain periods in a game in which the clock is stopped. This is at least the time until the first kick-off in each half. In playoff matches, the clock is also stopped between each goal and the next kick-off. Therefore, our robots do not send normal messages during such times, not even in preliminary matches. This allows us to set $t_{game} = 1200$ s.

3.3 Message Relevance

The approach so far would basically result in each robot sending a message every five seconds, but also adapting dynamically to phases without messages during preliminary games (in which the clock keeps running) and messages not sent by robots that are penalized for a while. However, whether a normal message is actually sent also depends on its relevance, i.e. whether it would contain information different enough from the previous message sent.

How "different enough" is defined depends on the kind of information. Behavior information such as the current role, tactic, set play, or pass target consists of discrete values, i.e. if one changes, it should be sent to the teammates. Position information such as the robot's own position, the ball position, or the kick target position consists of continuous values. Here, "different enough" means that it has changed significantly enough from its state sent previously both from the perspective of the sending robot as well as from the perspective of at least one of the receiving teammates. The further away a position is, the less precise it is usually observed. Also, the further away it is, the less important is its precision, because, for instance, kicking over long distances is also imprecise. As a result, the deviation accepted before a message is sent depends on the distance to the position in question. Another opportunity to save messages is to communicate information that allows to predict future positions for a while. The ball position is sent together with its speed. Using a friction model, this allows to predict the position where the ball will come to a halt. Robots also communicate how fast they are walking and where they intend to go. Thereby, teammates can predict their current position. Actually, the criterion for sending a message is whether the information sent previously would still suffice to predict the current position.

3.4 Team Play Under Limited Communication

Not all messages are sent, because the sliding budget might prevent it. Therefore, the team behavior must consider which information has actually reached the teammates, resulting in two different levels of strategic decisions: The ones that the behavior control would like to perform and the ones the teammates were actually informed about. Only the latter can actually be acted upon.

A very long-term variant of this problem is the coordination during the phases of the game without any communication. It is entirely based on the last information that was shared before the communication stopped. For instance, each robot decides about the kick-off position it will walk to based on the positions of all teammates at the time when the goal was scored.

During normal play, the robots act as a team, as described in Sect. 2. In addition, the robots orient themselves to face the ball at all times. This results

Table 4. Message delays due to budget limits in the German Open 2022 final

Player	Half	#Msg	$\frac{\text{Delay}}{\text{Msg}}$
1	1st	63	2.32 s
	2nd	84	1.48 s
2	1st	126	1.51 s
	2nd	107	2.09 s
3	1st	161	1.58 s
	2nd	162	2.13 s
4	1st	141	1.89 s
	2nd	129	1.91 s
5	1st	94	1.97 s
	2nd	129	2.33 s
Game		1196	1.91 s

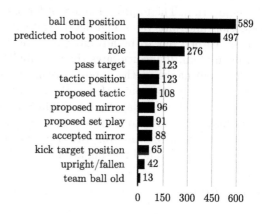

Fig. 3. Number of times a change was a reason for sending a message in the German Open 2022 final

in mostly sideways or diagonal walking, to ensure the head can still look in the walk direction to prevent collisions with other robots. Such behavior allows the robots to react to the movement of the ball. This way, a robot can intercept a rolling ball to receive a pass that could not be announced via communication.

3.5   Results

We analyzed the team communication of the RoboCup German Open 2022 final, which was very similar to the final at RoboCup 2022 (same teams, same score). On average, each message was delayed by 1.91 s (see Table 4), which was determined by summing up all the frames in which messages should have been sent, but were not allowed to due the dynamic limit, divided by the number of messages sent. Figure 3 shows the reasons for sending messages. Simulations conducted (50 playoff games each) show that the average score of our software playing against itself is 2.6:3.0 goals if the players of the second team send one message per second each (which would be illegal) and 3.2:2.6 goals if they always wait 5 s instead (which would just be legal). So our dynamic sending approach is better than a static one, but it cannot fully compensate the rule change in comparison to what was allowed before.

4   Conclusion

In this paper, we described some of the aspects that contributed to our success in the RoboCup 2022 competition. B-Human played seven games and scored a total of 48 goals while conceding 0. These accounted for 30% of all goals scored in this competition between 13 teams and a total of 38 games.

Our new behavior architecture enables more cooperative game play with less communication. The improved behavior for one vs. one situations maintains ball possession and integrates high-level strategic decisions. The dynamic passing strategy turned out to be advantageous and increased the performance of the B-Human team. It incorporates the new ability to play passes in any game situation to outmaneuver the opposing team. Experimental results show that this cooperative play style outperforms the previous behavior with improvements in key metrics such as the average goal difference and win rate. However, effective countermeasures have to be developed, such as marking opponents and intercepting their passes. Moreover, the accuracy of the kicks has to be improved in order to minimize failed passing attempts and losses of ball possession.

References

1. Anzer, G., Bauer, P.: Expected passes. Data Min. Knowl. Disc. **36**(1), 295–317 (2022)
2. Biswas, J., Mendoza, J.P., Zhu, D., Choi, B., Klee, S., Veloso, M.: Opponent-driven planning and execution for pass, attack, and defense in a multi-robot soccer team. In: Proceedings of the 2014 International Conference on Autonomous Agents and Multi-agent Systems, pp. 493–500 (2014)
3. Dick, U., Link, D., Brefeld, U.: Who can receive the pass? A computational model for quantifying availability in soccer. Data Min. Knowl. Discov. **36**(3), 987–1014 (2022)
4. Hasselbring, A.: Optimierung des Verhaltens von Fußballrobotern mittels simulierter Spiele und Gegnerverhaltensmodellen. Master's thesis, University of Bremen (2020)
5. Kaden, S., Mellmann, H., Scheunemann, M., Burkhard, H.D.: Voronoi based strategic positioning for robot soccer. In: Proceedings of the 22nd International Workshop on Concurrency, Specification and Programming, Warsaw, Poland, vol. 1032, pp. 271–282 (2013)
6. Kyrylov, V., Razykov, S.: Pareto-optimal offensive player positioning in simulated soccer. In: Visser, U., Ribeiro, F., Ohashi, T., Dellaert, F. (eds.) RoboCup 2007. LNCS (LNAI), vol. 5001, pp. 228–237. Springer, Heidelberg (2008). https://doi.org/10.1007/978-3-540-68847-1_20
7. Laue, T., Spiess, K., Röfer, T.: SimRobot – a general physical robot simulator and its application in RoboCup. In: Bredenfeld, A., Jacoff, A., Noda, I., Takahashi, Y. (eds.) RoboCup 2005. LNCS (LNAI), vol. 4020, pp. 173–183. Springer, Heidelberg (2006). https://doi.org/10.1007/11780519_16
8. Mendoza, J.P., et al.: Selectively reactive coordination for a team of robot soccer champions. In: Proceedings of the AAAI Conference on Artificial Intelligence, vol. 30, no. 1 (2016)
9. Reichenberg, P., Röfer, T.: Step adjustment for a robust humanoid walk. In: Alami, R., Biswas, J., Cakmak, M., Obst, O. (eds.) RoboCup 2021. LNCS (LNAI), vol. 13132, pp. 28–39. Springer, Cham (2022). https://doi.org/10.1007/978-3-030-98682-7_3
10. Röfer, T., et al.: B-Human 2019 – complex team play under natural lighting conditions. In: Chalup, S., Niemueller, T., Suthakorn, J., Williams, M.-A. (eds.) RoboCup 2019. LNCS (LNAI), vol. 11531, pp. 646–657. Springer, Cham (2019). https://doi.org/10.1007/978-3-030-35699-6_52

11. Röfer, T., Laue, T., Hasselbring, A., Richter-Klug, J., Röhrig, E.: B-Human 2017 – team tactics and robot skills in the standard platform league. In: Akiyama, H., Obst, O., Sammut, C., Tonidandel, F. (eds.) RoboCup 2017. LNCS (LNAI), vol. 11175, pp. 461–472. Springer, Cham (2018). https://doi.org/10.1007/978-3-030-00308-1_38

Winning the RoboCup Logistics League with Visual Servoing and Centralized Goal Reasoning

Tarik Viehmann[1]([✉]), Nicolas Limpert[2], Till Hofmann[1], Mike Henning[2], Alexander Ferrein[2], and Gerhard Lakemeyer[1]

[1] Knowledge-Based Systems Group, RWTH Aachen University, Aachen, Germany
viehmann@kbsg.rwth-aachen.de
[2] MASCOR Institute, FH Aachen University of Applied Sciences, Aachen, Germany

Abstract. The RoboCup Logistics League (RCLL) is a robotics competition in a production logistics scenario in the context of a Smart Factory. In the competition, a team of three robots needs to assemble products to fulfill various orders that are requested online during the game. This year, the Carologistics team was able to win the competition with a new approach to multi-agent coordination as well as significant changes to the robot's perception unit and a pragmatic network setup using the cellular network instead of WiFi. In this paper, we describe the major components of our approach with a focus on the changes compared to the last physical competition in 2019.

1 Introduction

The Carologistics RoboCup Team is a cooperation of the Knowledge-Based Systems Group (RWTH Aachen University) and the MASCOR Institute (FH Aachen University of Applied Sciences). The team was initiated in 2012 and consists of Doctoral, master's, and bachelor's students of both partners who bring in their specific strengths tackling the various aspects of the RoboCup Logistics League (RCLL): designing hardware modifications, developing functional software components, system integration, and high-level control of a group of mobile robots.

In previous competitions [6,7], we have pursued a distributed approach to multi-agent reasoning, where each robot acts on its own and coordinates with the other robots to resolve conflicts. This year, we have pursued a different strategy: Instead of having multiple agents each acting on its own, we now use one central goal reasoner that assigns tasks to each robot. This allows a more long-term strategy and avoids coordination overhead. Additionally, we have changed our approach to perception and manipulation. Instead of a pointcloud-matching approach that uses RGB/D data to iteratively determine an object's pose, we use a neural network to determine the bounding box of an object in an RGB image and then use closed-loop visual servoing to approach the object. Finally,

© The Author(s), under exclusive license to Springer Nature Switzerland AG 2023
A. Eguchi et al. (Eds.): RoboCup 2022, LNAI 13561, pp. 300–312, 2023
https://doi.org/10.1007/978-3-031-28469-4_25

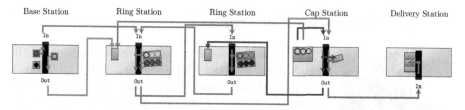

Fig. 1. Path for a product (green) of highest complexity, along with possible options to supply material for the ring assemblies (blue, brown and red). product in the RCLL (Color figure online).

we have taken first steps towards switching our navigation to ROS 2 and multi-agent path finding.

In the following, we summarize the RCLL in Sect. 2 and provide an overview of our system, starting with our robot platform in Sect. 3. In Sect. 4, we present our software architecture and continue by describing our advances towards multi-agent path planning in Sect. 5, before we explain our approach to perception and visual servoing in Sect. 6. In Sect. 7, we summarize our approach to high-level decision making and describe our new centralized approach to multi-agent coordination, before we conclude in Sect. 8.

2   The RoboCup Logistics League

The RoboCup Logistics League (RCLL) [12] is a RoboCup [10] competition with a focus on smart factories and production logistics. In the RCLL, a team of mobile robots has to fulfill dynamically generated orders by assembling workpieces. The robots operate and transport workpieces between static production machines to assemble the requested products or to supply the stations with material necessary to perform some assembly steps. The major challenges of the RCLL include typical robotics tasks such as localization, navigation, perception, and manipulation, with a particular focus on reasoning tasks such as planning, plan execution, and execution monitoring.

The game is controlled by a semi-automatic Referee Box (refbox) [18]. The refbox generates dynamic orders that consist of the desired product configuration and a requested delivery time window for the product, which must be manufactured by the robots of each team. Each requested product consists of a base piece (colored red, black, or silver), up to three rings (colored blue, green, orange, or yellow), and a cap (colored black or gray), resulting in 246 possible product configurations. The complexity of a product is determined by the number of required rings, where a C0 product with zero rings is a product of the lowest complexity, and a C3 product with three rings is a product of the highest complexity. Each team has an exclusive set of seven machines of five different types of Modular Production System (MPS) stations. To manufacture a requested product, the team has to execute a sequence of production steps by means of operating the MPS stations. An exemplary production is shown in Fig. 1.

3　The Carologistics Platform

The standard robot platform of this league is the Robotino by Festo Didactic [9]. The Robotino is developed for research and education and features omni-directional locomotion, a gyroscope and webcam, infrared distance sensors, and bumpers. The teams may equip the robot with additional sensors and computation devices as well as a gripper device for product handling. The Carologistics Robotino is shown in Fig. 2.

Sensors. We use one forward-facing and one tilted, backward-facing SICK TiM571 laser scanner for collision avoidance and self-localization. Using a second laser scanner in the back allows us to fully utilize the omni-directional locomotion of the Robotino. In addition to the laser scanners, we use a webcam for detecting the MPS identification tags, and a Creative BlasterX Senz3D camera for conveyor belt detection.

Fig. 2. The Carologistics Robotino

3.1　Gripper System

Our gripper system consists of three linear axes and a three-fingered gripper, as shown in Fig. 3. The three axes are driven by stepper motors, which allows movements with sub-millimeter accuracy. The axes are controlled by an Arduino, which in turn receives commands from the Robotino main computer.

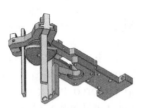

(a) The three linear axes driven by stepper motors　(b) The CAD model of the **three-fingered gripper**　(c) The complete gripper system

Fig. 3. The gripper system consisting of three linear axes and a self-centering gripper with three fingers

The gripper uses three fingers and grips the workpiece from above. This allows increased robustness and precision, as the workpiece is always centered between the three spring-loaded fingers, independent of positioning errors.

Since 2021, the laptop on top of the Robotino (cf. Fig. 2) was removed, as the Robotino 4 is capabale of running our full software stack without further need for additional computational power. As the laptop also served as an access point, initially, a small access point was mounted to ensure WiFi connectivity.

3.2   Cellular Network Setup via Tailscale

The challenging characteristics of tens of competing wireless networks communicating across the different leagues are an ever existing issue at RoboCup. The change of our hardware components in terms of the network equipment attached to the Robotinos rendered our communication platform virtually unusable due to tremendous paket loss among systems trying to communicate across the playing field. In addition, the change to a central goal reasoning approach increased the dependency on reliable communication among the participating machines.

Fig. 4. Smartphone for USB based LTE tethering to the robot

To address these issues and allow us to compete properly, we switched from a local WiFi connection to the cellular network of a generic local provider using a Long Term Evolution (LTE) network. However, according to the current rules of the RCLL[1] "Communication among robots and to off-board computing units is allowed only using WiFi". This rule was mainly intended to prohibit wired connections, so we approached the other teams and the TC to get approval for the usage of the cellular network during the competition.

Each robot has a direct connection to the internet by using a smartphone, which tethers its LTE connection to the robot without using WiFi (Fig. 4). As the robots expect a local network connection to each other, we equipped the VPN service Tailscale[2], which issues a static IP address to each robot and which is based on the WireGuard [4] network tunnel.

Albeit having some delay (100–200 ms), the UDP based connection was stable enough to reliably operate the robots and communicate to and from the central goal reasoning.

The authentication to join the Tailscale network is based on an existing identity provider (in our case we utilized our GitHub organization). In addition, the WireGuard tunnel encrypts the communication between the peers.

---

[1] See Sect. 7 of https://github.com/robocup-logistics/rcll-rulebook/releases/download/2022/rulebook2022.pdf.

[2] https://tailscale.com/.

A drawback of this solution is the dependency on the cellular network infrastructure on site, which at the venue of RoboCup Bangkok was no issue. Additionally, we had to be mindful of the data usage as the we had limited data available on the chosen prepaid plans. By only sparsely using network-based visualization tools (such as pointcloud or camera output streams), we had more than 50% of our 20 GB limit available by the end of the tournament, hence it turned out to be a feasible solution.

4 Architecture and Middleware

The software system of the Carologistics robots combines two different middlewares, Fawkes [13] and ROS [19]. This allows us to use software components from both systems. The overall system, however, is integrated using Fawkes. Adapter plugins connect the systems, for example to use ROS' 3D visualization capabilities. The overall software structure is inspired by the three-layer architecture paradigm [5], as shown in Fig. 5. It consists of a deliberative layer for high-level reasoning, a reactive execution layer for breaking down high-level commands and monitoring their execution, and a feedback control layer for hardware access and functional components. The communication between single components – implemented as *plugins* – is realized by a hybrid blackboard and messaging approach [13].

Recent work within Fawkes includes the support to couple the reasoning component CLIPS Executive (CX) with multiple reactive behaviour engines (cf. Sect. 7.1) of remote Fawkes instances. This enabled us to use Fawkes to build a centralized reasoner controlling the robots to fulfill the tasks of the RCLL (see Sect. 7). Now each Robotino runs a Fawkes instance without a reasoning unit along with a ROS-based

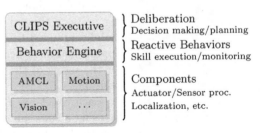

Fig. 5. Behavior Layer Separation [17]

navigation stack (cf. Sect. 5). Additionally, a central computer runs a Fawkes instance with the CX (see Sect. 7.2) that deliberates about the production strategy and sends commands to the behavior engines running on the robots.

Also, while the current setup offers bridging capabilities between Fawkes and ROS, as ROS 2 [22] becomes more prominent, we also implemented interfaces between Fawkes and ROS 2 to prepare for a future switch to ROS 2. Since the Carologistics are using Fedora as operating system on the Robotino platforms, which is not officially supported by ROS 2, we work on providing appropriate packages as we already do for ROS 1[3]. Moreover, as an entry point of ROS 2 into

---

[3] https://copr.fedorainfracloud.org/coprs/thofmann/ros/.

the RCLL, we are currently porting the Robotino hardware driver from Fawkes to ROS 2. The Fawkes driver directly uses the hardware interfaces instead of the Robotino REST API, which lacks reliable time stamps.

5   Towards Path Planning in ROS 2

Our current setup utilizes our navigation stack as described in [6]. As a first use case for ROS 2 we actively work towards a multi-agent path finding (MAPF) solution with the help of the ROS 2 Navigation framework [11]. With the MAPF approach, it is possible to handle narrow situations or intersection scenarios, which are well known problems for our current single-agent navigation solution.

However, as the work on the ROS 2 solution is still in active development and not yet ready for usage in competitions, we chose to deploy the ROS based navigation from previous years. Notably, the network middlware DDS[4] deployed in ROS 2 is quite complex and we could not configure it robustly, which sometimes caused faulty pose state estimations leading to unpredictable navigation behaviour.

6   Perception

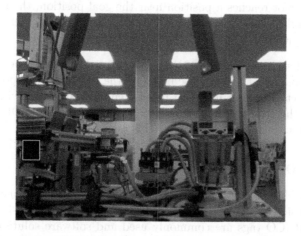

Fig. 6. Object detection with YOLO [23]. It detects objects of the three classes *conveyor belt* (green), *workpiece* (blue), and *slide* (red). (Color figure online)

Every production step in the RCLL comes down to a pick-and-place task on or from a narrow conveyor belt that is only a few millimeters wider than the workpiece itself. Since producing a medium-high complexity product can already involve 18 pick or place operations and a single manipulation error is likely

---

[4] https://www.dds-foundation.org/.

to result in total loss of the product, reliability (and therefore precision) is of paramount importance. In previous years [6], we have relied on a multi-stage procedure to detect conveyor belts of the target MPS stations. At its core, our previous approach used a model fitting approach based on the Iterative Closest Point (ICP) algorithm. It iteratively compared the current RGB/D pointcloud to a previously recorded model of the goal location (e.g., the conveyor belt) and computed a transformation from the current to the target position [2]. While this approach worked reliably, the iterative model matching of pointclouds made it comparably slow. Also, the approach relied on a good quality of the reference pointcloud, minor modifications to the machines often resulted in failed manipulation attempts.

For these reasons, we have replaced the pointcloud-based method by a simpler approach that only uses RGB camera images and point-based visual servoing (PBVS). It uses YOLOv4 [3,20] to detect objects in the image of the RGB camera, as shown in Fig. 6. The approach works in several stages [23]:

1. As long as the object of interest has not been detected near the expected position, the robot navigates to a pre-defined position near the expected goal location.
2. As soon as an object of the correct class has been detected in proximity to the expected position, the robot's base and its gripper are positioned simultaneously, using a closed-loop position-based visual servoing approach.
3. Once the robot reaches a position near the goal position, the robot's base is stopped while the PBVS task continues to position the gripper relative to the detected object.

The visual servoing task iteratively computes the distance between the current robot's pose and the goal pose based on the current object position. Therefore, the object detection needs to be fast enough to match with the control frequency of the robot. While YOLOv4 performed better, YOLOv4-tiny was sufficiently precise and fast enough for this task.

6.1   ARUCO Tag Detection

As of 2022 the rulebook of the RCLL requires ARUCO tags [24] in order to represent type and side of each machine. In comparison to the previously used ALVAR approach, ARUCO tags are commonly used and software solutions are widely available. We opted for the OpenCV based implementation[5] which required proper integration into Fawkes. During the development we encountered the need to actively calibrate the cameras for each robot to achieve a usable reported tag pose. The ALVAR-based solution did not require active calibration.

---

[5] https://docs.opencv.org/4.x/d5/dae/tutorial_aruco_detection.html.

7   Behavior Engine and High-Level Reasoning

In the following we describe the reactive and deliberative layers of the behavior components. In the reactive layer, the Lua-based behavior engine provides a set of skills. Those skills implement simple actions for the deliberative layer, which is realized by an agent based on the CX [16], a goal reasoning framework that supports multi-agent coordination.

7.1   Lua-Based Behavior Engine

In previous work we have developed the Lua-based Behavior Engine (BE) [14]. It serves as the reactive layer to interface between the low- and high-level systems. The BE is based on hybrid state machines (HSM). They can be depicted as a directed graph with nodes representing states for action execution, and/or monitoring of actuation, perception, and internal state. Edges denote jump conditions implemented as Boolean functions. For the active state of a state machine, all outgoing conditions are evaluated, typically at about 15 Hz. If a condition fires, the target node of the edge becomes the active state. A table of variables holds information like the world model, for example storing numeric values for object positions. It remedies typical problems of state machines like fast growing number of states or variable data passing from one state to another. Skills are implemented using the light-weight, extensible scripting language Lua.

7.2   Reasoning and Planning with the CLIPS Executive

We implemented an agent based on the CLIPS Executive (CX) [16], which uses a goal reasoning model [1] based on the goal lifecycle [21]. A goal describes objectives that the agent should pursue and can either *achieve* or *maintain* a condition or state. The program flow is determined by the *goal mode*, which describes the current progress of the goal. The mode transitions are determined by the goal lifecycle, which is depicted in Fig. 7. When a goal is created, it is first *formulated*, merely meaning that it may be relevant to consider. The goal reasoner may decide to *select* a goal, which is then *expanded* into one or multiple plans, either by using manually specified plans or automatic planners such as PDDL planners [15]. The reasoner then *commits* to one of those plans, which is *dispatched*, typically by executing skills of the behavior engine. Eventually, the goal is *finished* and the outcome is *evaluated* to determine the success of the goal.

7.3   Central Coordination

We utilize the CLIPS Executive framework to implement a central reasoner, which dispatches skill commands to the individual robots via the remote blackboard feature of Fawkes. In contrast to the distributed incremental approach pursued in the past [6,8], the central reasoner only maintains a single worldmodel, without the overhead of complex coordination and synchronization mechanisms required in the previous approach.

Setup. An off-field laptop runs a Fawkes instance with the CLIPS Executive and its dependencies. It connects to the blackboards of the remote Fawkes instances running on each Robotino over a TCP socket by subscribing as a reader to all necessary interfaces. This allows the central agent to read data from and send instructions to the robots. The most crucial communication channel is the Skiller interface, which is used to trigger skill execution and obtain feedback. Exploration tasks may require sensory feedback to locate machines based on their tags and laser feedback. The exploration results are then sent back to the navigator on the robots.

However, sending raw sensor data via the network can be a drawback of this setup compared to our previous distributed approach, where only processed worldmodel data was shared. This is especially critical in competitions where bandwidth and connection quality is suboptimal. To avoid this issue, the data could be pre-processed on the robot such that only the relevant information is sent, which is planned in the future.

Central Goal Reasoning. In contrast to our previous incremental approach [8], our new approach focusses on a long-term strategy, driven by a two-layer system: Decisions to commit to an order result in the creation of a goal tree with all necessary goals to build the requested product. Those decisions are made by filtering all available orders according to multiple criteria, such as estimated feasibility of attached time constraints, expected points and the workload required on each machine to assemble the product. Supportive steps, such as providing material to mount rings or caps are not part of the order-specific trees, but rather are maintained dynamically in a separate tree that contains all those tasks across all pursued orders and may perform optimizations based on the requested support tasks (e.g., providing a cap to a cap station yields a waste product at that cap station, which can be used as material at a ring station if any pursued order needs it, else it needs to be discarded). Essentially,

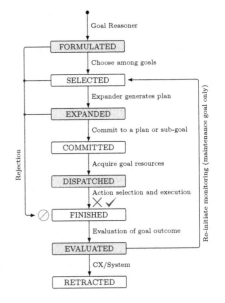

Fig. 7. The goal lifecycle with all possible goal modes [16].

the goal creation step defines the long-term strategy and goals for specific orders persist as long as the order is actively pursued. Figure 8 shows which goals are created if an order of complexity 1 and with a single material required to mount the ring is chosen.

The second layer consists of short-term decisions such as the distribution of robots to the respective goals, which is done lazily. Whenever a robot is idling, the reasoner evaluates the set of formulated goals that are currently executable for that robot and selects one among them. The selection is made by again filtering all the possible candidates. However, the criteria are much less complex and for now are based on a static priority of each goal (depending on the complexity of the belonging order), as well as constraints imposed through the corresponding goal tree (e.g., if goals must be executed in sequence).

Execution Monitoring. In order to become resilient to failures during the game, we handle execution errors in similar fashion as the distributed agent did: The damage is assessed, the worldmodel is updated accordingly and recovery methods are invoked if necessary (e.g., retrying failed actions or removing a goal tree, when the associated product is lost). In addition to the implications of a failed goal, the central agent needs to recognize a robot that not responding, de-allocate assigned tasks from it and react on successful maintenance. This is realized through a heartbeat signal, which is sent periodically by each robot. Lastly, the central agent itself may suffer a critical error that completely shuts it down. Even then the system is able to pick up on the work done so far by maintaining a persistent backup of the current worldmodel in a database and by restoring it, if necessary.

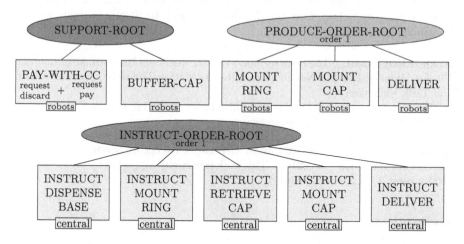

Fig. 8. Structure of goal trees for a single order. Ellipsis nodes are inner nodes, where blue ones always select the highest executable goal among them, while green ones only select the left-most child goal. Orange nodes denote the actual goals that are physically executed, either through the robots, or by the central instance itself, incase it only involves communication with the refbox (Color figure online).

8   Conclusion

In 2022, we have deployed a central agent based on the CLIPS Executive, which provides an explicit goal representation including plans and actions with their preconditions and effects. Focus was put on making explicit decisions about the orders that should be started. We replaced our established reasoner that grew over many past competitions, because we believe that its heavy focus on robustness made substantial tradeoffs to peak performance, which is not necessary anymore. This is mainly due to our matured gripping system. Unexpected side effects due to the dependency on a centralized reasoning server arrived in the form of WiFi issues, which we managed to overcome by utilizing cellular network-based communication with the help of a VPN and USB tethering from off-the-shelf smartphones.

Our perception setup is extended by a machine learning approach to detect workpieces, conveyor belts, and slides, which is used to approach the object with closed-loop visual servoing. This procedure turned out to be more robust and faster compared to our previous pointcloud-based approach.

A lot of effort was put into the integration of ROS 2 and towards multi-agent path planning to ensure fast and reliable navigation in tight and narrow environments. While not finished yet, important steps such as bridging Fawkes and ROS 2 are already implemented. We recon that completely exchanging most of our major components was an ambitious roadmap and the ROS 2 integration could not be finished yet, due to the amount of other tasks we worked on. Nevertheless, we believe that the decision to switch to ROS 2 is right and will benefit us in future competitions.

Acknowledgements. The team members in 2022 were Tom Hagdorn, Mike Henning, Nicolas Limpert, Hendrik Nessau, Simon Roder, Matteo Tschesche, Daniel Swoboda and Tarik Viehmann.

We gratefully acknowledge the financial support of RWTH Aachen University and FH Aachen University of Applied Sciences.

This work is partially supported by the German Research Foundation (DFG) under grants *GL-747/23-1, 2236/1* and under Germany's Excellence Strategy – EXC-2023 Internet of Production – 390621612. This work is partially supported by the EU ICT-48 2020 project TAILOR (No. 952215).

References

1. Aha, D.W.: Goal reasoning: foundations, emerging applications, and prospects. AI Mag. **39**(2) (2018)
2. Besl, P.J., McKay, N.D.: Method for registration of 3-D shapes. In: Sensor Fusion IV: Control Paradigms and Data Structures, vol. 1611, pp. 586–606. SPIE (1992)
3. Bochkovskiy, A., Wang, C.Y., Liao, H.Y.M.: YOLOv4: optimal speed and accuracy of object detection (2020). https://arxiv.org/abs/2004.10934
4. Donenfeld, J.A.: Wireguard: next generation kernel network tunnel. In: Proceedings of the Network and Distributed System Security Symposium (NDSS), pp. 1–12 (2017)

5. Gat, E.: Three-layer architectures. In: Kortenkamp, D., Bonasso, R.P., Murphy, R. (eds.) Artificial Intelligence and Mobile Robots, pp. 195–210. MIT Press (1998)
6. Hofmann, T., Limpert, N., Mataré, V., Ferrein, A., Lakemeyer, G.: Winning the RoboCup logistics league with fast navigation, precise manipulation, and robust goal reasoning. In: Chalup, S., Niemueller, T., Suthakorn, J., Williams, M.-A. (eds.) RoboCup 2019. LNCS (LNAI), vol. 11531, pp. 504–516. Springer, Cham (2019). https://doi.org/10.1007/978-3-030-35699-6_41
7. Hofmann, T., et al.: Enhancing software and hardware reliability for a successful participation in the RoboCup logistics league 2017. In: Akiyama, H., Obst, O., Sammut, C., Tonidandel, F. (eds.) RoboCup 2017. LNCS (LNAI), vol. 11175, pp. 486–497. Springer, Cham (2018). https://doi.org/10.1007/978-3-030-00308-1_40
8. Hofmann, T., Viehmann, T., Gomaa, M., Habering, D., Niemueller, T., Lakemeyer, G.: Multi-agent goal reasoning with the CLIPS executive in the Robocup logistics league. In: Proceedings of the 13th International Conference on Agents and Artificial Intelligence (ICAART) (2021). https://doi.org/10.5220/0010252600800091
9. Karras, U., Pensky, D., Rojas, O.: Mobile robotics in education and research of logistics. In: Workshop on Metrics and Methodologies for Autonomous Robot Teams in Logistics, IROS 2011 (2011)
10. Kitano, H., Asada, M., Kuniyoshi, Y., Noda, I., Osawa, E.: RoboCup: the robot world cup initiative. In: Proceedings 1st International Conference on Autonomous Agents (1997)
11. Macenski, S., Martín, F., White, R., Clavero, J.G.: The marathon 2: a navigation system. arXiv preprint arXiv:2003.00368 (2020)
12. Niemueller, T., Ewert, D., Reuter, S., Ferrein, A., Jeschke, S., Lakemeyer, G.: RoboCup logistics league sponsored by Festo: a competitive factory automation testbed. In: Jeschke, S., Isenhardt, I., Hees, F., Henning, K. (eds.) Automation, Communication and Cybernetics in Science and Engineering 2015/2016, pp. 605–618. Springer, Cham (2016). https://doi.org/10.1007/978-3-319-42620-4_45
13. Niemueller, T., Ferrein, A., Beck, D., Lakemeyer, G.: Design principles of the component-based robot software framework Fawkes. In: Ando, N., Balakirsky, S., Hemker, T., Reggiani, M., von Stryk, O. (eds.) SIMPAR 2010. LNCS (LNAI), vol. 6472, pp. 300–311. Springer, Heidelberg (2010). https://doi.org/10.1007/978-3-642-17319-6_29
14. Niemüller, T., Ferrein, A., Lakemeyer, G.: A Lua-based behavior engine for controlling the humanoid Robot Nao. In: Baltes, J., Lagoudakis, M.G., Naruse, T., Ghidary, S.S. (eds.) RoboCup 2009. LNCS (LNAI), vol. 5949, pp. 240–251. Springer, Heidelberg (2010). https://doi.org/10.1007/978-3-642-11876-0_21
15. Niemueller, T., Hofmann, T., Lakemeyer, G.: CLIPS-based execution for PDDL planners. In: ICAPS Workshop on Integrated Planning, Acting and Execution (IntEx) (2018)
16. Niemueller, T., Hofmann, T., Lakemeyer, G.: Goal reasoning in the CLIPS Executive for integrated planning and execution. In: Proceedings of the 29th International Conference on Planning and Scheduling (ICAPS) (2019)
17. Niemueller, T., Lakemeyer, G., Ferrein, A.: Incremental task-level reasoning in a competitive factory automation scenario. In: Proceedings of AAAI Spring Symposium 2013 - Designing Intelligent Robots: Reintegrating AI (2013)
18. Niemueller, T., Zug, S., Schneider, S., Karras, U.: Knowledge-based instrumentation and control for competitive industry-inspired robotic domains. KI - Künstliche Intelligenz 30(3), 289–299 (2016)
19. Quigley, M., et al.: ROS: an open-source Robot Operating System. In: ICRA Workshop on Open Source Software (2009)

20. Redmon, J., Divvala, S., Girshick, R., Farhadi, A.: You only look once: unified, real-time object detection. In: Proceedings of the IEEE Conference on Computer Vision and Pattern Recognition, pp. 779–788 (2016)
21. Roberts, M., et al.: Iterative goal refinement for robotics. In: Working Notes of the Planning and Robotics Workshop at ICAPS (2014)
22. Thomas, D., Woodall, W., Fernandez, E.: Next-generation ROS: building on DDS. In: ROSCon Chicago 2014. Open Robotics, Mountain View, CA (2014). https://doi.org/10.36288/ROSCon2014-900183
23. Tschesche, M.: Whole-body manipulation on mobile robots using parallel position-based visual servoing. Master's thesis, RWTH Aachen University (2022). https://kbsg.rwth-aachen.de/theses/tschesche2022.pdf
24. Wubben, J., et al.: Accurate landing of unmanned aerial vehicles using ground pattern recognition. Electronics 8(12), 1532 (2019)

FC Portugal: RoboCup 2022 3D Simulation League and Technical Challenge Champions

Miguel Abreu[1](✉)(iD), Mohammadreza Kasaei[2](iD), Luís Paulo Reis[1](iD), and Nuno Lau[3](iD)

[1] LIACC/LASI/FEUP, Artificial Intelligence and Computer Science Lab, Faculty of Engineering, University of Porto, Porto, Portugal
{m.abreu,lpreis}@fe.up.pt
[2] School of Informatics, University of Edinburgh, Edinburgh, UK
m.kasaei@ed.ac.uk
[3] IEETA/LASI/DETI, University of Aveiro, 3810-193 Aveiro, Portugal
nunolau@ua.pt

Abstract. FC Portugal, a team from the universities of Porto and Aveiro, won the main competition of the 2022 RoboCup 3D Simulation League, with 17 wins, 1 tie and no losses. During the course of the competition, the team scored 84 goals while conceding only 2. FC Portugal also won the 2022 RoboCup 3D Simulation League Technical Challenge, accumulating the maximum amount of points by ending first in its both events: the Free/Scientific Challenge, and the Fat Proxy Challenge. The team presented in this year's competition was rebuilt from the ground up since the last RoboCup. No previous code was used or adapted, with the exception of the 6D pose estimation algorithm, and the get-up behaviors, which were re-optimized. This paper describes the team's new architecture and development approach. Key strategy elements include team coordination, role management, formation, communication, skill management and path planning. New lower-level skills were based on a deterministic analytic model and a shallow neural network that learned residual dynamics through reinforcement learning. This process, together with an overlapped learning approach, improved seamless transitions, learning time, and the behavior in terms of efficiency and stability. In comparison with the previous team, the omnidirectional walk is more stable and went from $0.70\,\mathrm{m/s}$ to $0.90\,\mathrm{m/s}$, the long kick from $15\,\mathrm{m}$ to $19\,\mathrm{m}$, and the new close-control dribble reaches up to $1.41\,\mathrm{m/s}$.

1 Introduction

Historically, FC Portugal has contributed to the simulation league (2D and 3D) in numerous ways, including competitive methodologies and server improvements.[1] In the last years, the team has been focused on developing low-level

---

[1] For previous contributions concerning coaching, visual debugging, team coordination, sim-to-real, optimization algorithms and frameworks please refer to https://tdp.robocup.org/tdp/2022-tdp-fcportugal3d-robocupsoccer-simulation-3d/.

© The Author(s), under exclusive license to Springer Nature Switzerland AG 2023
A. Eguchi et al. (Eds.): RoboCup 2022, LNAI 13561, pp. 313–324, 2023
https://doi.org/10.1007/978-3-031-28469-4_26

skills and methodologies that leverage model knowledge to design more efficient reinforcement learning (RL) techniques [1–3,7–10,13,14]. In 2019, it introduced the first running behavior in the league, which was learned from scratch using RL. After that, the development focus was on leveraging the robot's symmetry and analytical models to improve the learning efficiency, producing human-like skills in less time. However, due to the intensive use of high-level optimization algorithms, the low-level C++ code of the team grew more convoluted with a network of interrelated skills and the addition of environment frameworks to bridge the gap between C++ and Python. Maintenance complexity along with lack of compatibility of some libraries with modern Linux distributions led to a turning point in 2021.

After RoboCup 2021, we decided to rebuild all the code from the ground up in Python, without using or adapting previous code, with the exception of the 6D pose estimation algorithm [4], and the get-up behaviors, which were re-optimized. The new code is compatible with most data science libraries and machine learning repositories, allowing for fast development of new behaviors and tactics. Due to hardware improvements in recent years, developing a team in Python is no longer a major concern in terms of computational efficiency. However, some computationally demanding modules were written in C++ to ensure the agent is always synchronized with the server.

2   3D Simulation League

The RoboCup 3D simulation league uses SimSpark [15] as its physical multiagent simulator, which is based on the Open Dynamics Engine library. The league's environment is a 30 m by 20 m soccer field containing several landmarks that can be used for self-localization: goalposts, corner flags and lines. Each team consists of 11 humanoid robots modeled after the NAO robot. Agents get internal data (joints, accelerometer, gyroscope and foot pressure sensors) with a 1-step delay every 0.02 s and visual data (restricted to a 120° vision cone) every 0.06 s. Agents can send messages to teammates every 0.04 s (see Sect. 3.3 for further details). There are 5 humanoid robot types with 22 to 24 controllable joints, and slightly different physical characteristics. Each team must use at least 3 different types during an official game.[2]

3   Team Description

An overview of the agent and server can be seen in Fig. 1. The white and green modules are implemented in Python and C++, respectively. The agent takes advantage of Python's development speed and compatibility with major data science libraries, and C++'s performance for time-sensitive modules.

The server is responsible for updating the soccer environment, based on the actions received from all agents. It adds noise to the world state in the form

---

[2] The official rules can be found at https://ssim.robocup.org/3d-simulation/3d-rules/.

of a calibration error that is constant and affects the robot's vision sensor, a perception error that follows a normal distribution and affects the coordinates of visible objects according to their distance, and a rounding error introduced by a lossy transmitter, where all numbers (including coordinates, joint angles and sensor readings) are rounded to the nearest hundredth.

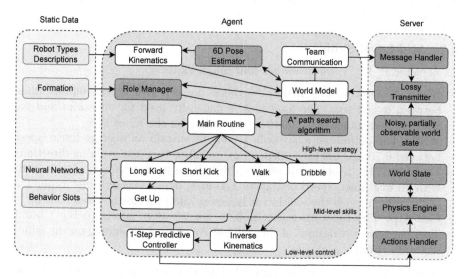

Fig. 1. Overview of the internal decision flow of agent and server, and their external interaction. White modules are implemented in Python while green modules are implemented in C++. Gray blocks represent static data. (Color figure online)

As aforementioned, the agent receives internal data from the server every 20 ms, and visual data every 60 ms. The latter is a noisy partial view of the world state, which is fed to a 6D pose estimator to extract the robot's localization and orientation in a three-dimensional space. The algorithm leverages the known noise models to maximize the probability of the current map of perceived objects, by iteratively adjusting the robot's 6D pose. For an extensive description of this process please refer to Abreu et al. [4]. Forward kinematics is then used to estimate the pose of every body part for a given robot type. This self-awareness ability in conjunction with team communication allows each agent to have a reliable representation of the world state. The following paragraphs will describe the main components of the agent.

3.1 Low-Level Control

The low-level control consists of an Inverse Kinematics module, used by the Walk and Dribble, and a 1-step predictive controller, which is indirectly used by the same skills, and directly used by the Long/Short Kick and Get Up. The Inverse Kinematics module simplifies locomotion by abstracting the joint information from the skill. The Long/Short Kick and Get Up do not use this module, as it did not improve the final behavior performance. Since the server sends the observations with a 1-step delay, the current robot state can only be estimated by combining the previous observations with the last sent actions. This technique is nearly optimal when the actuators work below their maximum torque specifications, although the lossy transmission protocol impedes actual optimality.

3.2   Mid-Level Skills

As depicted in Fig. 1, the team has four major skills:

- **Kick:** A short kick is mainly used for passes, with a controllable range between 3 and 9 m. The long kick is generally used for shooting, and has an average distance of 17 to 19 m, depending on the robot type;
- **Walk:** The omnidirectional walk is able to sustain an average linear speed between 0.70 and 0.90 m/s, depending on robot type and walking direction;
- **Dribble:** The dribble skill pushes the ball forward, retaining close control with an avg. max. speed of 1.25 to 1.41 m/s, depending on robot type. The walk skill can push the ball but it is slower and provides no close control;
- **Get Up:** The robot uses this skill to get up after falling to the ground. There are three variations of the skill per robot type depending on the falling direction (front, back, side).

The model architecture used for all skills that rely on neural networks can be seen in Fig. 2. An underlying base model is used to guide the optimization at an early stage, expediting the learning process and improving the quality and human-likeness of the final behavior. For the Walk and Dribble skills, the underlying model is based on a Linear Inverted Pendulum (LIP) solution, which generates a cyclic walk-in-place behavior. The kick skills are built upon a hand-tuned base model which is divided into two sections: back swing and forward acceleration. The state of the base model is fed to the neural network as a single integer variable. The state of the robot comprises its internal sensors, position and velocity of joints, head height and a step counter. Depending on the skill, the target denotes a set of variables that encode small variations of the main objective, such as direction and distance. The relative position of the ball is always required except for the Walk skill.

The output of the shallow neural network (single hidden layer with 64 neurons) is added to the output of the base model to generate a target position for each feet and hand. As shown in Fig. 1, only the Walk and Dribble skills generate relative positions, thus requiring the Inverse Kinematics module to obtain target joint angles. The optimization is performed by the Proximal Policy Optimization algorithm [12], extended with a symmetry loss.

3.3   Team Communication

According to official rules, team communication is restricted to messages of 20 characters, taken from the ASCII subset [0x21, 0x7E], except for normal brackets

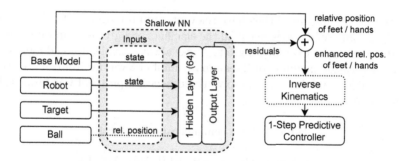

Fig. 2. Model architecture for the Long/Short Kick, Walk, and Dribble skills

{0x28,0x29}. However, due to server limitations, some characters create undefined behaviors, such as single (0x27) or double (0x22) quotation marks, backslash (0x5C), and semicolon (0x3B) at the beginning of messages. Therefore, FC Portugal uses 88 alternatives for the first character and 89 alternatives for the remaining 19 characters, since the semicolon is included. This gives about 9.6×10^{38} combinations per message.

The team uses three consecutive messages to send all the desired information. Accordingly, there are three distinct groups of variables, as seen in Table 1. Group A contains the position of the ball, and the position and state of teammates number 10 and 11, and opponents from number 7 to 11. The state is a Boolean indicating whether the agent has fallen. The position of a teammate has a precision of 10 cm and $x \in [-16, 16]$, $y \in [-11, 11]$, yielding $321 * 221 = 70941$ combinations. Considering the position and state of the robot, each teammate represents 141882 combinations. The opponents' position has the same range but lower precision (16 cm in x and 20 cm in y), generating $201 * 111 * 2 = 44622$ (including the state). Finally, the position of the ball has a precision of 10 cm and $x \in [-15, 15]$, $y \in [-10, 10]$, resulting in $301 * 201 = 60501$ combinations. Group B contains information about 7 teammates, and group C combines 2 teammates and 7 opponents.

Table 1. Description of message groups for the team communication protocol

Group	Teammates	Opponents	Ball	Combinations
A	10,11	7–11	Yes	$141882^2 * 44622^5 * 60501 = 2.2e38$
B	1–7	None	No	$141882^7 = 1.2e36$
C	8,9	1–6	No	$141882^2 * 44622^6 = 1.6e38$

The design of each group followed two principles: maximize the amount of information per message, and gather entities that are usually seen together on the field. Every agent can send messages at a given time step. However, the server only broadcasts one of the messages to the whole team. Therefore, agents should only try to communicate if they have relevant and up-to-date data. A naive rule would be to only send a message if all the elements of a given group were recently seen. However, as an example, if group A was always visible, it could monopolize the communication medium, or if teammate number 5 was far from all the others, group B would never be sent. To solve these issues, a protocol was proposed, as seen in Table 2.

Table 2. Team communication protocol

Time step	Round	Group	Max. RLP	Max. MP	Ball
0.00 s	1	A	0	0	is visible
0.04 s		B			–
0.08 s		C			–
0.12 s	2	A	1	–	is visible
0.16 s		B			–
0.20 s		C			–
0.24 s	3	A	2	–	is visible
0.28 s		B			–
0.32 s		C			–
0.36 s	1	...	...	...	...

Message groups are synchronized with the game time, which is the provided to all players by the server. There are three rounds of messages that are repeated every 0.36 s. Each round contains messages from all groups in a sequential order: A, B, C. Round 1 is the strictest, since an agent must see all the elements of the current group before it is allowed to broadcast. This means that the maximum number of recently lost players (RLP) and missing players (MP) is zero. The former includes player not seen in the last 0.36 s and the latter encompasses players that have not been seen in the last 3 s. Rounds 2 and 3 allow any number of missing players, but apply restrictions to the number of recently lost players. During a soccer match, missing players are typically far from the field or have crashed and left the server. In both cases, it is generally safe to ignore them. The main purpose of having rounds with different restrictions is to force high quality updates in round 1, and, if that is currently not possible, allow some losses in rounds 2 and 3. If a group is fully visible by any agent, all teammates will receive new information every 0.12 s, even if there are missing players; if one of the elements was recently lost, all teammates will be informed twice every 0.36 s; and if no agent has recently lost less than 2 elements from the current group, all teammates will get a broadcast once every 0.36 s.

3.4 Role Manager

The role manager is responsible for dynamically assigning roles to each player. Figure 3 shows an overview of this process. The formation, shown on the left, indicates the desired position of each role, not the current position of each team-mate. The red circles denote the actual position of the opponents. This example shows the formation when the team is attacking, i.e. the ball is closer to our team. The displayed roles are: goalkeeper (GK), central back (CB), man-marking (MM), left and right support (LS, RS), left and right front (LF, RF). The support roles are always close to the active player (AP) to assist when ball possession is lost, while the front roles are always closer to the opponent's goal to receive long passes. The man-marking roles mark the closest opponents to our goal in a sticky way, to avoid frequent switching.

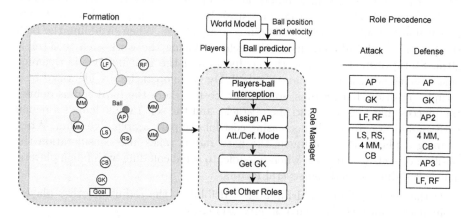

Fig. 3. Overview of the role manager. The formation (left) represents the desired position of each role, depending on the ball and opponents. The decision flow (middle) describes how the role manager assigns roles according to their precedence (right). (Color figure online)

The decision flow of the role manager is shown on the middle of Fig. 3. The world model provides the position and velocity of all players and ball. The ball predictor yields a sequence of ball positions for future time steps, until the ball stops completely. This information is combined with the maximum walking speed to estimate when and where each player would intercept the ball. At this point, it is possible to answer which player should go to the ball, and whether the opponent will get there faster. If the latter is true, the defense mode is activated, changing the role precedence, as indicated on the right of Fig. 3.

The roles with highest priority are the AP followed by the GK. Therefore, if the closest player to the ball is the goalkeeper, it becomes the AP, and the GK role will be assigned to another player. If the team is attacking, the next roles are LF and RF, to create space and supporting angles for the AP. Then, the low priority roles are assigned: LS, RS, 4 MM and CB. This assignment is an optimization problem where the objective is to reduce the distance between the player and the assigned role. However, the man-marking roles have different weights, according to the distance of the marked opponent to our goal. In general, the optimization algorithm prioritizes dangerous opponents, and gives less importance to the CB, in relation to the LS and RS roles. If the team is defending, after the AP and GK comes the Active Player 2 (AP2), and later the Active Player 3 (AP3). The position of these roles is the ball position. Three player will try to intercept it until at least one of them is closer than the closest opponent. In defense mode, there are no support roles, and the front roles are the least important.

3.5   Path Planning

The path search algorithm is based on A* [6] and divides the soccer field into 70941 nodes (32 m × 22 m, 10 cm grid size). It is only employed when the player is not able to walk to the objective in a straight line. When dribbling the ball, the player is only allowed inside the field, reducing the work area to 30 m × 20 m. Static obstacles include the goalposts and the goal net, while dynamic obstacles include the players and the ball. Dynamic obstacles are identified by a 4-tuple ⟨position, hard radius, soft radius, cost⟩, where the hard radius defines the inaccessible region around the obstacle, and the soft radius defines a region with an additional radially decreasing cost for the path search algorithm. When defining a player as an obstacle, the hard radius takes into consideration the position of each arm and leg, when possible, to avoid colliding with fallen players. The soft radius depends on several factors:

- Is the agent walking or dribbling? If dribbling, the radius of other players increases, since the maneuverability decreases;
- The radius of an opponent decreases when closer to the ball, to allow tackling;
- The radius of a teammate increases if it was assigned a more important role.

Regarding the last factor, the role precedence introduced in Fig. 3 is not only used to assign roles to players. It is also important to ensure some additional space is given to higher priority roles. As an example, when the team is defending and there are 3 active players, AP2 will give extra space to AP (through a larger soft radius), and AP3 will get away from AP and AP2. This also applies to all the other roles. Players are not considered obstacles if their distance is over 4 m or they have not been seen (or perceived through team communication) for longer than 0.5 s.

In order to achieve path stability, the initial path position is obtained by estimating the position of the robot after 0.5 s, considering the current velocity as a constant. To extract the current target for walking or dribbling, the number of considered path segments depends on the current speed of the robot. This technique generates a smooth path without losing too much accuracy.

This module was implemented in C++ due to its computational complexity. To ensure that no simulation cycle is lost, the path planning algorithm runs until the end or until a timeout (5 ms) expires. In the latter case, the best current path is returned. Therefore, while running asynchronously with the server, the performance of the team depends on the computational power of the host machine, but not to the point where it misses a significant amount of cycles, except for abnormally slow host machines.

3.6   Main Routine

The main routine is divided into 5 major steps as depicted in Fig. 4. The agent starts by selecting a general intention from: get up, dribble or push the ball, kick to pass or shoot, move to some point, or beam before a kickoff. Some situations require a specific intention: the active player is required to kick the ball during kickoff and other play modes, fallen players must get up, inactive players must move to strategic positions, etc. However, the active player faces more complex scenarios where it must decide whether to dribble, push or kick the ball.

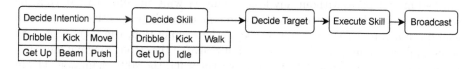

Fig. 4. Major steps of the agent's main routine

To reach a verdict, the algorithm scores multiple passes along a grid of feasible options, and shooting alternatives if the goal is reachable. The passing score considers several factors as shown in Fig. 5: alignment between kicker, ball and target (1); opponents' distance within a kick obstruction radius of 0.8 m, centered at 0.3 m from the ball in the target direction (2); distance of all players to the kick path, considering a kick angle error, α, which depends on the robot type and kick skill (3); target distance difference between the closest standing teammate and the closest opponent (4); alignment between receiver, ball and opponent goal (5); difference between the distance ball-goal before and after the kick, to ensure field progression. Shooting takes into account factors 1–3. During a game, if shooting is likely to be successful, the decision is immediate. Otherwise, passing is only preferred if it allows a faster game progression than dribbling. When there is no space to kick or dribble, the robot will push the ball by walking towards it.

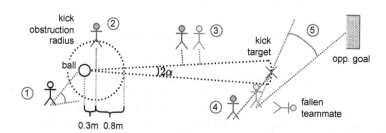

Fig. 5. Passing score factors: convenience for kicker (1), kick obstructions (2), ball obstructions (3), distance of both teams to target (4), convenience for the receiver (5).

The second step is to decide the current skill. As an example, if the agent decides to kick or dribble, it must first walk until it is in the correct position to use the desired skills. Then the agent must compute a target for the selected skill, if required, e.g., a target position and orientation while walking. Finally, the skill is executed, by applying the respective model and feeding the output to the low-level controller. An extra step broadcasts visual information to other teammates if the conditions are met, as explained in Sect. 3.3.

4   Results

The results of the main competition are summarized in Fig. 6. FC Portugal finished in first place without any loss, having 17 wins and only 1 tie during the seeding round.[3] It managed to end in first place in the respective group of all round-robin rounds. During the course of the competition, the team scored 84 goals while conceding only 2. In the final, it defeated the strong magmaOffenburg team by 6–1, a team that defeated UT Austin Villa, the 2021 champion [11], in the semi-finals by 3–0.

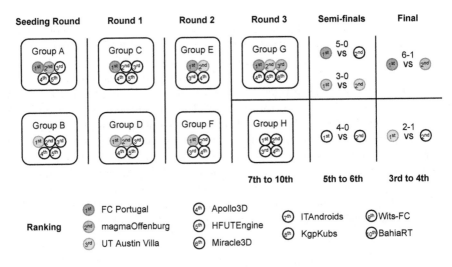

Fig. 6. Main competition results

In addition to the main competition, there was a Technical Challenge, which was composed of two events: the Free/Scientific Challenge, where competing teams presented their research work in the context of the 3D simulation league; and the Fat Proxy Challenge, which was played in a single round-robin tournament using the magmaFatProxy [5]. The purpose of the proxy is to provide the same skills to all teams, preventing the agent from controlling the joints of the robot, thus making the competition exclusively about high-level strategies.

Table 3 shows the scores obtained in the Free/Scientific and Fat Proxy challenges by each participating team. The Technical Challenge is won by the team

---

[3] Official results can be found at https://cloud.robocup.org/s/ifX7TDsaHpCFWWH.

that accumulates more points in both events. In each case, the number of awarded points is $25 - 20 * (rank - 1)/(number\_of\_participants - 1)$.

Table 3. Technical Challenge results based on the number of points accumulated in the Free/Scientific Challenge and the Fat Proxy Challenge

Team	Overall		Free/Scientific		Fat proxy	
	Rank	Points	Rank	Points	Rank	Points
FC Portugal	**1**	**50**	**1**	**25**	**1**	**25**
magmaOffenburg	2	25	3	5	2	20
BahiaRT	3	20	2	15	5	5
UT Austin Villa	4	15	–	–	3	15
Wits-FC	5	10	–	–	4	10

FC Portugal won both the Free/Scientific Challenge and the Fat Proxy Challenge, leading to a victory in the overall Technical Challenge with 50 points. The presented scientific contribution introduced the dribble skill—the league's first close control dribble behavior, reaching speeds of up to 1.41 m/s. In the Fat Proxy Challenge, the team registered 4 wins, no ties and no losses, 21 scored goals and 3 conceded.

5   Conclusion

FC Portugal has developed numerous skills and methodologies concerning the NAO humanoid robot and the simulation league in general. Currently, the team has a very robust code base developed from scratch after RoboCup 2021, which led to a victory in the 2022 RoboCup 3D simulation league, as well as the Free/Scientific Challenge, the Fat Proxy Challenge, and consequently, the Technical Challenge. In the main competition it registered 17 wins, 1 tie, 0 losses, 84 goals scored, only 2 conceded, and in the Fat Proxy Challenge 4 wins, 0 ties, 0 losses, 21 goals scored, and 3 conceded.

An integrated learning approach guaranteed that skills such as the omnidirectional walk, dribble and kick can attain high performance but also smooth transitions, without falling or requiring intermediate steps. Despite the considerable gain in performance and competitiveness, in comparison with previous years, there are still many improvement opportunities. Future research directions include high-level multi-agent coordination strategies, opponent modeling, goalkeeper skills, omnidirectional kicks, optimization algorithms, and more.

Acknowledgment. The first author is supported by FCT—Foundation for Science and Technology under grant SFRH/BD/139926/2018. The work was also partially funded by COMPETE 2020 and FCT, under projects UIDB/00027/2020 (LIACC) and UIDB/00127/2020 (IEETA).

References

1. Abdolmaleki, A., Simões, D., Lau, N., Reis, L.P., Neumann, G.: Learning a humanoid kick with controlled distance. In: Behnke, S., Sheh, R., Sariel, S., Lee, D.D. (eds.) RoboCup 2016. LNCS (LNAI), vol. 9776, pp. 45–57. Springer, Cham (2017). https://doi.org/10.1007/978-3-319-68792-6_4
2. Abreu, M., Lau, N., Sousa, A., Reis, L.P.: Learning low level skills from scratch for humanoid robot soccer using deep reinforcement learning. In: 2019 IEEE International Conference on Autonomous Robot Systems and Competitions (ICARSC), pp. 256–263. IEEE (2019)
3. Abreu, M., Reis, L.P., Lau, N.: Learning to run faster in a humanoid robot soccer environment through reinforcement learning. In: Chalup, S., Niemueller, T., Suthakorn, J., Williams, M.-A. (eds.) RoboCup 2019. LNCS (LNAI), vol. 11531, pp. 3–15. Springer, Cham (2019). https://doi.org/10.1007/978-3-030-35699-6_1
4. Abreu, M., Silva, T., Teixeira, H., Reis, L.P., Lau, N.: 6D localization and kicking for humanoid robotic soccer. J. Intell. Robot. Syst. **102**(2), 1–25 (2021)
5. Amelia, E., et al.: magmaFatProxy (2022). https://github.com/magmaOffenburg/magmaFatProxy
6. Hart, P., Nilsson, N., Raphael, B.: A formal basis for the heuristic determination of minimum cost paths. IEEE Trans. Syst. Sci. Cybern. **4**(2), 100–107 (1968)
7. Kasaei, M., Abreu, M., Lau, N., Pereira, A., Reis, L.P.: Learning hybrid locomotion skills - learn to exploit residual dynamics and modulate model-based gait control. arXiv preprint arXiv:2011.13798 (2020)
8. Kasaei, M., Abreu, M., Lau, N., Pereira, A., Reis, L.P.: A CPG-based agile and versatile locomotion framework using proximal symmetry loss. arXiv preprint arXiv:2103.00928 (2021)
9. Kasaei, M., Abreu, M., Lau, N., Pereira, A., Reis, L.P.: Robust biped locomotion using deep reinforcement learning on top of an analytical control approach. Robot. Auton. Syst. **146**, 103900 (2021)
10. Kasaei, S.M., Simões, D., Lau, N., Pereira, A.: A hybrid ZMP-CPG based walk engine for biped robots. In: Ollero, A., Sanfeliu, A., Montano, L., Lau, N., Cardeira, C. (eds.) ROBOT 2017. AISC, vol. 694, pp. 743–755. Springer, Cham (2018). https://doi.org/10.1007/978-3-319-70836-2_61
11. MacAlpine, P., Liu, B., Macke, W., Wang, C., Stone, P.: UT Austin Villa: RoboCup 2021 3D simulation league competition champions. In: Alami, R., Biswas, J., Cakmak, M., Obst, O. (eds.) RoboCup 2021. LNCS (LNAI), vol. 13132, pp. 314–326. Springer, Cham (2022). https://doi.org/10.1007/978-3-030-98682-7_26
12. Schulman, J., Wolski, F., Dhariwal, P., Radford, A., Klimov, O.: Proximal policy optimization algorithms. arXiv preprint arXiv:1707.06347 (2017)
13. Simões, D., Amaro, P., Silva, T., Lau, N., Reis, L.P.: Learning low-level behaviors and high-level strategies in humanoid soccer. In: Silva, M.F., Luís Lima, J., Reis, L.P., Sanfeliu, A., Tardioli, D. (eds.) ROBOT 2019. AISC, vol. 1093, pp. 537–548. Springer, Cham (2020). https://doi.org/10.1007/978-3-030-36150-1_44
14. Teixeira, H., Silva, T., Abreu, M., Reis, L.P.: Humanoid robot kick in motion ability for playing robotic soccer. In: 2020 IEEE International Conference on Autonomous Robot Systems and Competitions (ICARSC), pp. 34–39. IEEE (2020)
15. Xu, Y., Vatankhah, H.: SimSpark: an open source robot simulator developed by the RoboCup community. In: Behnke, S., Veloso, M., Visser, A., Xiong, R. (eds.) RoboCup 2013. LNCS (LNAI), vol. 8371, pp. 632–639. Springer, Heidelberg (2014). https://doi.org/10.1007/978-3-662-44468-9_59

RoboFEI@Home: Winning Team of the RoboCup@Home Open Platform League 2022

Guilherme Nicolau Marostica⬛, Nicolas Alan Grotti Meireles Aguiar⬛,
Fagner de Assis Moura Pimentel[(✉)]⬛, and Plinio Thomaz Aquino-Junior⬛

FEI University Center, Av. Humberto de Alencar Castelo Branco 3972-B - Assunção,
09850-901 São Bernardo do Campo, São Paulo, Brazil
{fpimentel,plinio.aquino}@fei.edu.br

Abstract. For the first time, the HERA robot won the RoboCup@Home in the Open Platform League in Bangkok, Thailand. This robot was designed and developed by the RoboFEI@Home team, considering all mechanical, electronic, and computational aspects. It is an Open League platform capable of performing autonomous tasks in home environments, in addition to human-robot interaction, collaborating with people who share the same environment. In this edition of the competition, the platform presented advances in the methods of interacting with people and social navigation. Interaction with people and objects is supported by image segmentation processes, enhancing environment perceptions and people recognition during tasks.

Keywords: RoboCup@Home · Open platform league · Domestic service robotics

1 Introduction

The RoboCup 2022 edition in Thailand was the first in-person edition after the pandemic. That was the first time the RoboFEI@Hone team won first place in the RoboCup@Home Open Platform League. We are a passionate team for RoboCup@Home, participatory and active since 2016, in our first participation in the world competition. Participation in local competitions helps to promote the league in South America, being a powerful ecosystem of knowledge exchange and preparation. In the Brazilian Robotics Competition (partner RoboCup Brazil) five consecutive titles were won.

In this edition of RoboCup in Thailand, a well-designed area was found, comprising a living room, kitchen, bedroom and office, containing 2 entrances/exits

This study was financed in part by the Coordenação de Aperfeiçoamento de Pessoal de Nível Superior - Brasil (CAPES) - Finance Code 001; Acknowledgment to FEI (Fundação Educacional Inaciana Pe. Saboia de Medeiros) and a special acknowledgment to all members of the RoboFEI@Home team.

ⓒ The Author(s), under exclusive license to Springer Nature Switzerland AG 2023
A. Eguchi et al. (Eds.): RoboCup 2022, LNAI 13561, pp. 325–336, 2023
https://doi.org/10.1007/978-3-031-28469-4_27

from the arena. There were several pieces of furniture in each environment, allowing you to perform the tasks provided for in the rulebook. Using this environment, it was possible to perform tasks with new features of computer vision, new methodology with social navigation, and new software architecture that integrates the various packages.

1.1   The RoboFEI@Home Team

The RoboFEI@HOME team started its activities in 2015 using the PeopleBot platform to perform domestic tasks. Research on human-robot interaction has been intensified with master's and doctoral projects. Research is carried out in different contexts considering human behavior, user modeling, interaction design, social navigation, among others [2,9,13]. The mechanics and electronics of the robot were completely redesigned after difficulties in purchasing spare parts from the robotic platform on the market.

The mechanical, electronic, and computational design considered an economical platform for maintenance, but it can perform many domestic activities such as: social and safe navigation, object manipulation, interaction with people and appliances, and command recognition based on gestures or voice [1].

The RoboFEI@Home team seeks to be in constant evolution to always develop new technologies in the domestic assistant area. The main researchers developed by our team in these years are focused on making the HERA robot even more autonomous.

2   Hardware

The Hera robot has an omnidirectional base, which makes it possible to move in any direction, making it a great differential for movement in places with restricted navigation. For manipulation, has a robotic arm of 6 DOF, composed of Dynamixels servo motors, being controlled through an OpenCM 9.04 board, with a gripper revolute using flexible filaments for better grip objects. Has attached a Logitech 1080p camera to its end effector, allowing it to perform a wide variety of tasks. For the computer vision part, it used a Microsoft Kinect, with the RGB camera and Depth for integration with the system. Counting on a servo motor in its joint to adjust the tilt of the camera according to the need. The robot head, has an Apple Ipad 2, and 2 RODE VideoMic GO directional microphones, which make up our audio system, and a MATRIX CreatorTM at the top of the head, with the main purpose of using this board is to perform directional voice recognition.

The navigation system contains a Hokuyo UTM-30LX-EW sensor, capable of detecting obstacles in the environment, and Asus Xtion, used to detect obstacles that are difficult to recognize. For processing, has a Zotac Mini-PC with core i5, 7500T, 16 Gb RAM, with Ubuntu 20.04, ROS Noetic, and an Nvidia Jetson AGX Xavier, to compensate for the system's graphics processing. The robot's power supply consists of 5000 mAh Lipo battery packs, modularly connected within the robotic platform (Fig. 1).

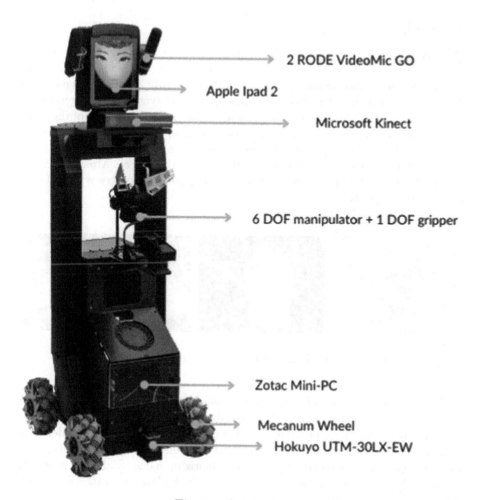

Fig. 1. Robot hardware

3 RoboCup@Home

The RoboCup@Home league aims to develop service robots for home applications. It is the largest annual international autonomous service robot competition. A domestics task group of varying themes is used to assess the skills and performance of robots.

The main skills required by robots are Human-Robot Interaction and Cooperation, Navigation and Mapping in dynamic environments, Computer Vision and Object Recognition under natural light conditions, Object Manipulation,

Adaptive Behaviors, Behavioral Integration, Environmental Intelligence, Standardization, and Integration of Systems [11].

4 Skills

4.1 Robot Vision

The object detection system consists of Efficientdet-d0 [6], using TensorFlow 2.0. To create this dataset of objects, have been used synthetic data generation [5] to save time in tagging the images and create a larger amount of data. An efficient background removal method based on Deep Salient Object Detection [10], is an algorithm to detect the most noticeable and important object in an image, returning a binary mask of the image with the object. After generating the mask, the object image can be used to compose new images with different backgrounds (Fig. 2).

Fig. 2. Efficient background removal

Integrating a system of vision and manipulation of specific objects, with image segmentation techniques using color extraction to perform the fine adjustment in the object manipulation, not being necessary to train a model from scratch to recognize a single object, reducing the time spent in the training. The robot has people recognition, capable of memorizing names and faces using the library dlib [3] that can identify a landmark, which allows guaranteeing a wide variety of tasks with people.

4.2 Voice Recognition

The team decided to use Google's Speech Recognition API. For this, a ROS package was developed that operates through a set of APIs. They are online tools that work directly on Ubuntu. In addition, a comparison is made with generic sentences using the Hamming distance to recognize sentence variations.

This API was created from methods that facilitate the code adaptation to a given environment, creating a new use of word choices in speech.

In competition, the team is using the MATRIX CreatorTM [4], a board with sensors, wireless communication and an FPGA. The main objective of using this board is to perform directional voice recognition, thus being able to recognize where the operator is talking to the robot from.

The Raspberry Pi connected to the MATRIX is used for communication with the core of our robot. The Raspberry is responsible for reading the information from the various sensors on the board and sending this information to the main system.

4.3 Manipulator

The manipulator has a number of degrees of freedom (DOF) contained in a human arm, aiming to obtain a great similarity with real movements using the anthropomorphic principle. From this, a study of human anatomy and kinesiology began, more specifically in the skeleton of the free portion of the upper limbs, namely: arm, forearm, carpus, metacarpal. It was noticed that the main movements are extension and flexion.

A new change in the manipulator is the new materials we are using, for parts with more complex shapes we use 3D printing and for flat parts we are using carbon fiber, resulting in greater resistance with less weight and smaller dimension.

In the manipulation system, we used the Dynamixel Workbench package for direct kinematics control when we need simpler movements. When we need trajectory planning and deeper precision, we use Moveit with inverse kinematics. For a more optimized and safer manipulation, we use the octomap integrated with the manipulation system (Fig. 3). With this, we can have the perception of the environment through the vision, considered in the robotic arm trajectory planning, allowing a safety movement.

Fig. 3. Sensor interpretation using octomap in manipulation system.

4.4 Robot Navigation

An autonomous robot, to be able to navigate alone, needs the ability to map where it is, define its position in space and decide the best possible route.

For this to be possible, sensors that capture external environments are used, and this information is transformed into interpretable data so that the robot chooses the best route. When the robot is in an unknown location, it must map the environment, where it is located, and at the same time define its position in space. This technique is known as Simultaneous Location and Mapping (SLAM). In navigation, the robot has the ability to choose the best possible route and avoid possible obstacles using parameters where the smallest path error is corrected instantly.

5   Implementation Highlights in Tasks

This section presents the main challenges encountered in implementing the tasks, and how they were resolved. The implementation of these tasks was considering the main strengths and weaknesses of the robot, and how they can be improved for the next competition. Aiming to generalize the advances obtained, and facilitate the implementation of new tasks.

5.1   Stage I

In the first stage, the robot needs to perform simple tasks in a domestic environment. The HERA robot scored 800 points in this stage.

Take Out the Garbage: This task consists of removing the garbage from the recycle bin (300 score points) and placing it in a predefined location. The challenge of this task was the positioning of the robot for manipulation. The garbage location was predefined, however, it could vary within an area, thus making it difficult to accurately position the robot. To solve this problem, we trained a neural network with Framework TensorFlow 2 and Efficient-net [6] to detect garbage, with this detection we use the depth camera to capture the position of the object around the robot. When the robot arrived at the garbage position, it used a Logitech camera coupled to an end effector to center the object with image segmentation where we removed the HSV [14] of the object to visualize the binary mask. After centering, the robot approached the garbage and picked it up with the claw.

Receptionist: The purpose of this task was to receive people at a party and take them to the Host in a pre-defined area. However, the place where the Host and the other guests were seated was dynamic. The robot had to introduce the new guest to the rest of the party, pointing them at their positions, worth 250 points for each person, in addition to talking about physical characteristics and their favorite drinks worth 150 points.

The biggest challenge for the success of this task was finding an empty seat for the new guest to sit. To solve this problem, we used a calculation performing a triangulation of the position of the seated people detected within the useful area of the party, thus, having an estimate of their positioning within the room. With this, we were able to predict which seat would be empty for the new guest to be able to sit.

To brag about the task it is necessary to perform the delivery of a guest in real-time. For this we use a model of disclosure of body recognition, with U-2-Net [10] then remove the body from the image and that we remove only the recognition of points of the body afterward, to extract only the points of the body and thus, an image of the person in 3 parts: head, trunk, and legs. For each part performed, by a resource acquisition process, we use neural to detect, networks and masks. In the lower ones, it performs a classification of pieces and color pieces.

5.2   Stage II

In the second stage, the robot needs to perform more complex tasks in a domestic environment. The HERA robot scored 700 points in this stage.

Stick for the Rules: This task had a higher level of difficulty, the robot had to be able to monitor the four rooms of the house with 5 people walking. Inside the house, there were four rules that the robot would have to identify when they were being broken and correct the violators (100 points for each infraction detected). Among them, prohibited room, banned the use of shoes inside the house, prohibited throwing garbage on the floor and all people should have drinks in hand.

However, the difficulty encountered was detecting a guest with a drink in his hand, and thus, directing the guest to the bar. For this, we use two neural networks with TensorFlow to detect a person and another for drinks. Thus performing an IOU (Intersection Over Union) heuristic and bounding box relation to join the two detections and create a complex object. With that, the robot approached the guest and directed him to the bar.

To monitor people in the forbidden room (worth 100 points to detect the person and clarify which rule is being broken), we used a neural network trained to perform Person Recognition, and with the point cloud, we obtained the coordinate of the person in relation to the map of the house, being able to identify if she really was in the forbidden room and then take the necessary measures.

5.3   Final

The final task consisted of carrying out an emergency care approach in a home environment. This theme was proposed by the organizing committee of the competition. Our team chose to perform home emergency assistance, in which the robot made the connection between the patient and the hospital, informing and following instructions from the medical team.

For this, the robot needed to identify the act of a fall inside the house and confirm the accident. Then, establishing a connection via Telegram with the hospital, opening the triage process, sending all the data of the injured resident, and ending with a photo of the accident. With the communication established, the judges of the task were able to make the decisions by interacting with a tablet that was given to them for the Human-Robot Interface. With the type of medicine chosen by them and the specific milligram, the robot went to the shelf, collected the exact medicine requested and delivered it to the patient, while the doctor did not arrive.

Upon the arrival of the doctor, the robot received him and took him to the injured patient. Due to the competition being in Thailand, the robot explained everything that happened in the local language, for a better understanding of the doctor. And so, the robot ended his assistance task successfully.

6   Current Research

6.1   Social Navigation

The main focus of this research is on the people's comfort in spatial interactions with a social robot. This research had as motivation the difficulty found when dealing with the social robot's navigation in a safe, natural and social way, making the robot's presence comfortable for the people interacting around it.

Initially, simulated experiments presented in [9] were carried out, then 20 volunteers were invited to participate in the real experiments. The characteristics of the volunteers varied in terms of age, gender, previous experience with a robot and previous knowledge of robotics. The real experiments were carried out following the project with a Certificate of Presentation of Ethical Appreciation number 43096121.7.0000.5508 presented to the ethics committee in research in Brazil. In this research, all safety protocols related to the pandemic of the new coronavirus (Covid-19) indicated by competents institutions were followed.

For the real experiments, two types of spatial interaction between people and the robot were applied. In the first type, the robot navigated through the environment passing through some specific points. Between each point, there were people performing a certain action (standing, moving, interacting with other people or objects). In the second type, the robot approached a person or a group of people in a certain location. From these experiments, an ontology was developed, initially proposed in [2] where it was possible to determine the type of navigation that the robot performs, social distancing and how to approach people in a socially accepted and comfortable way.

The ontology was used in the robot to build the semantic maps as proposed by [7] in the form of social navigation layers as proposed by [8].

The robot receives objects information, people and the relationships between these entities. 3 layers of cost maps are created representing objects, people or people formations and relationships as areas of interaction. Then, the expansion of obstacles is performed in each of the elements of the classes of objects, people and formations. The robot's radius is used to expand objects and areas where interactions take place, and proxemics are used to expand the area of people and formations. The cost maps are used by ROS Navigation Stack to plan the robot's trajectories.

The robot also receives an identification of a place or a person existing in its knowledge base and performs the navigation to this place in a social way, respecting the social norms and rules based on the ontology. The robot receives the destination name and Navigation type and returns a robot pose at the target. Checks if the destination is a location or a person. If is a location, the destination is set to the location itself, if it is a person, the destination is defined as a location close to the person within its field of vision, respecting the proxemics and within a possible trajectory for the robot, so the robot send a new destination to the navigation system. During navigation, constant changes are made to the trajectory to prevent it from passing through people's personal space.

The robot performs reasoning on ontological information to perform social navigation respecting social rules and norms. The type of formation group is not classified here, however, the positioning of the formation members and the best approach can be found based on the guidance of the closest person and cost maps. Thus, it is possible to generalize the solution to any formation of groups.

The ontology identifies the type of navigation (location, person, walking side by side, guide, follow). Calculates the robot's destination coordinates depending on the navigation type. It publishes information about people, formations, objects and the interactions between them. Then from the navigation type and proxemics it returns the approach pose.

The type of social navigation the robot is currently performing can be To-Local, To-Person, Side-to-Side, Guiding or Following navigation. The type of navigation determines the robot's destination, the angle relative to the destination, safe distance from the destination, orientation relative to the destination as seen in the Table 1 and in the Fig. 4 which presents a robot positioning depending on the type of navigation performed. In Fig. 6, the cost maps of people and areas of interaction are shown (in yellow it is possible to check the possible approach points that a robot can use, these points are inferred using the ontology).

Table 1. Robot navigation types

Navigation type	Robot destination	Approach angle relative to the destination	Distance from destination	Angle of approach relative to destination
To-location	Location	0	0	0
To-person	Person	0	Defined by proxemic	180
Side-by-side	Person	90	Defined by proxemic	0
Guiding	Person	0	Defined by proxemic	0
Following	Person	180	Defined by proxemic	0

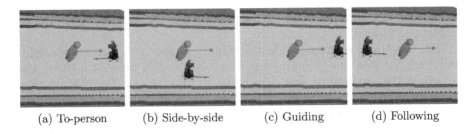

(a) To-person   (b) Side-by-side   (c) Guiding   (d) Following

Fig. 4. Types of navigation when a destination is a person.

In this study, was observed that elements such as the appearance of the robot and noise produced by the robot stand out compared to elements of spatial interaction causing discomfort in the human being. It was observed that the volunteer's previous experience influences the way in which social norms are accepted. Volunteers with previous experience of the robot's capabilities usually reveal some points of discomfort such as appearance, noise and even sudden interventions in people's interactions. With this, it was observed that the existence of previous experience in people with the capabilities of the robot has a great influence on the comfort of these people.

At the end of this study, a computational solution was obtained that allows a mobile social robot to be able to interact properly in spatial terms in a social environment, reducing the feeling of discomfort for the human being during this type of interaction. This study is designed to address current scientific and social challenges. Having potential for a positive impact both in the academic environment and in the daily life of the common citizen.

6.2   Dynamic Power Management

The constant development of our robotic platform substantially increased its power consumption. As a consequence, our first solution was to increase the our robot. After that, the next step was to migrate to higher energetic density batteries, which solved our problem but wasn't a good solution due to its prices so we started research into Dynamic Power Management [15]. The Dynamic Power Management (DPM) is a method developed with the purpose of optimizing available energy sources. DPM proposes to optimize energy use through the control of the energy used by the system's modules, made by the idleness exploitation: If a device (or components of a device) is idle, its energy consumption should be reduced as much as possible to save for when it will be needed. There are several ways to implement the DPM, but the first step is to detect with accuracy the idleness in the system's module to quickly deactivate and force it into an energetic dissipation state in which the wasted energy is as low as possible. To develop the research and implement the most suitable DPM [15]. In our system, it was necessary to conduct a lot of tests, which consisted into analyse the operating current of each module as a function of time, while the robot performs a task. The current was acquired using a ACS712 sensor and the data was sent

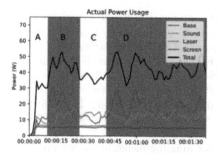

Fig. 5. Consumption of each module and total consumption.

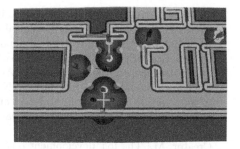

Fig. 6. People approach points.

to a computer through an Arduino. Then the operating current of each module was multiplied by its operating voltage to obtain the energy consumption of each module, generating the graph shown in Fig. 5.

Analyzing the graph, the Average Consumption per Second (Cps), Execution Time (Texec) and the Total Consumption (Ct) were calculated. From the data obtained, simulations were made considering different DPM methods, and the one closest to the ideal was the predictive method with pre-wake up, saving more than 35% of wasted energy, increasing the running time substantially and by consequence the batteries lifespan [15]. Our team implemented this DPM module on HERA utilizing an electronic relay module to deactivate the module that is in idleness and is conducting new research into implementing a battery bank to supply energy to the entire robot and make the management even more efficient.

7   Summary

This paper describes the main strategies and technologies used on robot HERA to win first place in the RoboCup@Home 2022 (Open Platform League), focusing on the organization of tests and rapid implementations of technologies for validation. In addition, we modulate our strategy to be easily adapted to different environments and situations. With this, we hope that the strategy presented can be applied to different contexts and replicated by other groups.

References

1. Aquino Junior, P.T., et al.: Hera: home environment robot assistant. In: Fei, C.U. (ed.) Proceedings of the II Brazilian Humanoid Robot Workshop (BRAHUR) and III Brazilian Workshop on Service Robotics (BRASERO), Sao Bernardo do Campo, Sao Paulo, Brasil, vol. 1, pp. 68–73. IEEE South Brazil Robotics & Automation Society Chapter (2019)

2. de Assis Moura Pimentel, F., Aquino, P.T.: Proposal of a new model for social navigation based on extraction of social contexts from ontology in service robots. In: 2019 Latin American Robotics Symposium (LARS), 2019 Brazilian Symposium on Robotics (SBR) and 2019 Workshop on Robotics in Education (WRE), pp. 144–149 (2019). https://doi.org/10.1109/LARS-SBR-WRE48964.2019.00033

3. Boyko, N., Basystiuk, O., Shakhovska, N.: Performance evaluation and comparison of software for face recognition, based on dlib and opencv library. In: 2018 IEEE Second International Conference on Data Stream Mining & Processing (DSMP), pp. 478–482. IEEE (2018)

4. Creator, M.: The IoT development board for building incredibly smart products. https://www.matrix.one/products/creator

5. Dwibedi, D., Misra, I., Hebert, M.: Cut, paste and learn: surprisingly easy synthesis for instance detection. In: Proceedings of the IEEE International Conference on Computer Vision, pp. 1301–1310 (2017)

6. Koonce, B.: EfficientNet. In: Koonce, B., et al. (eds.) Convolutional Neural Networks with Swift for Tensorflow, pp. 109–123. Apress, Berkeley (2021). https://doi.org/10.1007/978-1-4842-6168-2_10

7. Kostavelis, I., Giakoumis, D., Malassiotis, S., Tzovaras, D.: Human aware robot navigation in semantically annotated domestic environments. In: Antona, M., Stephanidis, C. (eds.) UAHCI 2016. LNCS, vol. 9738, pp. 414–423. Springer, Cham (2016). https://doi.org/10.1007/978-3-319-40244-4_40

8. Lu, D.V., Hershberger, D., Smart, W.D.: Layered costmaps for context-sensitive navigation. In: 2014 IEEE/RSJ International Conference on Intelligent Robots and Systems, pp. 709–715. IEEE (2014)

9. Pimentel, F., Aquino, P.: Performance evaluation of ROS local trajectory planning algorithms to social navigation. In: 2019 Latin American Robotics Symposium (LARS), 2019 Brazilian Symposium on Robotics (SBR) and 2019 Workshop on Robotics in Education (WRE), pp. 156–161 (2019). https://doi.org/10.1109/LARS-SBR-WRE48964.2019.00035

10. Qin, X., Zhang, Z., Huang, C., Dehghan, M., Zaiane, O.R., Jagersand, M.: U2-net: going deeper with nested u-structure for salient object detection. Pattern Recogn. **106**, 107404 (2020)

11. RoboCupHome: Robocup@home. https://athome.robocup.org/

12. Romano, V., Dutra, M.: Introdução a robótica industrial. Robótica Industrial: Aplicação na Indústria de Manufatura e de Processo, São Paulo: Edgard Blücher, pp. 1–19 (2002)

13. dos Santos, T.F., de Castro, D.G., Masiero, A.A., Aquino Junior, P.T.: Behavioral persona for human-robot interaction: a study based on pet robot. In: Kurosu, M. (ed.) HCI 2014. LNCS, vol. 8511, pp. 687–696. Springer, Cham (2014). https://doi.org/10.1007/978-3-319-07230-2_65

14. Sural, S., Qian, G., Pramanik, S.: Segmentation and histogram generation using the HSV color space for image retrieval. In: Proceedings of the International Conference on Image Processing, vol. 2, p. II. IEEE (2002)

15. Techi, R.D.C., Aquino, P.T.: Dynamic power management on a mobile robot. In: 2021 6th International Conference on Mechanical Engineering and Robotics Research (ICMERR), pp. 8–14. IEEE (2021)

Tech United Eindhoven Middle Size League Winner 2022

S. T. Kempers$^{(\boxtimes)}$, D. M. J. Hameeteman, R. M. Beumer, J. P. van der Stoel,
J. J. Olthuis, W. H. T. M. Aangenent, P. E. J. van Brakel, M. Briegel,
D. J. H. Bruijnen, R. van den Bogaert, E. Deniz, A. S. Deogan,
Y. G. M. Douven, T. J. van Gerwen, A. A. Kokkelmans, J. J. Kon,
W. J. P. Kuijpers, P. H. E. M. van Lith, H. C. T. van de Loo, K. J. Meessen,
Y. M. A. Nounou, E. J. Olucha Delgado, F. B. F. Schoenmakers, J. Selten,
P. Teurlings, E. D. T. Verhees, and M. J. G. van de Molengraft[iD]

Tech United Eindhoven, De Rondom 70, P.O. Box 513,
5600 MB Eindhoven, The Netherlands
techunited@tue.nl
https://www.techunited.nl/

Abstract. During the RoboCup 2022 tournament in Bangkok, Thailand, Tech United Eindhoven achieved the first place in the Middle Size League. This paper presents the work done leading up to the tournament. It elaborates on the new swerve drive platform (winner of the technical challenge) and the progress of making the strategy software more semantic (runner-up of the scientific challenge). Additionally, the implementations of the automatic substitution and of more dynamic passes are described. These developments have led to Tech United winning the RoboCup 2022 tournament, and will hopefully lead to more successful tournaments in the future.

Keywords: RoboCup soccer · Middle Size League · Multi-robot · Swerve drive · Semantic strategy

1 Introduction

Tech United Eindhoven represents the Eindhoven University of Technology in the Robocup competition. The team joined the Middle Size League (MSL) in 2006 and played in 13 finals of the world championship, winning them 6 times. The MSL team consists of 4 PhD, 7 MSc, 2 BSc, 6 former TU/e students, 7 TU/e staff members, and 1 member not related to TU/e. This paper describes the major scientific improvements of the Tech United soccer robots over the past year and elaborates on some of the main developments for future RoboCup tournaments. The paper starts with a description of the fifth generation soccer robot used during the RoboCup 2022 competition in Sect. 2. Additionally, some statistics of the robots during the tournament are given in Sect. 3. In Sect. 4 the developments on the swerve drive platform are described. Section 5 elaborates on how we will increase

© The Author(s), under exclusive license to Springer Nature Switzerland AG 2023
A. Eguchi et al. (Eds.): RoboCup 2022, LNAI 13561, pp. 337–348, 2023
https://doi.org/10.1007/978-3-031-28469-4_28

the level of semantics in our strategy. The work on the automatic substitution and on more dynamic passes is briefly described in Sect. 6. Finally, the paper is concluded in Sect. 7, which also presents our outlook for the coming years.

2   Robot Platform

The Tech United soccer robots are called TURTLEs, which is an acronym for Tech United Robocup Team: Limited Edition. Their development started in 2005, and through years of experience and numerous improvements they have evolved into the fifth generation TURTLE, shown in Fig. 1. A schematic representation of the robot design can be found in the work of Lopez et al. [8]. A detailed list of hardware specifications, along with CAD files of the base, upperbody, ball handling and shooting mechanism, is published on the ROP wiki[1].

Fig. 1. Fifth generation TURTLE robots, with the goalkeeper on the left-hand side. (Photo by Bart van Overbeeke)

The software controlling the robots consists of four modules: Vision, Worldmodel, Strategy, and Motion. These parts of the software communicate with each other through a real-time database (RtDB) designed by the CAMBADA team [1]. The Vision module processes the vision sensors data, such as omni-vision images, to obtain the locations of the ball, opponents, and the robot itself. This position information is fed into the Worldmodel. Here the vision data from all the team members is combined into a unified representation of the world. The Strategy module makes decisions based on the generated worldmodel using the Strategy, Tactics and Plays (STP) framework. More information on STP can be found in [7]. Finally, the Motion module translates the instructions of Strategy

---

[1] http://roboticopenplatform.org/wiki/TURTLE.

into low-level control commands for the robot's actuators. Further details on the software can be found in [11].

3   RoboCup 2022 Statistics

Six teams participated at RoboCup 2022 in the MSL: Barelang 63, ERSOW, IRIS, Falcons, Robot Club Toulon, and Tech United Eindhoven. Of these six teams, five participated in the soccer competition, which resulted in a total of thirty-three matches. Tech United played fourteen of these matches, during which they were able to score 172 times, while only conceding one goal (plus three regulatory goals). Compared to the previous tournament in 2019 where our robots scored on average 7.0 goals per match, this year they scored 12.3 goals per match. The robots passed a total of 364 times. Based on odometry, the robots collectively drove 61.7 km over the course of the tournament, of which 1.4 km was covered by the goalkeeper.

4   Swerve Drive Platform

The RoboCup MSL matches take place on flat, homogeneous surfaces and are primarily focused on strategic and autonomous multi-agent decision-making at high velocity. Nevertheless, humans are able to maneuver on a wide variety of terrains, where bumpy soil and tall (wet) grass are no challenge at all. Most teams have equipped their platforms with multiple (mostly three) omni-directional wheels, enabling the possibility to instantaneously translate in both forward and sideways direction. In addition, being able to quickly rotate around the robot's vertical axis is very effective in outplaying its opponent, like ball shielding and interception. These wheels have shown great performance in terms of maneuverability and flexibility, but require flat surfaces due to the limited radius of the small rollers on the perimeter. Besides, more uneven or slippery surfaces result in slipping motion or even getting stuck, hence the omni-directional wheels are not functional on (artificial) grass-like surfaces. To achieve the ultimate goal of RoboCup, namely to win against the winner of the human World Cup, an important aspect is being able to play soccer on more diverse and outdoor terrains.

An earlier attempt to improve the traction in the target direction has resulted in the development of an eight-wheeled platform [5], consisting of four wheel sets each having two hub-drive motors. The torque delivered by the wheels could directly be applied into the desired direction, thus increasing the overall accelerations and velocities. Furthermore, each pair of wheels could act as a differential drive using the friction between wheel and surface to generate a rotational movement around the pivot point. Indeed the results show great performance in terms of acceleration and agility. Nevertheless, the platform lacks performance when accelerating on uneven surfaces, since the evenly distributed friction between the wheels in each set and the ground needs to be guaranteed.

Our primary motivation is enabling the important step of the RoboCup MSL to play outdoor on regular fields and is able to achieve higher maximum accelerations and velocities in all directions. Besides, developing an open-source platform that is robust and cost-effective will provide an accessible option for all current teams.

4.1   Hardware Design

The swerve drive principle [4] seems rather promising when considering movements that are less dependent on the friction between wheel and surface, while these platforms are generally equipped with separate steering actuators. Besides, these platforms could take full advantage of their acceleration capabilities, since 'normal' wheels have greater contact surface (and thus friction) than the small rollers of omni-directional wheels.

We have chosen for a three-wheeled coaxial-drive configuration (rather than two- or four-wheeled) to comply with the MSL dimensions, while simultaneously considering the required space for shooting and ball handling mechanism (see Fig. 2a). Besides, even mass distribution and axisymmetric design were important reasons for choosing this configuration. The coaxial-drive controls the rotation and propulsion of each wheelset separately, which could be more easily manufactured and made dust- and water-tight, and hence is a more cost-effective solution.

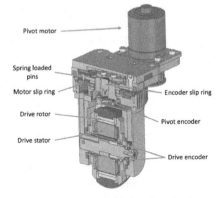
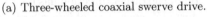

(a) Three-wheeled coaxial swerve drive.   (b) Cross-section of a single wheelset.

Fig. 2. Design and realisation of a robust, outdoor motion platform

Figure 2b shows the compact design of a single wheel-set, consisting of the drive motor and the pivot motor. Direct-drive in-wheel brushless DC (BLDC) motors are found in a wide variety of electrical scooters and skateboards nowadays and form the basis of the wheelset design. Since space is very limited, the wheel and drive motor are combined into one part. The inner-coils of the motor controls the outer magnets, which in turn is attached to the tire.

One major challenge is how to provide high power to the drive motor, while it rotates along the pivot axis. It is not preferable to limit the possible amount of rotations, therefore twisting wires through the center of the pivot is not feasible. Since the motor requires a considerable amount of current of at least 15A RMS, special slip rings have been developed. Three brass rings, one for each BLDC motor coil, are put together in a single disk, with the motor wires assembled to one side of the ring. Spring loaded pins are pushed against the other side of the ring, which bridges the gap between the rotational and stationary side.

A secondary motor is attached to the propulsion motor, responsible for rotating the wheel around its own axis. Again, a low-cost BLDC outer runner motor was chosen that is widely available within the consumer market. Since the pivot angle requires a high torque and a lower maximum velocity, the motor and the driven axle are connected by a belt transmission. For ease of assembly it was chosen to point the motor upwards, but could as well be pointing downwards in later iterations. The wheelset has been designed such that three identical sets could be created: one for each corner of the platform.

Next, the position and velocity of each motor needs to be obtained for accurate control of the overall platform. Most off-the-shelf high resolution encoders are either not accurate enough, built rather bulky or are expensive. Therefore, a small PCB for accurate 19-bit encoder positioning was developed for accurate encoder placement, including an external EEPROM for configuration settings and line driver for stable communication. A magnetic ring with a unique pattern is centered on the rotating side, while the encoder chip is placed off-center against the stationary side.

The pivot encoder was directly placed onto the outgoing axis to make sure the wheels are always properly aligned, even if a problem appears, such as slip on the driven belt. The encoder for the drive motor was placed directly along the pivot axis within the fork. The encoder cables are fed through a slip ring, centered along the pivot axis.

4.2   Software Design

First, each of the motor drivers needs to be configured by saving information about the motor, gear ratio, encoders and low-level control structure in the driver's Service Data Object (SDO) dictionary. Furthermore, the Process Data Object (PDO) map has to be assigned, such that the device knows which registers should be available for reading and writing real-time data. Configuration and communication between the master controller and the motor driver slaves make use of the Simple Open EtherCAT Master (or SOEM) library[2]. The library provides an application layer for reading and writing process data, keeping data synchronized and detecting and managing potential errors.

The propulsion and pivot motor of one wheelset have a different amount of pole pairs. Besides, the pivot motor is equipped with a hall-sensor and includes a gear ratio with the outgoing axis. Finally, the minimal PDO-map consists of:

---

[2] https://github.com/OpenEtherCATsociety/SOEM.

controlword, torque, velocity and position setpoint for the RxPDO and status-word, error code, torque, velocity and position value for the TxPDO. The full configuration file has been published on the ROP wiki[3].

The software architecture was designed such that parts of the robot could change without having to change the entire model. In this case, only the motion platform has changed when introducing the swerve drive, for which a new model was designed.

Receiving and sending information between the motor driver and Simulink is handled by creating read and write Simulink S-functions. These functions set RxPDO and get TxPDO values respectively based on the selected slave ID. Besides, a MATLAB function has been written for stepping through the Drive State Machine (CiA 402) by reading the current state using statusword and transitioning towards the next state with the controlword. Furthermore, when an error occurs and the fault state is entered, the function automatically resets and enables the driver again.

The motor drivers are already capable of performing cyclic synchronous position and cyclic synchronous velocity mode based on the low-level setpoints. What remains is to calculate the setpoint position for the pivot motor and setpoint velocity for the drive motor.

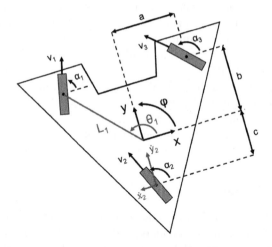

Fig. 3. Schematic representation of swerve drive platform.

Figure 3 shows a graphical representation of the swerve drive platform. Each wheelset has a position and center of rotation with respect to the platform's center of rotation, which is known by design. These positions are used to calculate the polar coordinate position of each wheelset, represented as arm L_i and angle θ_i with respect the platform's x-axis, as

$$L_i = \sqrt{x_i{}^2 + y_i{}^2}$$
$$\theta_i = \arctan2\left(y_i, x_i\right)$$

(1)

---

[3] http://roboticopenplatform.org/wiki/SwerveDrive.

where $i = 1, 2, 3$ correspond to the wheelset.

Next, the desired platform reference velocity $[\dot{x} \ \dot{y} \ \dot{\phi}]_{\text{ref}}^T$ could be used to calculate the velocity and orientation of each wheelset. First, the velocity vector of each wheelset is calculated as follows

$$
\begin{aligned}
\dot{x}_i &= \dot{x}_{\text{ref}} - \sin(\theta_i) \, v_{c,i} \\
\dot{y}_i &= \dot{y}_{\text{ref}} + \cos(\theta_i) \, v_{c,i}
\end{aligned}
\tag{2}
$$

in which the linear velocity from the angular reference is calculated as

$$
v_{c,i} = \dot{\phi}_{\text{ref}} L_i
\tag{3}
$$

Finally, the angular setpoint of each wheel is calculated as follows, which is used as the target position for the pivot motor.

$$
\alpha_i = \arctan2 \, (\dot{y}_i, \dot{x}_i)
\tag{4}
$$

The velocity setpoint for the propulsion motor is calculated as

$$
v_i = \sqrt{\dot{x}_i^2 + \dot{y}_i^2}
\tag{5}
$$

The calculations above were implemented in the model-based framework within MATLAB and Simulink, in order to create an easy to maintain and scalable design. The hardware communication takes place through EtherCAT, using the S-functions as mentioned above. The software design is compiled and runs real-time onto the robot platform.

4.3 Results

The swerve drive platform was presented for the first time during the MSL technical challenge at RoboCup 2022 and got first place. The results are very promising as the prototype showed great performance in terms of acceleration and robustness on uneven, bumpy terrain. Currently, the acceleration is limited by the motor drivers, which are not capable of delivering more than 30A peak. The goal for next year is to have a swerve drive platform based robot participating in the team and scoring its first goal.

5 Decision Making Through Semantic Regions in Robotic Soccer

In the game of soccer, strategy has a significant contribution to winning games. The effectiveness of a strategy is dependent on the capability of the players and the opposing strategy. It is therefore something that is adjusted often, requiring a proper definition. In human soccer this is developed over the years and comprises the formation, the type of build up, the width of play and many more qualitative notions [3]. In the context of the RoboCup MSL, the strategy and configuration of this has been solved in various ways [2,9,10]. Currently the Tech United team uses Skills, Tactics and Plays (STP) as the overall strategy framework [7]. Within STP ad-hoc decisions about which skill to deploy still have to be made to react properly to changes on the field, which is solved by the use of potential fields, called mu-fields. These fields include many parameters and are therefore hard to configure. This work aims at enhancing this configurability by incorporating semantic information about the skills in the world model of the robot. 2D regions are constructed that represent the affordance of a certain skill. Knowledge of the possible combination of skills and targets allows for explainable decision making, improving configurability.

5.1   Strategy Framework

As already mentioned, the STP framework is used to deploy the strategy. The global team plan consists of a list of plays that are selected based on the game situation. Within a play each robot gets a specific role, e.g., goal keeper, attacker, defender. Each role has a specific action or a sequence of actions to execute. An example of the first being in an attacking play and possession of the ball. The action is then described by 'advancing the game', in which the player needs to choose the appropriate skill and target ad-hoc based on the game situation (position of ball, peers and opponents). The potential field used for this is a grid field where for each cell a number of cost-functions are calculated giving a score to that specific location, which is visualized in Fig. 4a. Assisting players create such a field to select to position themselves on the field strategically and communicate this to the player in possession of the ball. The player in possession of the ball creates a field for dribbling, and calculates its current scoring position. It then compares all skills and selects the skill and target with the highest score. The downside of using potential fields is that it has a lot of parameters, in the case of our TURTLEs up to 16 parameters. Only regarding the region to where a skill is actually possible would make all involved parameters redundant.

5.2   Semantic Representation of Soccer

Figure 4b visualizes the semantic regions for an arbitrary game situation. To create these, first the different skills of a soccer robot are defined. In case of attacking game play we make a distinction between a shot, pass (forwards and backwards) and dribble (forwards and backwards). Hereafter, constraints are defined restricting the region of a skill. Shots are constrained by opponents and peers between the ball and goal, while a dribble is constrained by opponents within the dribbling region. The dribbling region is a confined region where a robot is allowed to dribble from the position were it got the ball and is set by the competition rules. A pass has the following constraints:

- The region in which a pass can be received is constrained by the estimated pass time.
- Opponents between the ball and the pass receiving region limit this region.

– Target positions closer to an opponent than the receiver are not reachable.

These geometrical calculations are executed by the use of the Shapely Python package[4] and result in a world model as visualised in Fig. 4b. The resulting skill regions are evaluated by a simple decision tree structure explained in Sect. 5.3. Assisting players use a similar semantic representation to decide on their target for strategic positioning.

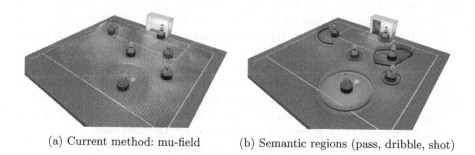

(a) Current method: mu-field   (b) Semantic regions (pass, dribble, shot)

Fig. 4. Visualization of decision making methods for soccer robots

5.3   Decision Making

Now that a semantic representation of possible skills (and targets) exist, this enables the decision process to be based on a tree structure. A few risk and reward parameters are used to evaluate the actions. For now the following parameters are used:

– Scoring position: distance to the optimal position for making a shot.
– Pressure on the ball: the area of the region is a measure for the amount of pressure on the ball by opponents.
– Duration of the action: the longer the action takes, the more time the opponent has to react or intercept.

Actions are evaluated in the following order. From the semantic map it is determined if a shot is possible. If possible and the shooting position score is above a certain threshold, a shot is taken. If not, the possible forward passes are evaluated. If the reward and risk are above and below a certain threshold a pass is given. If not, forward dribbles are evaluated. Hereafter backward passes and dribbles are considered. In case none of them satisfies the thresholds, values are compared and the best possible action is taken. Note that this can also be staying idle.

---

[4] Shapely, a Python package for computational geometry - https://shapely. readthedocs.io/en/stable/manual.html.

6   Miscellaneous

6.1   Automatic Substitution

Last year's addition of autonomous substitutions is one of the advancements towards an even more autonomous game. Without the autonomous substitution, a human team member has to remove the robot from the field and during the next game stoppage after a repair period of 20 s, a robot may be placed at the sideline again. In the new situation, the humans only indicate the number of the robot to be substituted. Once that robot leaves the field and enters the Team Technical Area (TTA) next to the field autonomously, then the other robot is immediately allowed to move from the TTA to the sideline and participate when the game continues. In this way, there is no longer a disadvantage of temporarily playing with one less player. If the robot is unable to autonomously leave the field, a manual substitution has to be performed and a 20 s penalty is applied.

One of the main challenges for the implementation of the autonomous substitution on the TURTLEs is their localization outside the field lines. The position of robots in the TTA is about 1.5 m away from the sideline of the field. There our original localization method [6], which was built upon the assumption that the robot is in the field or at least very close to it, was not sufficiently robust anymore. This issue was solved by initializing the robots within the field before the game starts, and moving the substitute robot from there into the TTA. Even if the position with respect to the field lines cannot be found anymore using the omnivision camera data, the fused encoder and IMU data still provide an accurate enough update of the location estimate until the substitute enters the field when it is allowed to.

6.2   Through Balls

Instead of only using static passes, i.e., a robot passes the ball to the position another robot is already located, the gameplay becomes a lot more dynamic by through balls. In general, a through ball is a pass between opponents to receive the ball behind them, as shown in Fig. 5. When we refer to through balls, we also mean passes into open space that are not in between defenders. The addition of such passes does not only drastically increase the amount of opportunities to give a pass, it also enables the pass receiver to already cover a significant distance within the time the ball is on its way. This accelerates the gameplay during ball possession, making it more difficult for the opponent to defend.

Through balls are implemented in our software by first letting the pass giver and pass receiver agree on a position to give the pass to. Whereas for static passes the only possible position is the current position of the pass receiver, now all positions on the field are considered as possible targets. The best target within this large set is found by applying a multi-objective optimization, for instance taking into account the distance of the receiver to the target, whether there are opponents blocking the path of the ball and how close to the opponent goal the pass target is. This optimization is now performed by the mu-fields, but can in

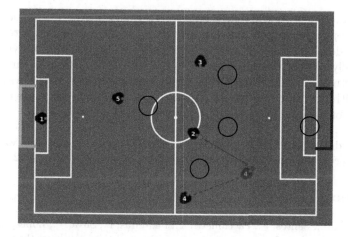

Fig. 5. Example of a through ball from robot 2 to robot 4. The black circles represent opponents.

the future be based on semantic regions, both explained in Sect. 5. As long as the decision to pass is not considered to be better than dribbling or shooting, the players without the ball will continue moving around the field. Once the pass giver and receiver agree on a good target for a pass, the robot in possession of the ball will kick it towards the agreed target and the other robot moves towards the target to intercept the ball.

7   Conclusion

In this paper we have described the major scientific developments of team Tech United Eindhoven in preparation of RoboCup 2022, and the results we have achieved during the tournament. Not all developments contributed actively to the result, but will lead to improvements of our soccer robots in future tournaments.

We have elaborated on our work on a new platform that utilizes the swerve drive principle. This platform won the Technical Challenge of RoboCup 2022 and will hopefully participate in its first matches at RoboCup 2023. The new design reduces the dependency on friction between the wheels and soccer field, therefore making the robot rely less on a flat and homogeneous field and hence more future-proof.

We also discussed how we are transforming our potential fields, used in the decision making, by more semantic models. This gives us more insight in the decision making process, which in turn allows for better configuration of the overall strategy.

The automatic substitution will make the MSL even more autonomous. The challenges and solutions for implementing this were briefly discussed. By utilizing the through balls our robots can now perform, the game play will be more dynamic and fluent. It will allow the pass receivers to cover much more distance during the pass, making it extra dangerous for static opponents. Altogether we think our developments will contribute to an even higher level of dynamic and scientifically challenging robot soccer. The latter, of course, while maintaining the attractiveness of our competition for a general audience. In this way we hope to stay with the top of the MSL for many more years and contribute to the long term goal of beating the human world champion in soccer in 2050.

Acknowledgement. The development of the swerve drive platform was part of RoboCup Federation and MathWorks Support for Research Projects 2022.

References

1. Almeida, L., Santos, F., Facchinetti, T., Pedreiras, P., Silva, V., Lopes, L.S.: Coordinating distributed autonomous agents with a real-time database: the CAMBADA project. In: Aykanat, C., Dayar, T., Körpeoğlu, İ (eds.) ISCIS 2004. LNCS, vol. 3280, pp. 876–886. Springer, Heidelberg (2004). https://doi.org/10.1007/978-3-540-30182-0_88
2. Antonioni, E., Suriani, V., Riccio, F., Nardi, D.: Game strategies for physical robot soccer players: a survey. IEEE Trans. Games **13**(4), 342–357 (2021). https://doi.org/10.1109/TG.2021.3075065
3. FIFA: Futsal coaching manual (2019)
4. Holmberg, R., Slater, J.C.: Powered caster wheel module for use on omnidirectional drive systems. uS Patent 6,491,127 (2002). https://www.google.it/patents/US4741207
5. Houtman, W., et al.: Tech United Eindhoven middle-size league winner 2019. In: Chalup, S., Niemueller, T., Suthakorn, J., Williams, M.-A. (eds.) RoboCup 2019. LNCS (LNAI), vol. 11531, pp. 517–528. Springer, Cham (2019). https://doi.org/10.1007/978-3-030-35699-6_42
6. Kon, J., Houtman, W., Kuijpers, W., van de Molengraft, M.: Pose and velocity estimation for soccer robots. Student Undergraduate Res. E-J. **4** (2018). https://doi.org/10.25609/sure.v4.2840
7. de Koning, L., Mendoza, J.P., Veloso, M., van de Molengraft, R.: Skills, tactics and plays for distributed multi-robot control in adversarial environments. In: Akiyama, H., Obst, O., Sammut, C., Tonidandel, F. (eds.) RoboCup 2017. LNCS (LNAI), vol. 11175, pp. 277–289. Springer, Cham (2018). https://doi.org/10.1007/978-3-030-00308-1_23
8. Lopez Martinez, C., et al.: Tech United Eindhoven team description (2014). https://www.techunited.nl/media/files/TDP2014.pdf
9. Neves, A.J.R., Amaral, F., Dias, R., Silva, J., Lau, N.: A new approach for dynamic strategic positioning in RoboCup middle-size league. In: Pereira, F., Machado, P., Costa, E., Cardoso, A. (eds.) EPIA 2015. LNCS (LNAI), vol. 9273, pp. 433–444. Springer, Cham (2015). https://doi.org/10.1007/978-3-319-23485-4_43
10. Reis, L.P., Lau, N., Oliveira, E.C.: Situation based strategic positioning for coordinating a team of homogeneous agents. In: BRSDMAS 2000. LNCS (LNAI), vol. 2103, pp. 175–197. Springer, Heidelberg (2001). https://doi.org/10.1007/3-540-44568-4_11
11. Schoenmakers, F., et al.: Tech United Eindhoven team description (2017). https://www.techunited.nl/media/images/Publications/TDP_2017.pdf

Author Index

© The Editor(s) (if applicable) and The Author(s), under exclusive license
to Springer Nature Switzerland AG 2023
A. Eguchi et al. (Eds.): RoboCup 2022, LNAI 13561, pp. 349–351, 2023
https://doi.org/10.1007/978-3-031-28469-4

Printed in the United States
by Baker & Taylor Publisher Services

Printed in the United States
by Baker & Taylor Publisher Services